LATEST APARTMENT BUILDING

Dalian University of Technology Press

Latest Apartment Building

Edited by
Hi-design International Publishing Co., Ltd.

Published by
Dalian University of Technology Press

Address: Section B, Sic-Tech Building, No.80 Software
Park Road, Ganjingzi District, Dalian, China
Tel: +86 411 84709043
Fax: +86 411 84709246
E-mail: designbooks_dutp@yahoo.cn
URL: http://www.dutp.cn

ISBN 978-7-5611-6362-7

In the nineteenth century the German architect Gottfried Semper formulated in his theoretical writings the myth that the origin of architecture was in the natural and primitive act of covering the body with woven. From this primordial act began a symbolic string in which the hearth, where the fire is lit, became the heart of the house, around which people gathered. A platform guaranteed settlement and isolation of natural terrain, a substructure holding walls and ceilings and a coating of textile materials of different origin formed the spatial enclosure and provided protection from the elements.

There is no doubt that architecture has evolved greatly in recent years, reaching high levels of sophistication and technical complexity. In an age where almost everything is possible, we must not lose sight of Semper descriptions which talk about the essence of dwelling, the anthropological origins of the home, in short, what each of us rooted in the subconscious more intimate when we think of our house. Therefore, if to make any kind of architecture is to acquire a high responsibility to society, to make "homes" should be the highest of responsibilities. Not surprisingly, we project the places where people live, is sheltered, and somewhat protected from the surrounding environment in an intuitive act. This area is identified with the privacy and intimacy.

Paradoxically, building homes is also working with communal spaces. In the intimate act of inhabiting, the domestic space overlaps the public space and the social act, the relationship with the community, where you share space, duties and obligations. Providing and enhancing this social relations is essential in architecture, especially in housing projects, as an act of civic engagement, and as intrinsic function unavoidable when dealing with collective housing. The perfect balance and coupling of public and private, the domestic and social, is essential for the smooth operation of this type of architecture.

However, we have witnessed a new revolution of capitalism and information in the age of globalization. And the question arises in what way it will affect the design of housing, in the domestic and social environment. If we will tend, perhaps, like in the utopian work of Toyo Ito *PAO for the Tokyo Nomad Women*, to merge the two fields, expanding the boundaries of our home throughout the city, without possessions. We won´t need a library, a kitchen or wardrobe. Nothing more than a few technological devices that shall act as interfaces on the technological metropolis: an information console (communication), a dressing table (embellishment) and a chair (relaxation) are the artifacts into which the house has drifted. A new triad that will replace the Vitruvius "Utilitas, Firmitas and Venustas".

María José Pizarro, Oscar Rueda

Contents

Master architect: Nihon Sekkei, Inc
Design architect: Jun Mitsui & Associates Inc. Architects
Façade design collaborator: Studio ABK,
Landscape design: Ai-shokubutsu Landscape Planning Office
Interior design: Garde Co., Ltd.
Location: 2-10, Higashigotanda, Shinagawa-ku, Tokyo, Japan
Area: 62188.55m²
Photographer: Norio Koike, Nacasa & Partners

Park Tower Gransky

Park Tower Gransky is a project which has sophisticated design to add consistency and variety to the whole town and to realize a rich residential space and beautiful cityscape, based on the master plan of the redevelopment area in Higashigotanda district. In this area, as a 150m tall high-rise, this tower has to create an overall cityscape that works with other future high-rises in the adjacent Kitashinagawa district. During the exterior design process, in order to add various inspirations to produce a sophisticated cityscape, the designers collaborated with other foreign designers; first, the designers set up the cascade-shaped design along Meguro River adjacent to the site and visualized the landscape with the water in Meguro River. The designers then made more than 20 kinds of patterns using different tile combinations and applied them to the exterior wall, which produced a rich expression like a delicately woven fabric. Glossy luster tiles were scattered across the facade to reflect natural light thus adding to the appearance of the tower from different angles and viewpoints. In order to match the Jounan Gozan Mountains, the designers picked more earth colored and dark colored metal frames, which also enforces the imagery of the cascading water.

For the design of the top of the tower, the designers used the spire shape so the tower appears as if it's shooting from the ground towards the sky, and added the grid design to create a contrast. Also to establish a more luxurious space on the rooftop, roof garden and a Jacuzzi bath was added as part of the residents' amenity; it offers a spectacular view of the city and is part of the hospitality towards the residents. The designers carefully designed every detail for each material, and created a rich and warm space by using delicate craftsmanship. Overall, for this tower design the designers believe they have achieved alignment and harmony with natural environment in this area.

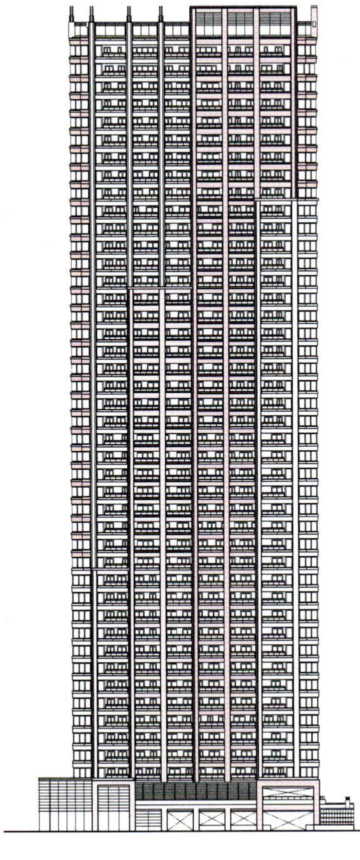

South elevation plans

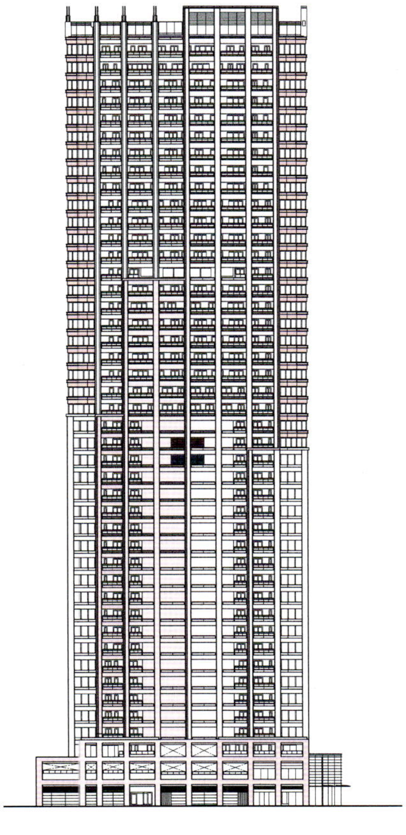

North elevation plans

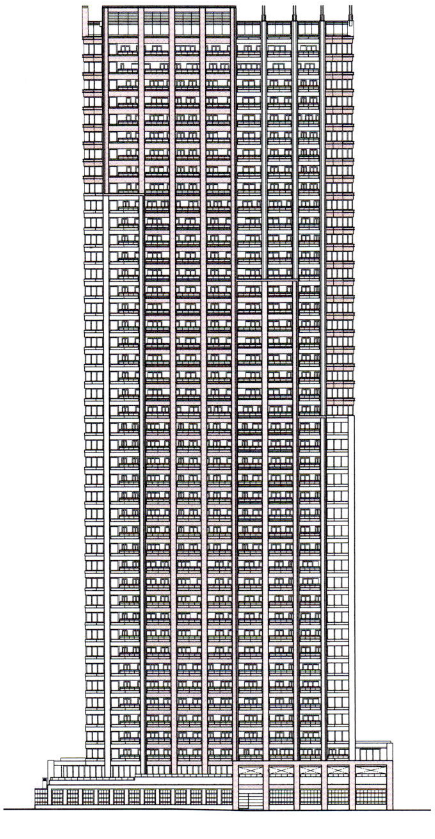

East elevation plans

West elevation plans

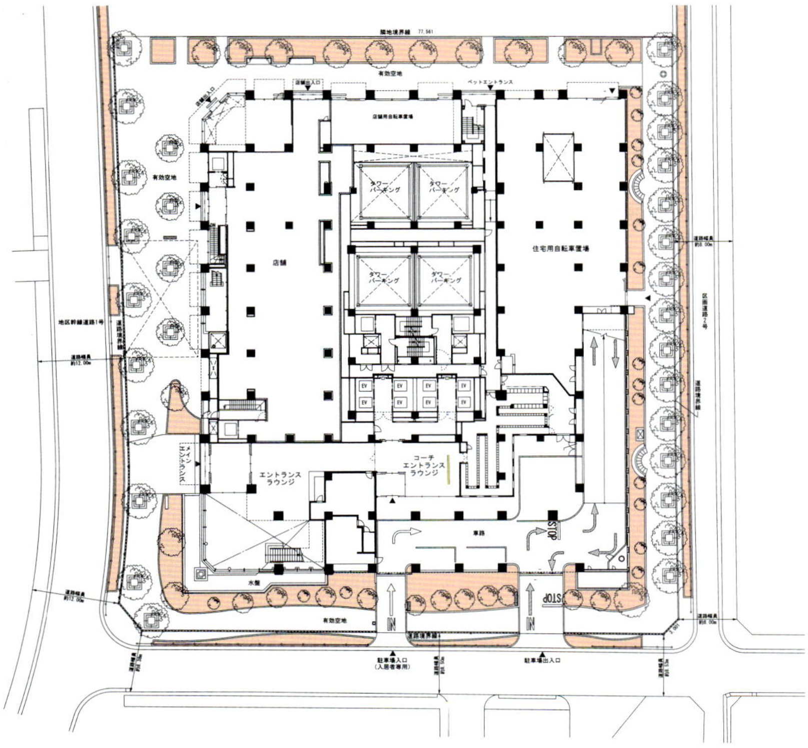

site plan 1F

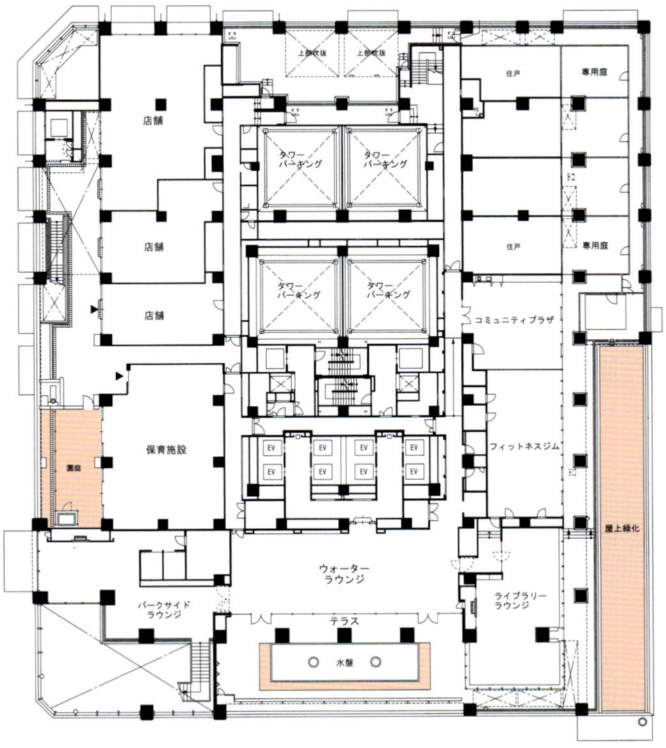

floor plan 2F

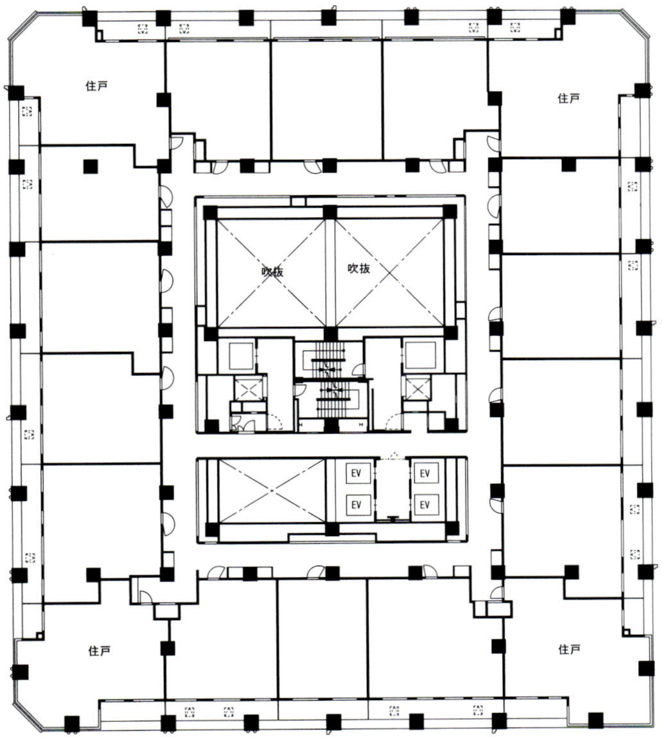

floor plan 32-34F

Architect: Stefan Behnisch, Martin Haas, David Cook
Firm: Behnisch Architekten
Location: Hamburg
Area: 12500m²
Photographer: Roland Halbe

Marco Polo Residential Tower

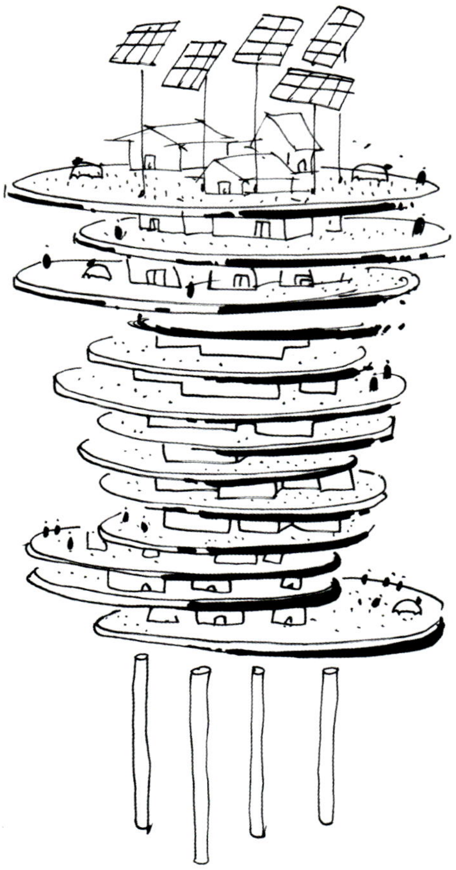

The 55m high tower punctuates the end of the route from the inner city out to the new attractions, the Cruise Ship Terminal and the Promenade on Strandkai. In its form and arrangement, it is a unique and remarkable sculptured building, adding to Hamburg's silhouette on the Elbe. Its 17 above ground levels, each turned a few degrees on their axis, allow all 58 apartments spectacular views over the harbor and the city. Generous perimeter terraces and balconies extend the living areas out into a soft play of lines, and lend the tower its distinctive image. A high degree of flexibility is offered with the possibility of increasing areas by connecting neighbouring apartments. External variations in appearance are reflected in the interiors, in that no level, or apartment, is quite like any other. Choices can range from two apartments to large penthouses with maisonette characteristics. Load-bearing structural elements and necessary fixed services have all been reduced as much as possible, so that the residents rtment one has an uninterrupted view over an open plan living room landscape, through generously sized glass panels, to the outside world and Hamburg's roofscape. The apartments are sold design-ready. Clients can design their new home, with the help of an interior architect, according to their own taste. The concept for residential spaces proposed by the Behnisch Architekten emphasizes natural light and views. The Marco Polo Residential Tower brings together high-class living accommodation and a holistic ecological building concept. The recessed façades are protected from direct sun by the overhanging terraces above so that additional sunshades are not necessary. Vacuum collectors on the roof, using a heat exchanger, turn heat into a cooling system for the apartments. Innovative sound insulated air louvres in the sleeping areas make natural ventilation possible without increased noise pollution from outside.

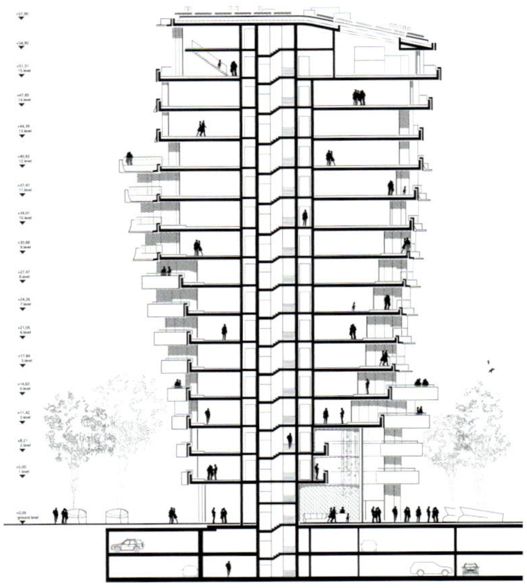

1m SECTION SOUTH-WEST

1m NORTH ELEVATION

1m EAST ELEVATION

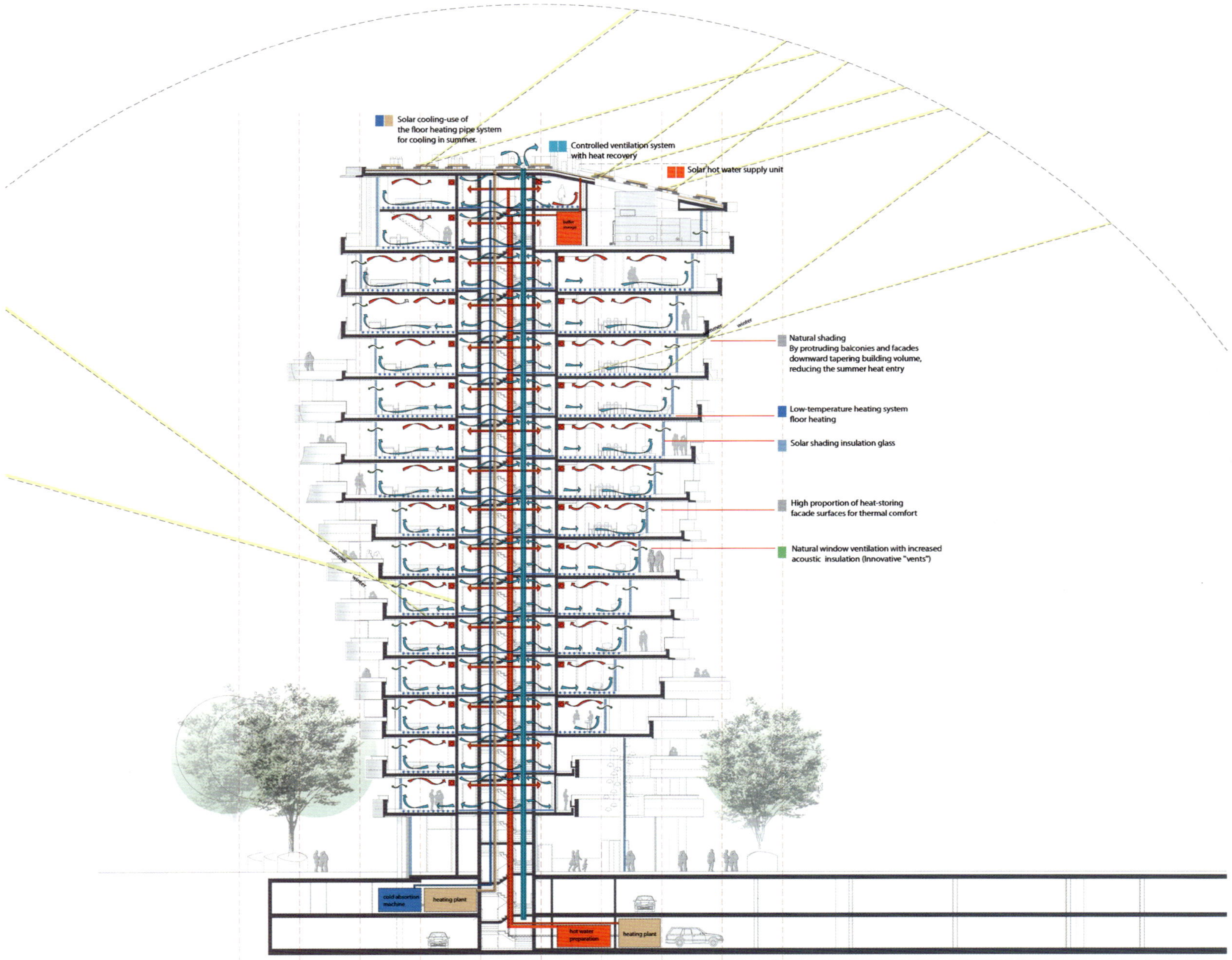

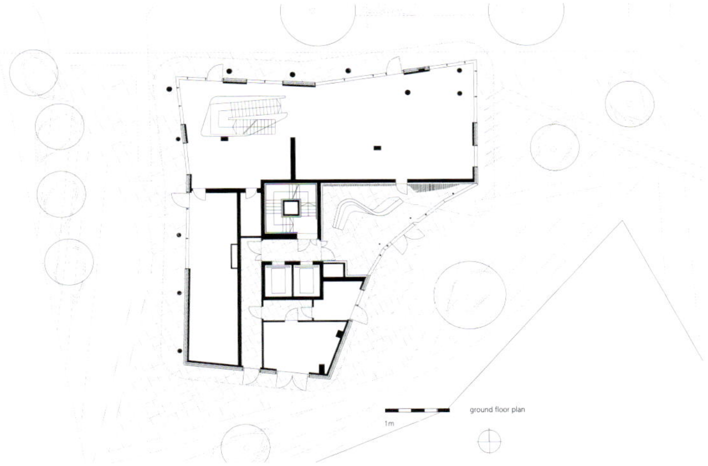
1m
ground floor plan

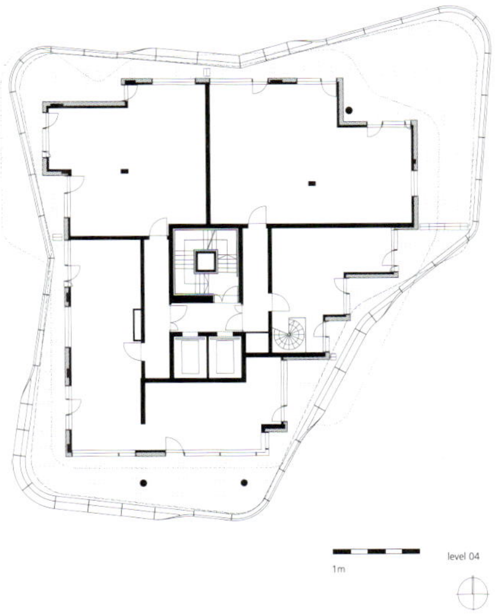
1m
level 04

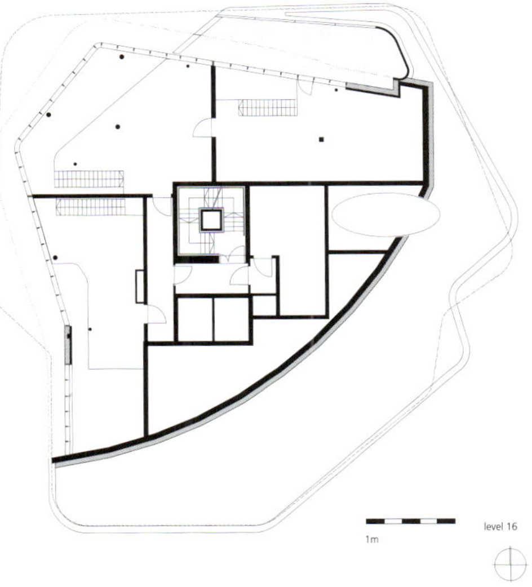
1m
level 16

Architect: Josef Weichenberger Architects
Firm: Josef Weichenberger Architects
Location: Sensengasse 1-3, Vienna, 9., Alsergrund, Austria
Area: 7217m²
Materials: mixed construction, large concrete slabs, site-mixed concrete
Photographer: Lisa Rastl, Ditz Fejer

Wohngarten Sensengasse

The proposed project responds to the described situation with a linear building structure along the Sensengasse and a free-standing, "point-house" positioned in an extended park.
The linear construction, composed of three connected individual buildings, creates a membrane-like break between the park area and the urban space. In doing so, particular attention has been paid to permeability with paths and visual connections in the direction of the park on the one hand and the partly two-floor ground floor area forms part of the open urban space.
These spaces with their views upon the surrounding greenery create a "green break" in association with the untouched and completely preserved tree avenue in the Sensengasse which fuse the façade of the structure and the airy opening of the "vertical green" together with the park to form one organic whole. The row of trees along the Sensengasse is taken up into the arrangement of the individual structures, so that the project reacts with sensitivity to the existing tree stock. The green area does not end at the perimeter of the property. Instead, it interweaves with the planted greenery of the old hospital, the sports ground and the school.
Through the combination of a sensible development together with an eventful layout of paths through the area, it was possible to maintain the quality of this urban recreational area. The free-standing point house marks the construction's boundary to the northwest and stands, as a result of the now clearly defined park perimeter, literally in the middle of greenery.

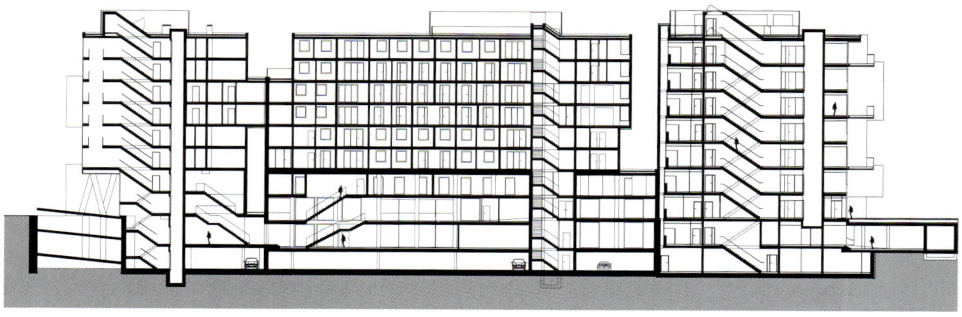

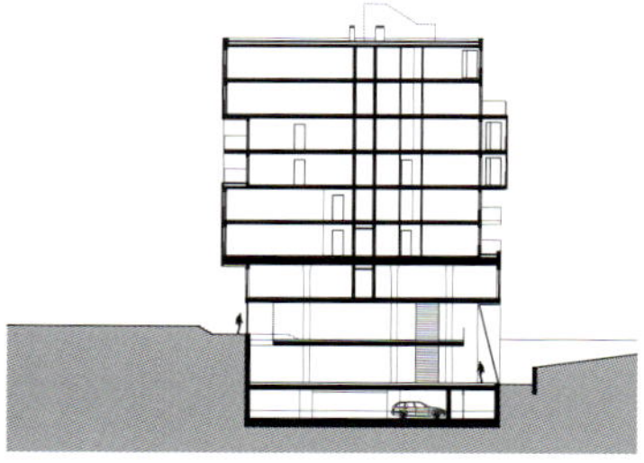

SECTION 2_2

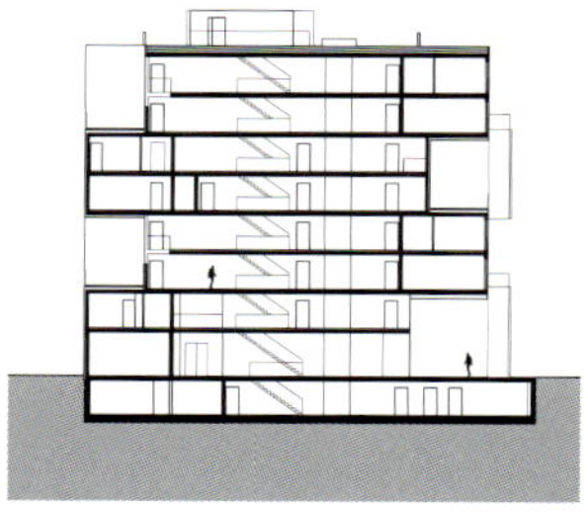

SECTION 3_3

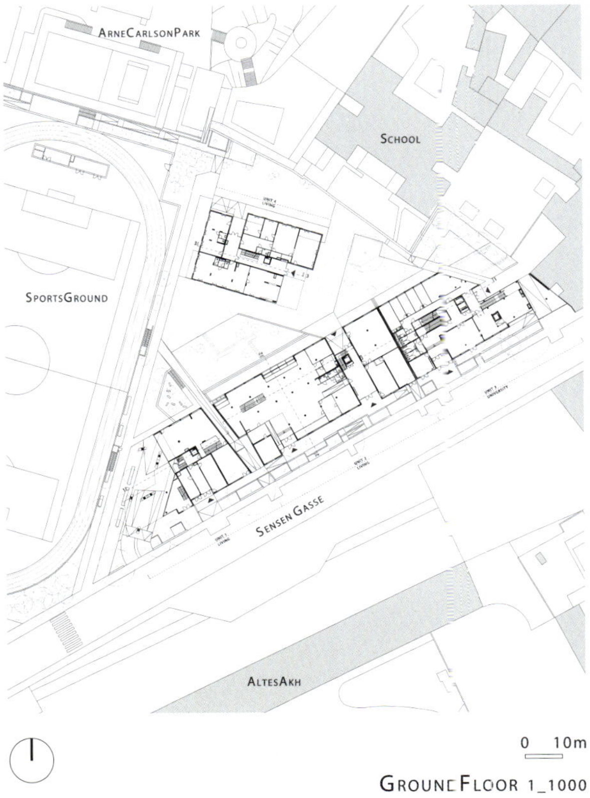

GroundFloor 1_1000

1_Steel piping d=42,2/2 galvanized or niro
set in 2 levels
2_EPS-plates sealed full-surfaced according to construction physics
3_U-profile or sword according to static
4_Fitting anchors in supporting rails
5_Ground plate on EPDM-base
thermically separated
6_Rosette
7_FSG-E30-Glass railings
8_Steel piping closed and welded
9_Pre-cast concrete segment
10_Weather groove
11_Sheet steel angle, 4 mm
12_Window sill in incline
protective steel sheet, 4mm
13_Sheet steel angle, EI90
3 mm, galvanized (into the **reveal**)

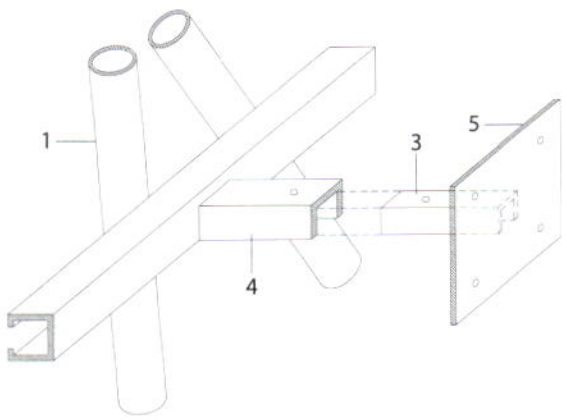

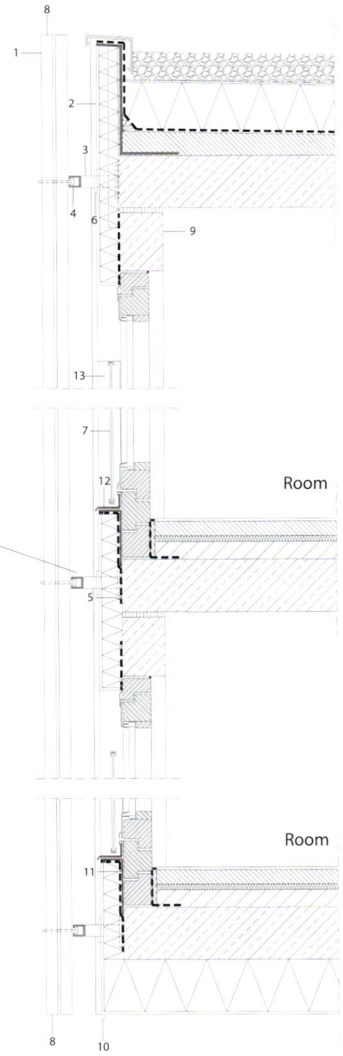

Detail 1_10
Façade element transection bay

1_Steel piping d=42,2/2 galvanized or niro
set in 2 levels
2_EPS-plate sealed full-surfaced
according to construction physics
3_U-profile or sword according to static
4_Fitting anchors in supporting rails
5_Ground plate on EPDM-base
thermically separated
6_Rosette
7_FSG-Glass railing
8_Steel piping closed and welded
9_Pre-cast concrete segment
10_Weather grooving
11_Sheet steel angle 4 mm
12_Window sill in incline
protective steel sheet, 4mm
13_Sheet steel angle, EI90
3 mm, galvanized (into the reveal)
14_Flat steel angle according to static
15_Steel-U-profile galvanized and laminated
16_Waterspoud
plumbing integrated
17_Shrouding galvanized
18_Concrete slabs in gravel bed
19_U-profile with plates at the railing stands,
galvanized and laminated
20_Railing stands steel square pipe 50/50
galvanized and laminated
21_Railing stands
plumbing integrated
22_Steel-profile
bolted down with countersunk screws

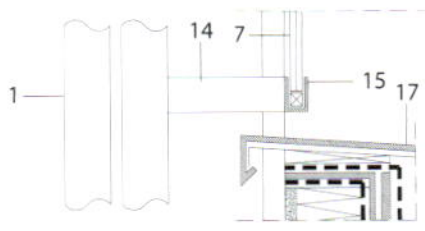

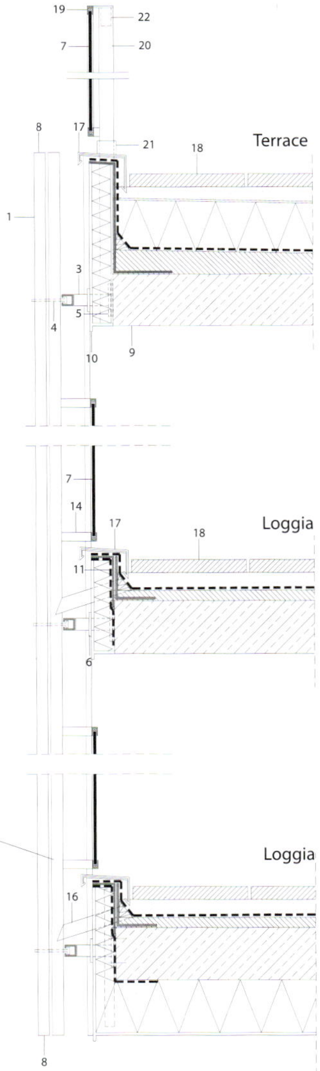

Detail 1_10
Façade element transection bay

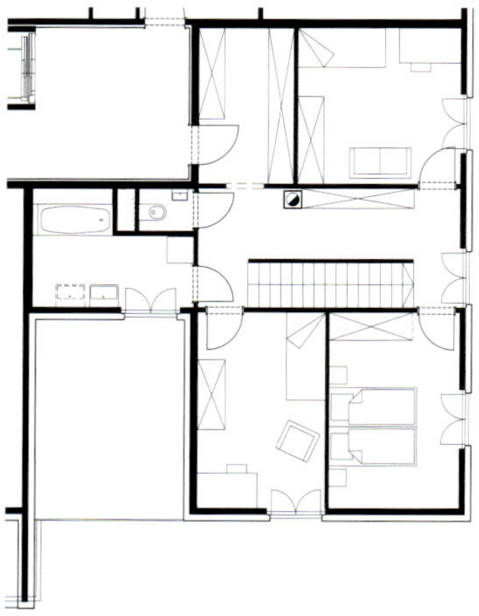

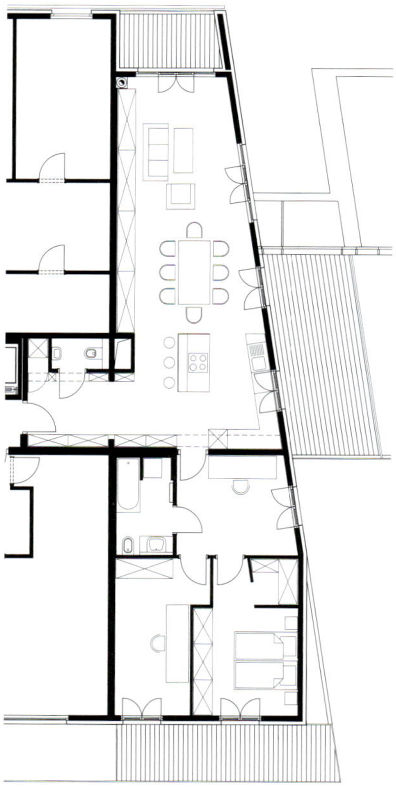

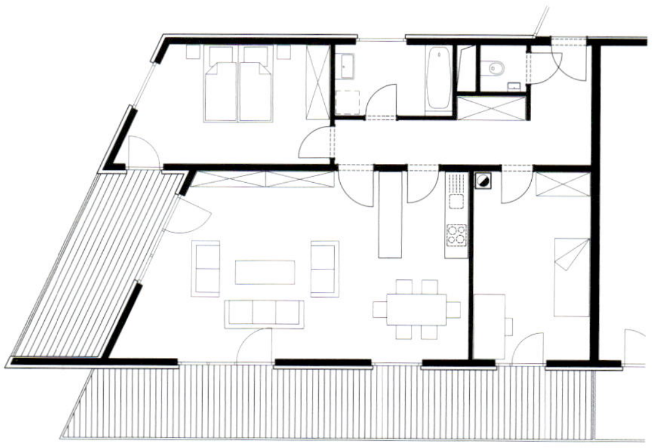

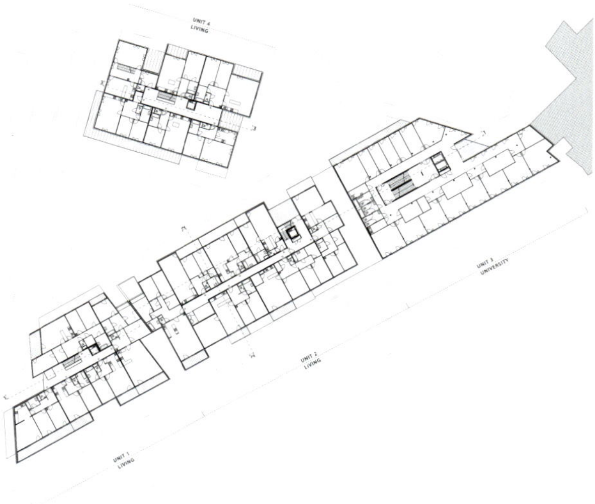

Architect: Jakob+Macfarlane architects
Firm: Jakob+Macfarlane architects
Engineering: BET Choulet
Acoustic: J.-P. Lamoureux
Structure: Batiserf
Landscape: Cap Paysage
Location: site of the ancient hospital Hérold, Paris 19ème arrondissement
Area: 6600m²
Materials: steel concrete, volcanic stone, metal façade
Photographer: Nicolas Borel

Herold Social Housing

The programme set by the client for the 100 apartments project calls for the design of apartments of varying size, with the ground floor being reserved to accommodate apartments for handicapped people.

The site planning involved the creation of three separate buildings on fragments of the existing site. The fragmentation of the site resulted from various factors such as site views, Hausmanian setback rules, the preservation of old-growth trees and a stringent site development code. After studying the requirements for development and the difficulty they imposed, the architects decided to conceive the buildings in a way that accentuates the existing urban condition and that of the site, creating a relation between the site and its developmental limitations. The project is conceived as a large-scale urban 3D matrix, to which applied a careful 'cutting' and 'coring' process. The finished project is a result of both existing urban and ecological factors, which register as conceptual starting points and determining factors in the creation of the design of new urban space.

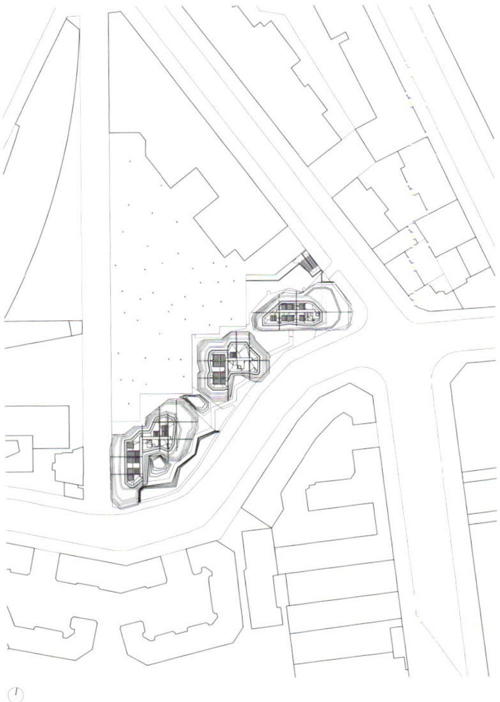

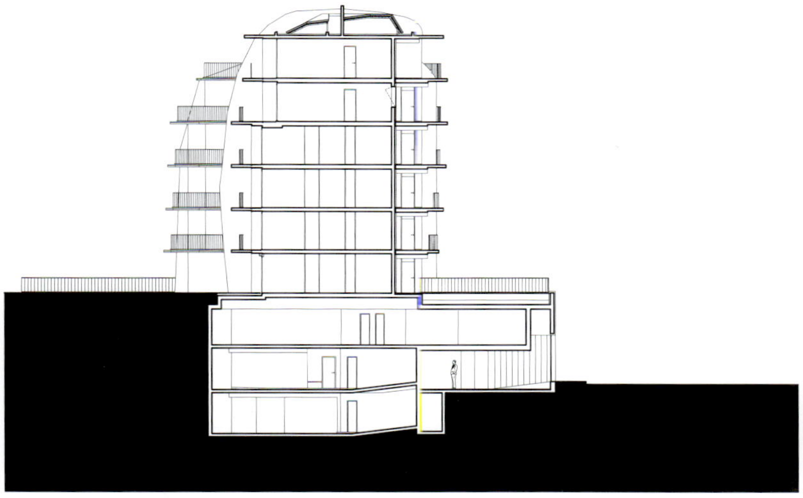

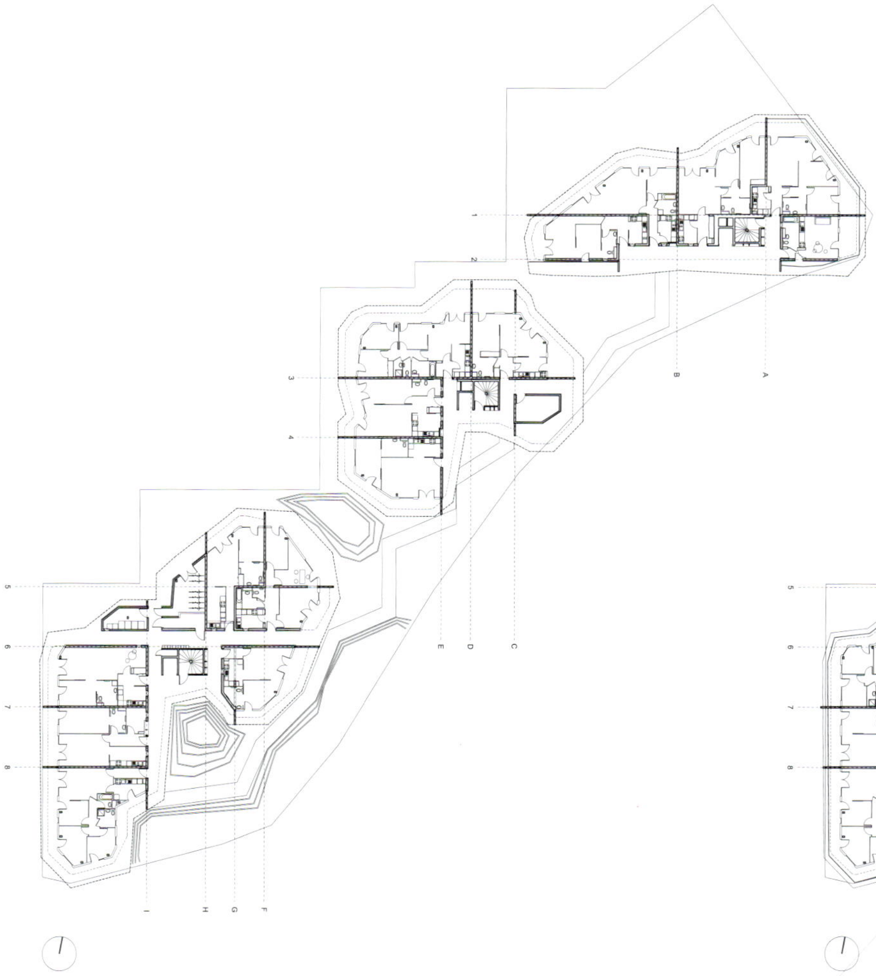

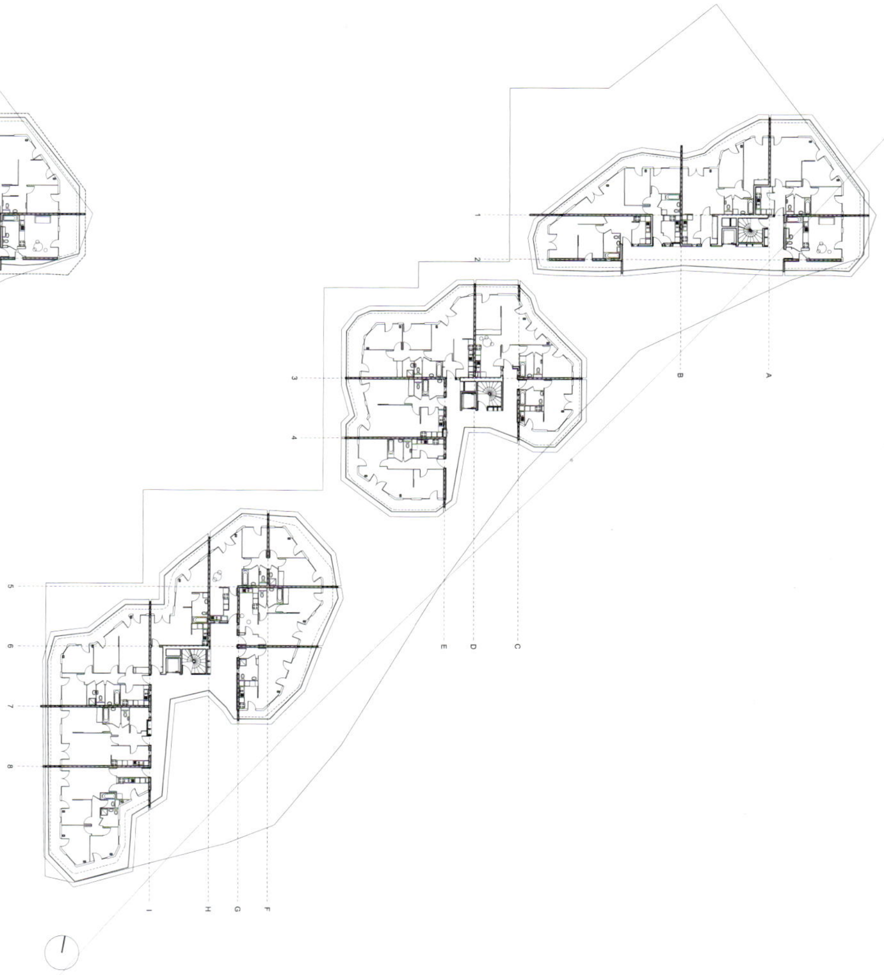

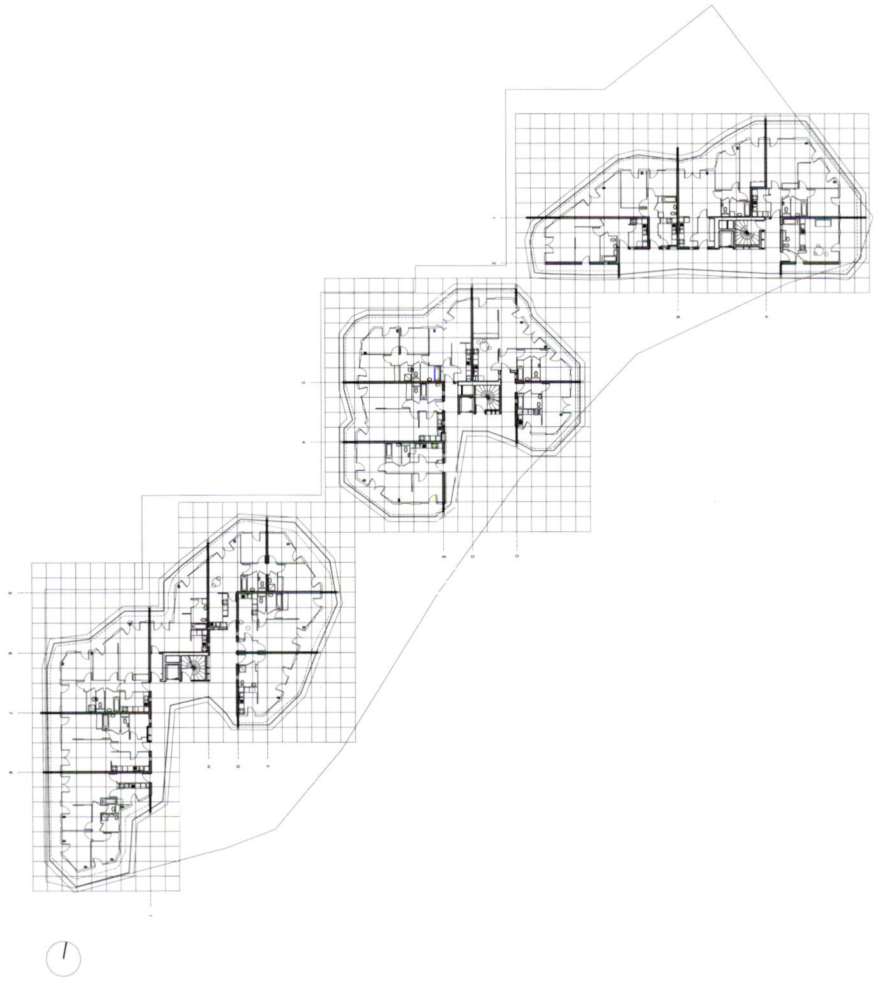

Architect: SAKO Architects
Firm: SAKO Architects
Location: Beijing, China
Area: 100360m²

MOSAIC in Beijing

The motif of MOSAIC has been used everywhere. Paper snowstorm flies up, light, weak, transient, but clear, full of actual feelings of existence. Office was changed into one room apartment, and store would be rent to tenants. It says, mosaic is also meaning a collection of small capital.
The facade of 4th story store will be occupied by inside lighting ads, therefore it achieves a goal of lifeline in where tenants could show their existence to outside people. And the same, motif is hiding behind as well. There are 2 moving lines being added into store. Main entrance seems a large opening mouth to the outside, and people can access directly from B1 to 3rd floor. People's moving and other various view are divided, amplified, or radiated through complicated reflecting faces.
The motif of mosaic is running through sign, landscape as well as interior.

0m 30m

Landscape 1/800

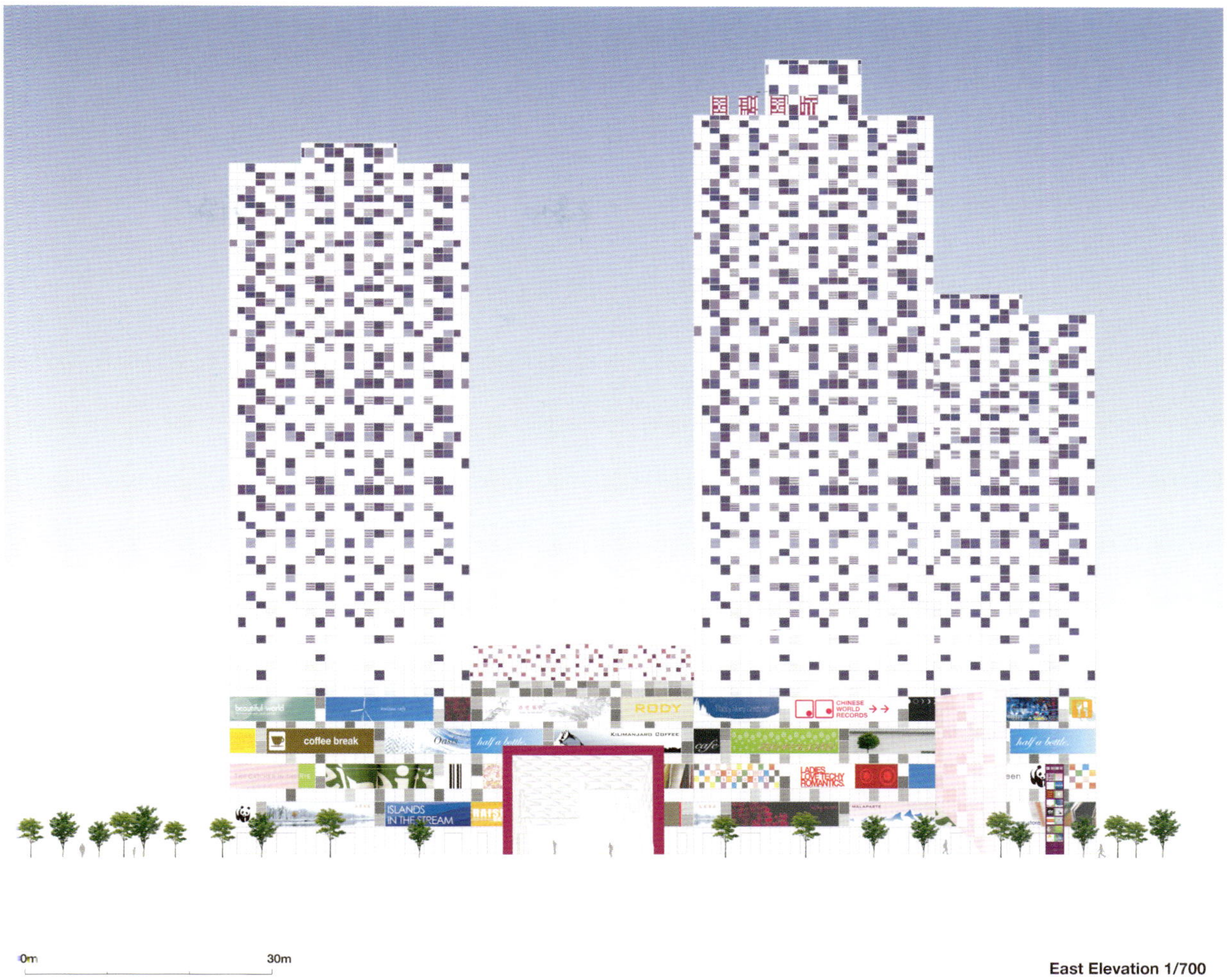

East Elevation 1/700

1019

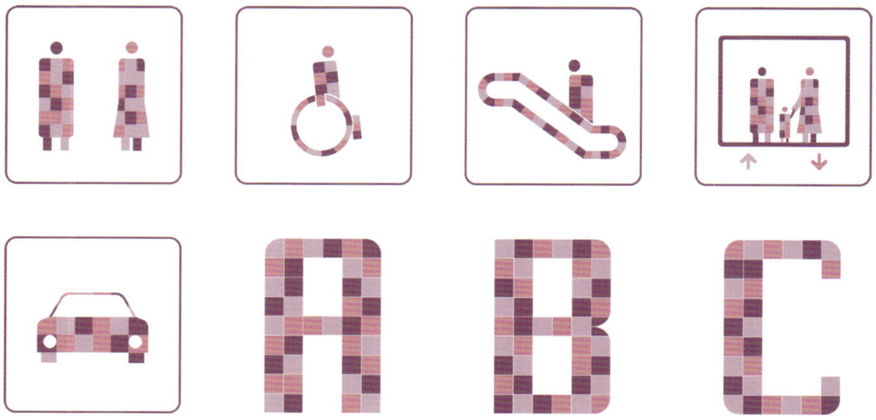
A
B
C

B 07

Architect: Rok Oman, Spela Videcnik
Firm: OFIS architects
Location: Bohinjska Bistrica, Slovenia
Area: 7500m²
Photographer: Tomaž Gregorič

Shopping Roof Apartments

The initial task from the client was to build a new shopping mall on the plot of the existing one.
Furthermore the new project proposed use of the shopping roof for additional volume as new apartments.
The wooden mass is located centrally in the village of Bohinjska Bistrica in the alpine area of Lake Bohinj. The village has beautiful views towards the mountains but unfortunately it was architecturally damaged in the 1960s with a textile factory and housing blocks, that were build around them. The textile factory was later closed down and the plot later occupied the old shopping mall and today the new one. The side views therefore are not very nice.
The organization of the housing and the envelope of the apartments open towards mountain views and the sun. Therefore the front, wooden facade is mostly transparent with panoramic windows. The side facade is closed, the windows open towards the balcony that is cut into the volume. From side windows also view open to the mountains.
The stepped volume of the building follows the silhouette of surrounding landscape. On top of the shopping mall apartments are set in the form of stepped L-volume. From the west strong wind and snow arrive the facade which is opened only towards enclosed balconies and its material is gray slate which is designed as a vertical roof. L-shape volume encloses inner communal garden that is the roof of the shopping mall.
The front and courtyard facade is warm and opened, made of wooden verticals with different rhythm.

POKLJUKA
PL. LIPANCA
2050
PL. KONJŠČICA
Rudno Polje
Mrzli Studenec
SPORT HOTEL
PL. PRAPROTNICA
PL. ZAJAMNIKI
KOPRIVNIK
GORJUŠE
Pojelje
USKOVNICA
NOMENJ
Sv. Martin
SREDNJA VAS
BOHINJSKA ČEŠNJICA
Jereka
BLED
Sv. Pavel
STARA FUŽINA
Senožeta
Bitnje
APART HOTEL TRIGLAV
BOHINJSKA BISTRICA
Brod
Sv. Janez
JEZERO
SLOM
Savica
Kamnje
Sv. Nikolaj
BELLEVUE
Polje
Žlan
Nemški Rovt
RIBČEV LAZ
Laški rovt
ski center KOBLA
Ravne
Črna prst 1844
Poljanski vrh 1897
DOM Z. JELINČIČA
ŽELEZNIKI
ski center SORIŠKA PLANINA

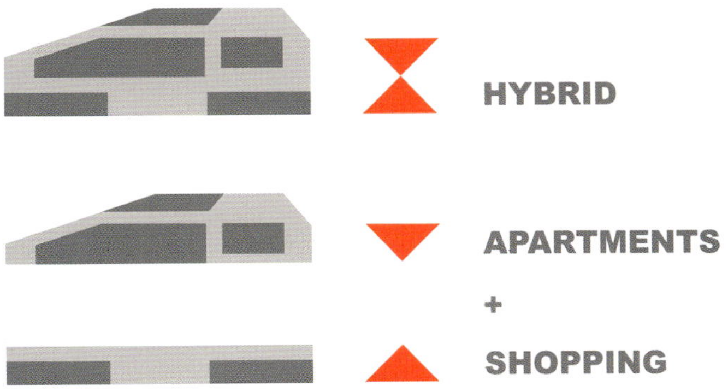
HYBRID
APARTMENTS
+
SHOPPING

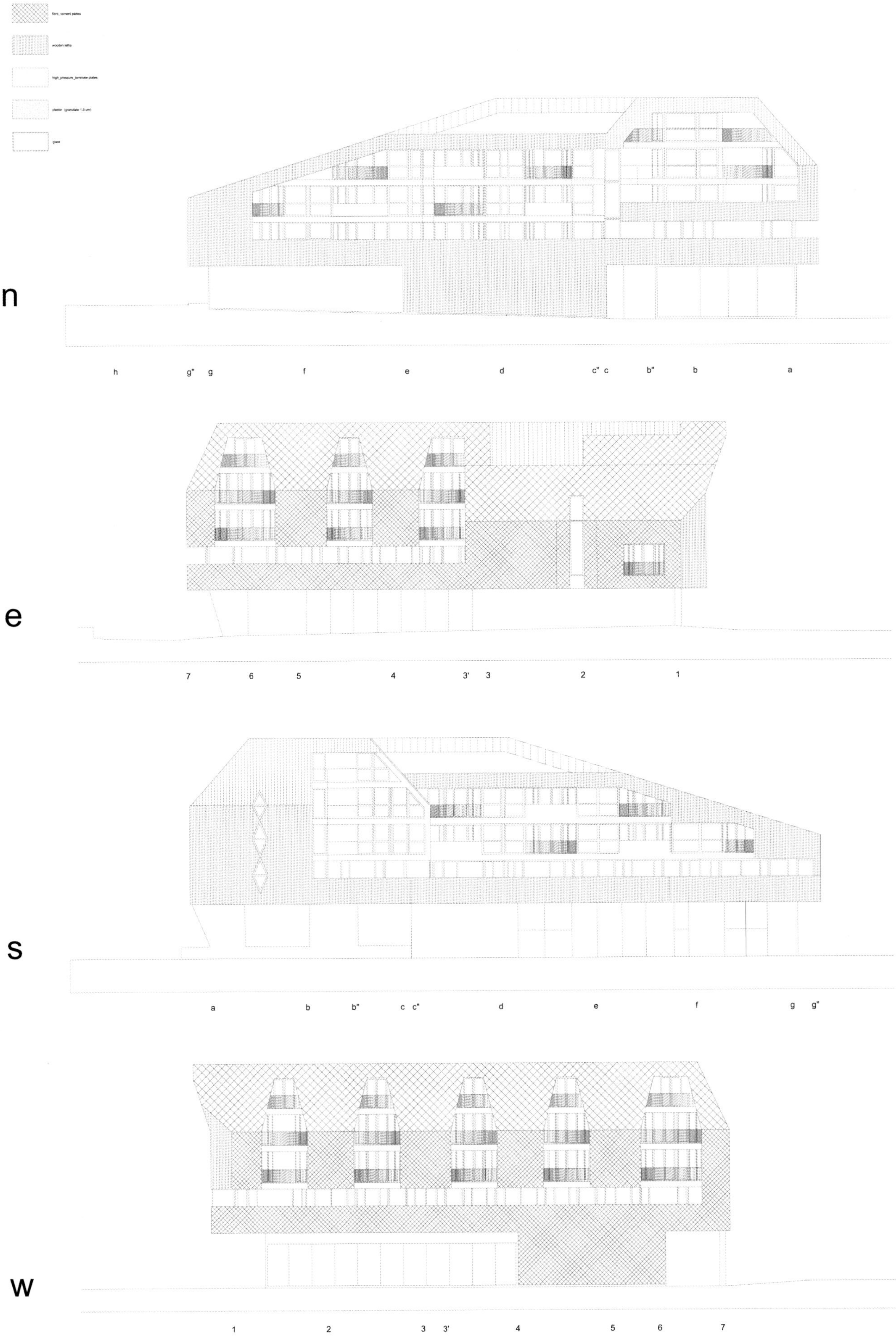
fibre_cement plates
wooden laths
high_pressure_laminate plates
plaster (granulate 1.5 cm)
glass
n
h g" g f e d c" c b" b a
e
7 6 5 4 3' 3 2 1
s
a b b" c c" d e f g g"
w
1 2 3 3' 4 5 6 7

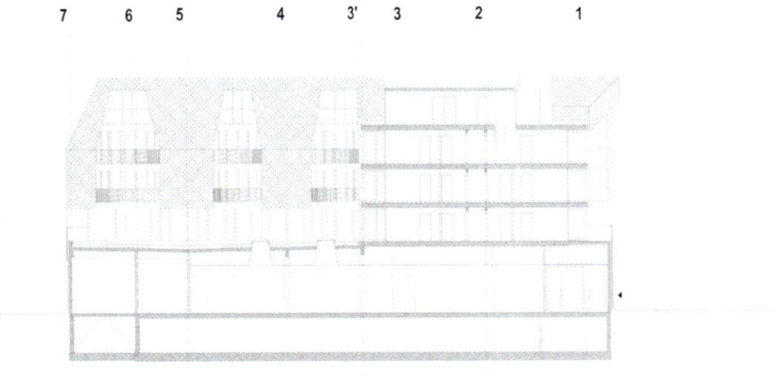
7 6 5 4 3' 3 2 1

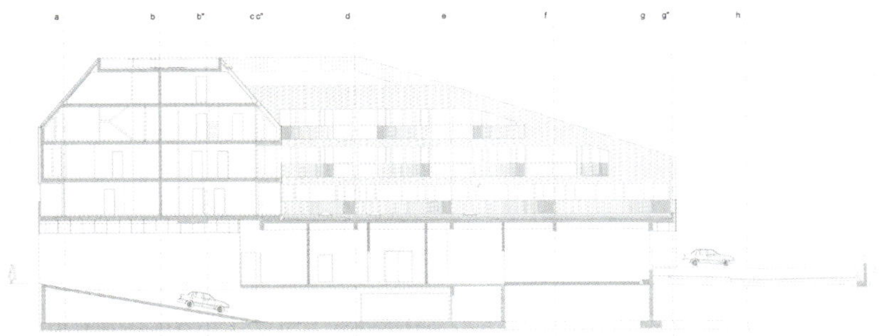

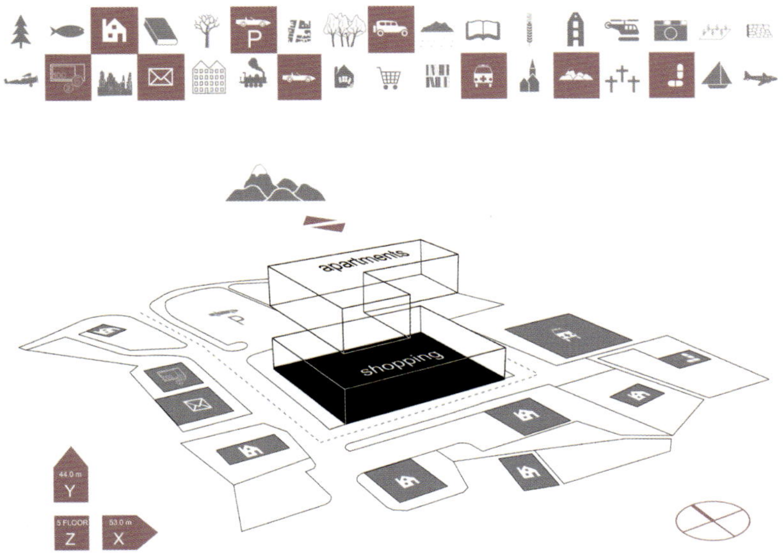
apartments
shopping
P
Y
Z
X

Df-2 DETAJL BALKONOV S KOVINSKO OGRAJO - FASADA JUG

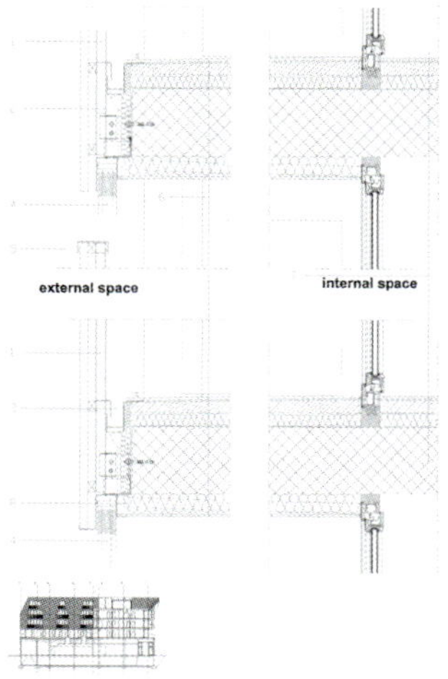

1 -steel structure to support fence, profiles 40x40mm

2 -thermal insulation 10.0cm
-hydro insulation
-ventilated cavity 5cm
-steel support structure
-larch timber 30x40 mm

4 -standard PVC window frame with double glazing
space for external shading sistem

5 -steel structure to support fence
-facade - larch
-sheet metal cover

6 balcony
-stone tiles 1,0cm
-glue 0,5cm
-concrete estrih 5,5-4,0cm
-thermal insulation 5,0cm
-polyurethane sheeting 0,05cm
-raiforced concrete slab 26,cm
-thermal insulation 8,0cm
-facade plaster

2 -sheet metal rainwater channel

7 -parquet 1,0 cm
-glue 0,5 cm
-concrete layer 6,0 cm
-polyurethane sheeting 0,02 cm
-sound insulation 5,5cm
-reinforced concrete slab 26,0 cm

8 -thermal insulation
-plaster

Df-2 DETAJL FASADE IN STREHE - ZAHODNA FASADA

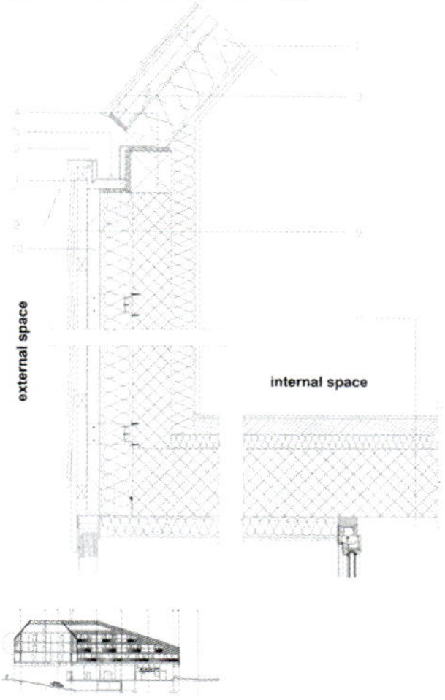

1 -timber beam 12/16 cm

3 roof construction
-cement composite sheets 400x400 mm
-timber structure, 5/4cm 4,0cm
-ventilated cavity, 5/5cm 5,0cm
-rainproof foil
-timber beam, 14/18cm 18,0cm
-thermal insulation 18,0cm
-vapour barrier
-construction for plasterboard
-plasterboard, fire resistant 2x1,5cm

4 -timber beam 16/16

5 -gutter, heated

6 -steel construction

7 -timber board

8 -ventilated cavity

9 Facade
-cement composite sheets 400x400
-timber structure 40/60 4,0cm
-rainproof foil
-thermal insulation 10,0cm
-reinforced concret wall 15,0cm
-plaster

10 -vertical outflow

11 -parquet 1,0 cm
-glue 0,5 cm
-concrete estrih 6,0 cm
-polyurethane sheeting 0,02 cm
-sound insulation 5,5cm
-reinforced concrete slab 26,0 cm

Df-1 DETAJL BALKONOV S KOVINSKO OGRAJO IN FASADE - ZAHODNA FASADA

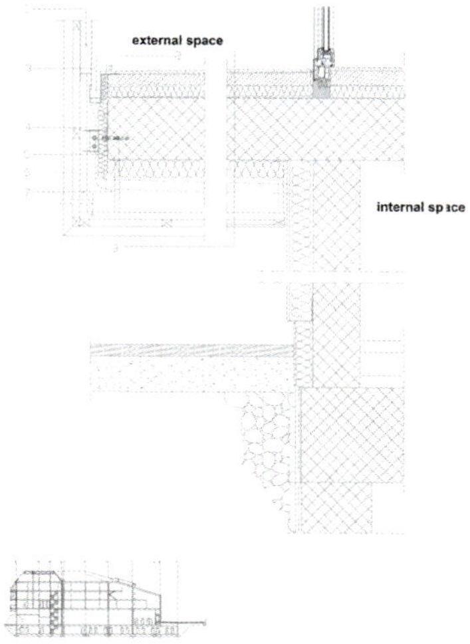

1 -steel fence structure
-timber vertical 2/5cm
-cement composite sheets

2 -rainwater gutter

3 -outflow

4 -steel profile

5 -thermal insulation (XPS)

6 -screw

7 -steel construction

8 balcony
-stone tiles 1,0cm
-glue 0,5cm
-concrete estrih 5,5-4,0cm
-thermal insulation 5,0cm
-polyurethane sheeting 0,05cm
-reinforced concrete slab 26,cm
-thermal insulation 8,0cm
-plaster 1,0cm

Df-10 DETAJL OKNA - FASADA JUG

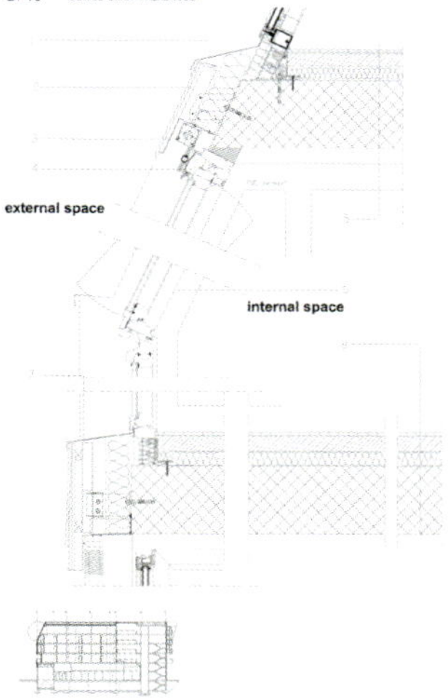

1 -metal sheet

2 -thermal insulation 10.0cm
-hydroinsulation foil
-ventilated cavity 5cm
-steel profile fixed in reinforced concrete slab
-cement composite sheets 400x400 mm

3 -external shading sistem

5 -parquet 1,0 cm
-glue 0,5 cm
-concrete estrih 6,0 cm
-polyurethane sheeting 0,02 cm
-sound insulation 5,5cm
-reinforced concrete slab 26,0 cm

6 -roof window

7 -metal sheets

8 -parquet 1,0 cm
-glue 0,5 cm
-concrete estrih 6,0 cm
-polyurethane sheeting 0,02 cm
-sound insulation 5,5cm
-reinforced concrete slab 26,0 cm

Df-3 DETAJL BALKONOV S KOVINSKO OGRAJO - ZAHODNA FASADA

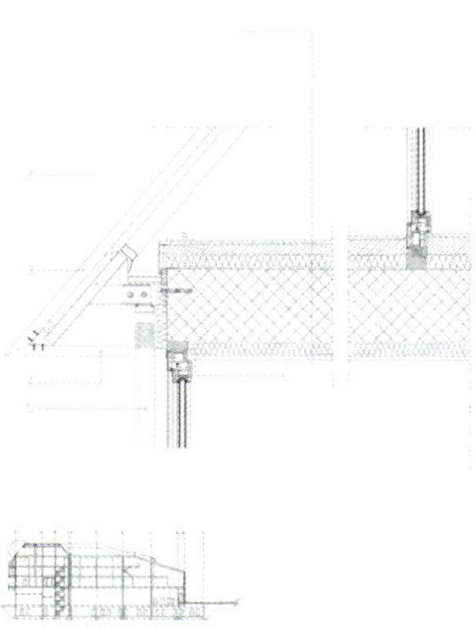

1 balcony
-stone tiles 1,0 cm
-glue 0,5 cm
-concrete estrih 5.5-3,0 cm
-thermal insulation 5,0 cm
-polyurethane sheeting 0,05 cm
-reinforced concrete slab 26,0 cm
-thermal insulation 8,0cm
-plasterboard 1.2cm

2 -steel construction for larch timbers

3 -timber

4 -timber ceiling

5 -external shading sistem

6 -filled with propen

Df-4 DETAJL STREHE - ZAHODNA FASADA

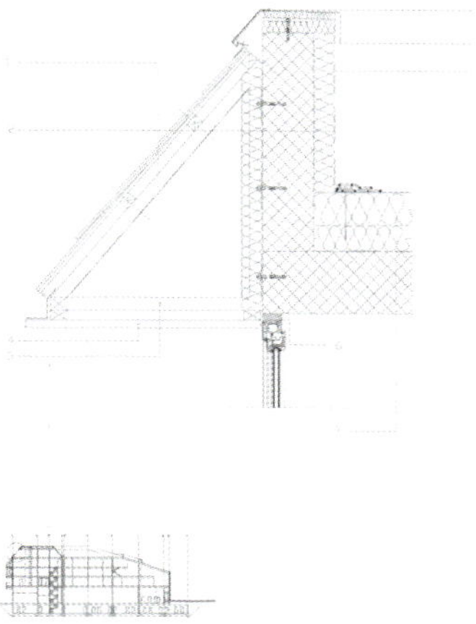

1 -cement composite sheets 1,3cm
-timber construction, 5/4cm
-timber construction, 5/8cm
-polyurethane sheeting
-timber beam, 14/18cm 14,0cm

2 -thermal insulation 8,0cm
-reinforced concrete wall 20,0cm
-thermal insulation 8,0cm
-rainproof foil (Sica)

4 -ceeling timber horizontals 30x40

5 -steel construction for celling

6 -filled with propen

7 -hydroinsulation
-thermal insulation 20,0cm
-vapour barrier
-reinforced concrete slab 24,0cm

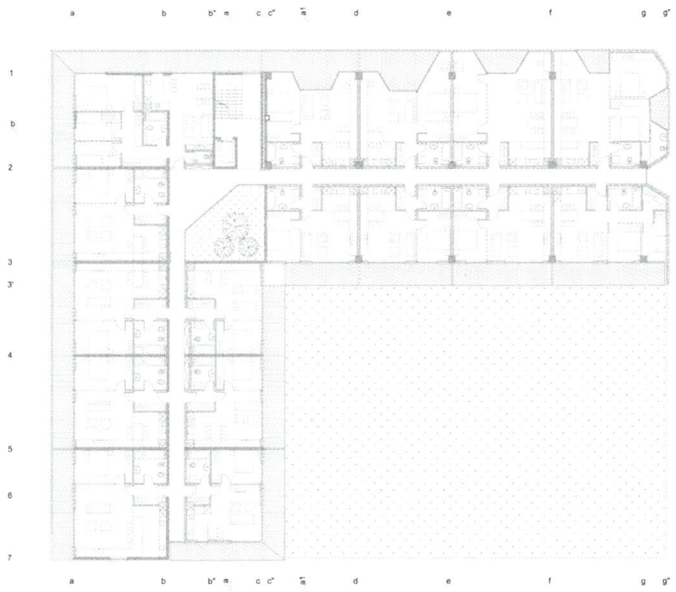

1

3

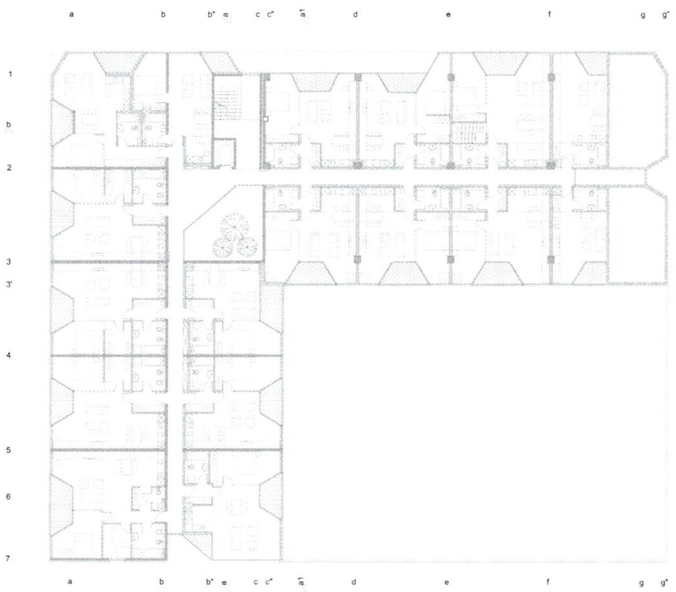

2

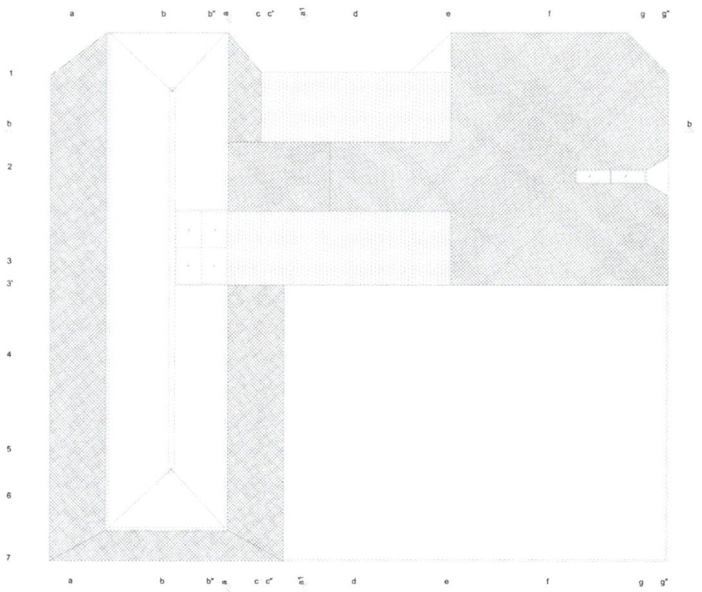

R

Architect: Marlies Rohmer
Firm: Architectural office Marlies Rohmer
Location: De Uithof, Utrecht
Area: 18600m²
Photographer: Scagliola / Brakkee, René de Wit, AkzoNobel

Smarties

The building of student housing on the Utrecht University site turned De Uithof into a fully-fledged campus, simultanously tackling the chronic housing shortage for young people in this city. The new complex of 380 independent and clustered rooms manifests itself as a solitary mass in the 'strip of objects' of the OMA master plan (which also includes the Unnik tower and the Educatorium). A four-storey tall concrete 'leg', which projects by eight metres, shelters a range of communal and commercial spaces.

The facade consists of a grid of multicoloured aluminium panels in which the windows are omitted. Seen from a distance, the colours blend into a grey, scaly skin. The closer you come, the more it appears as a colourful honeycomb for the bright young students – people's 'smarties' – from all over the world.

Beneath this lively skin, the interior of the mass fosters encounter and interaction at every scale. Party rooms and niches along the staircases and corridors make a microcosm in which young love can flourish and lifelong friendships can develop.

De Uithof used to be a monotonous cluster of university buildings on the city margins of Utrecht. The separate buildings bore little relation to one another. Rem Koolhaas (OMA) and Art Zaaier devised a concept which would give De Uithof more coherence and character. Since then the object has been to achieve greater contrast between the built–up zones and open areas. The existing landscape qualities must be accentuated, but this must be balanced by compact building in functional clusters to enhance the level of urban interaction and eliminate the feeling of an urban desert.

The central section consists of the Kasbah zone, a dense strip in individual buildings are closely spaced, and the 'Object' zone, in which the buildings are of a similar scale but grouped separately around a walkable public strip. A hallmark of the central area is that the buildings also communicate internally through a continuous walkway at first floor level.

Hart voor de klant
7848

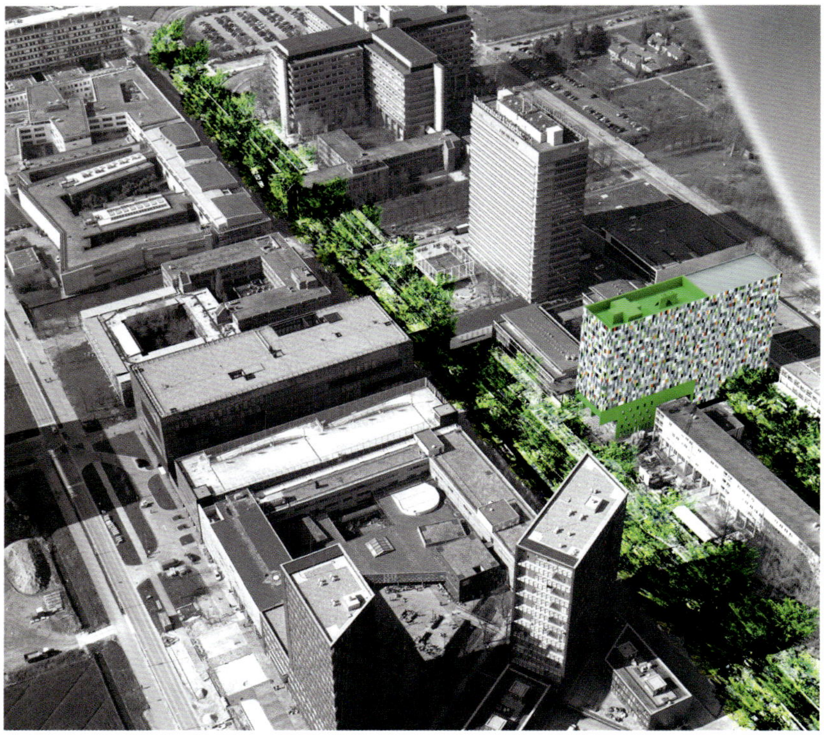

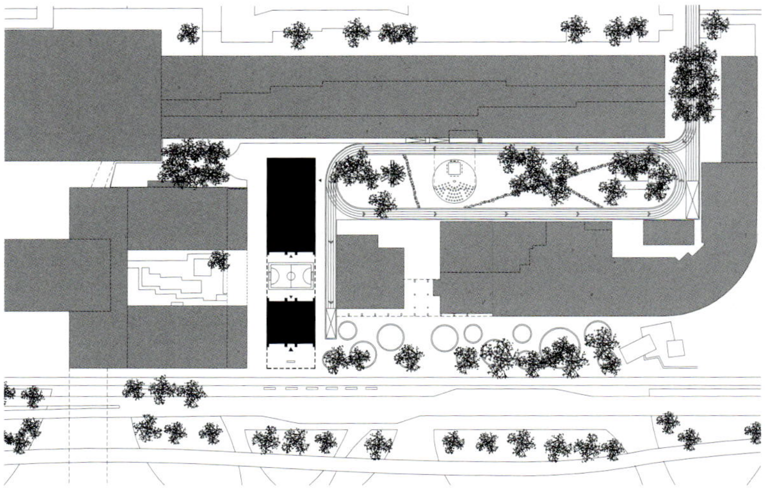

FLEXIBILITEIT

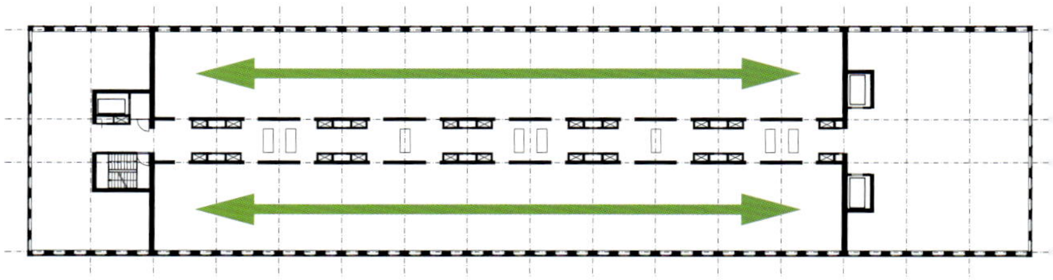

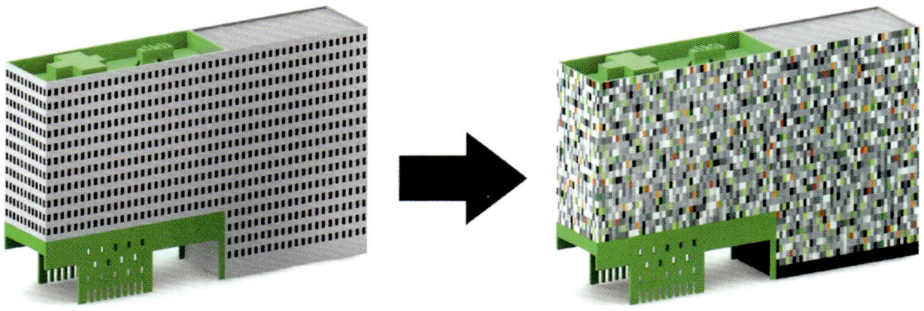

1510 RAMEN 'VALLEN WEG'

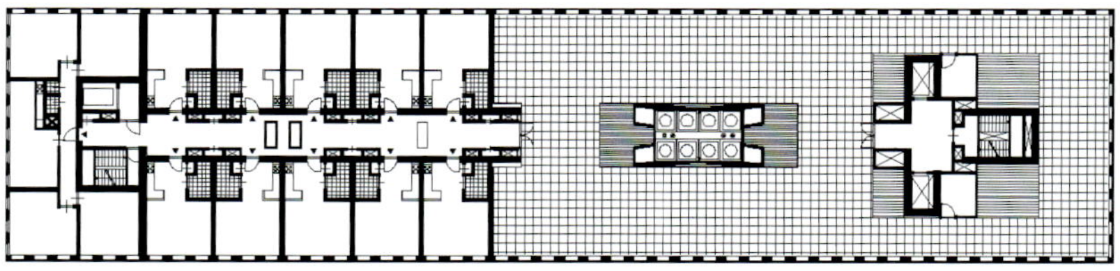

DAKTERRAS

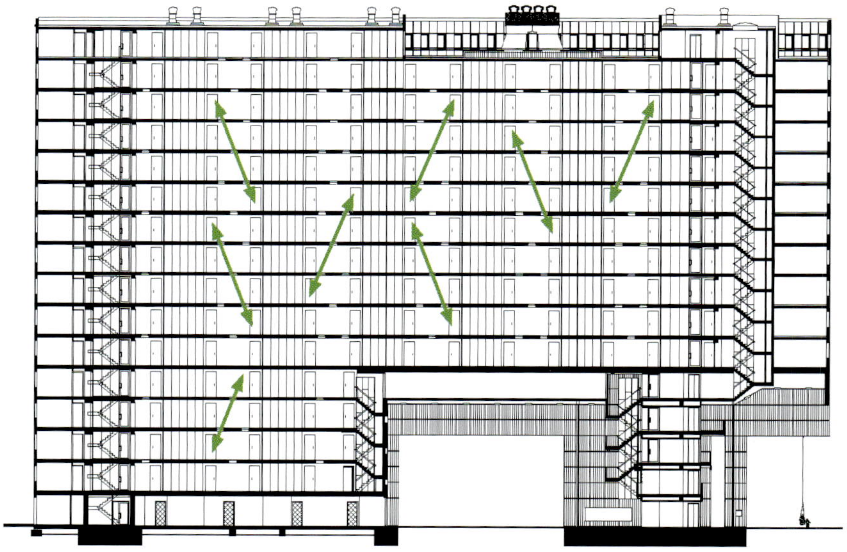

DOORSNEDE: GLAZEN VLOERDELEN VERBINDEN DE GANGEN ONDERLING

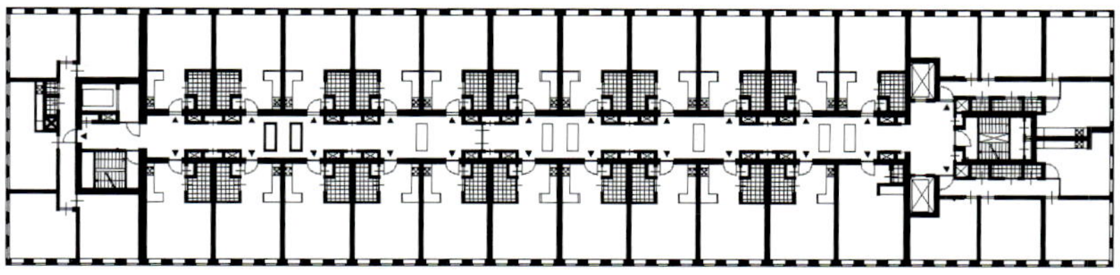

STANDAARDPLATTEGROND

11
226
221
11

Architect: Andrej Kalamar
Firm: Studio Kalamar
Location: Murska Sobota, Slovenia
Area: 6370m²
Materials: reinforced concrete, ceramic tiles, gravel
Photographer: Miran Kambič, Studio Kalamar

Cvetkova Apartments

Apartment buildings are located in a redeveloped downtown area that has recently been converted from derelict industrial to public and residential use. Despite its central position in the city, the site is removed from main roads and downtown bustle. Next to the new music school on the north of the city block, three apartment buildings rise from a green surface. Their positions reflect the heterogeneous surroundings; variety of directions influences the varying orientations of the volumes. As there are no parallel facades, each apartment can establish an individual character without obstruction. Lower floors comprise four apartments each, top floor is divided into three apartment units, all with flexible floor plans. As a result, long facade surfaces ensure each apartment abundant natural light. The smooth facades with enclosed loggias enfold the volumes into a shimmering ceramic coat, varying in colour tone and reflectivity.
Transparent fence around the complex prevents exclusion from surrounding city life while allowing unsupervised children's play. Gently sloped green areas provide various opportunities for that, and a small playground for the youngest children is located in the eastern corner of the site.
Parking and storage spaces are arranged below ground, serving as a base for all three volumes. Special attention was accorded to energy use of the complex. Each building has a separate boiler room, and cooling is arranged individually. The compact facade and energy supply scheme both stimulate economical consumption and thus reduced energy use.

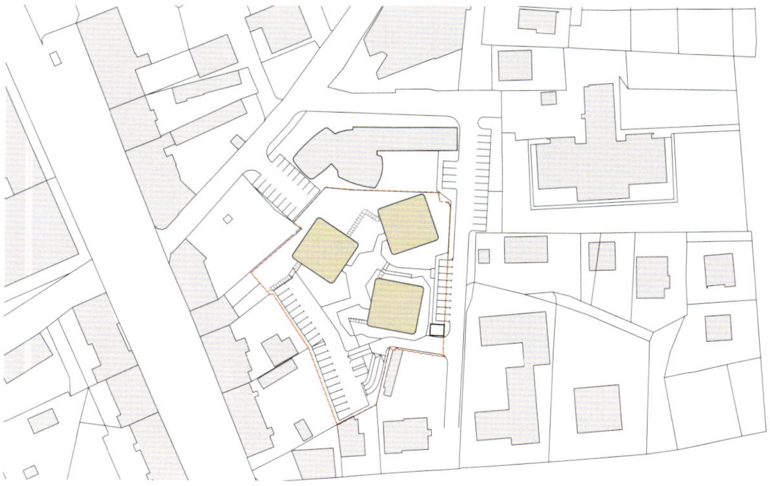

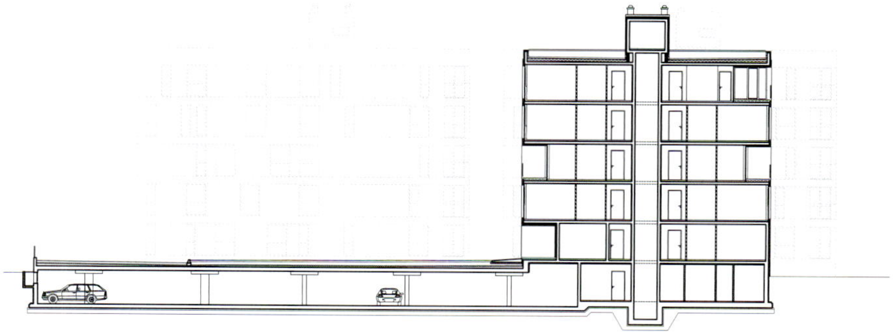

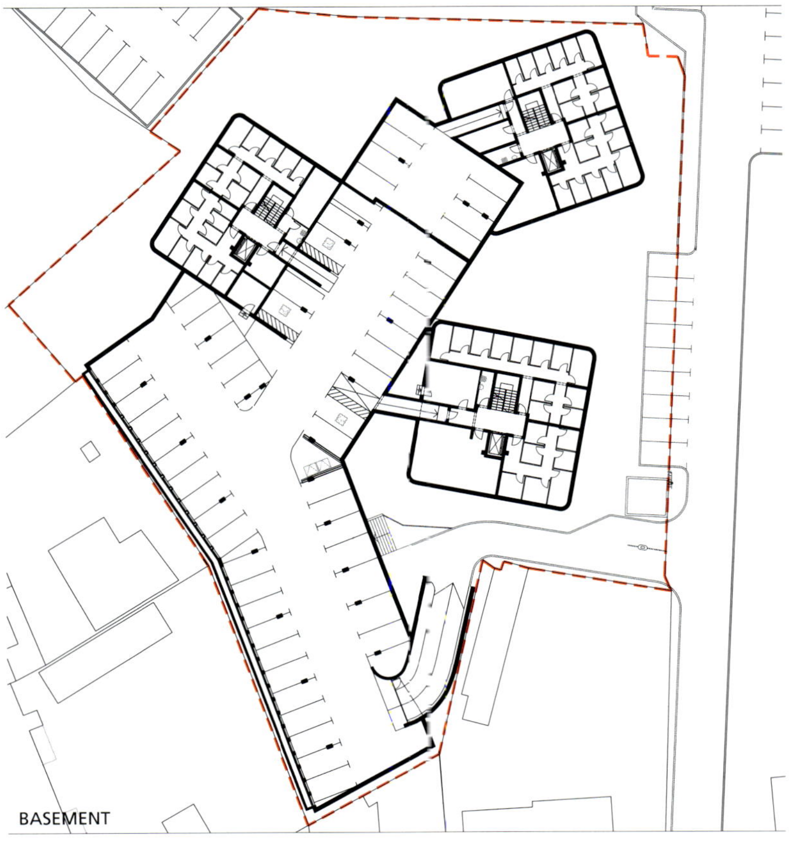
BASEMENT

GROUND FLOOR

FIRST FLOOR

GROUND FLOOR

Architect: Jaime Magén
Firm: Magén Arquitectos
Collaborator architect: Francisco J. Magén.
Developer: Sociedad Municipal Zaragoza Vivienda
Quantity surveyor: Fernando Bardavío, Felisa Sánchez, Miguel A. Romero y Adrián Zorraquino
Construction company: Construcciones Galve y Gil.
Site manager: Zorba Shabaro
Location: Calle Biescas 15-29, Zaragoza
Area: 9194m²
Photographer: Roland Halbe www.rolandhalbe.com

68 Social Housing

The orientation of the building supports the idea of contrast between the Northern closed exterior and the uneven and open to the South interior. The housing units participate in this double orientation, with the bedrooms to the North and the daytime rooms towards the South, facing the garden space. The bathrooms ocupy a central position that allows a double circulation around them, as well as the extension of the lines of sight in the housing, giving a sensation of openess despite its reduced surface.

On the exterior the building presents a closed configuration that seeks a discreet integration within the immediate environment. These façades follow the regulated alignements and are configured from the repetition in series of the openings. The fencings are built with white bricks and the base on the ground floor is made of prefabricated concrete panels.

The entrance halls are located in the places where the building delineates its height to adapt to the slope. These are of wide dimensions and connect the street with the interior garden space which gives access to the dwellings and creates a sequence between the public and private spaces. The differences of heights are solved by soft entrance ramps to the doorways.

In the interior space the building is characterized by some uneven walls defined by a poligonal. The seen structure with concrete slabs and panels configures a middle space between the housing and the exterior in front of the daytime rooms. The fencings of these spaces and doorways are characterized by its finishing with black stony mono-coating. The different directives of the façade and the variable positions of the sliding aluminium panels give movement to the façade and a changing aspect throughout the day.

The configuration of the building favours good thermal conditions: the closed northern façade isolates from the noisy environment and adds thermal inertia, while towards the South, the aluminium sheets, the slabs of the terraces and the deciduous grove allow the sun´s exposure in the winter while in the summer it is alleviated.

23

WEST ELEVATION - PADRE HUARTE STREET

1/400

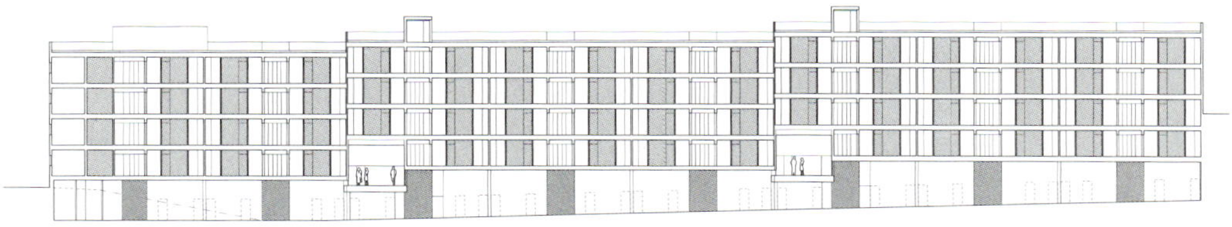

SECTION L1

1/400

SOUTH ELEVATION - INNER SEMI-PUBLIC SPACE

1/400

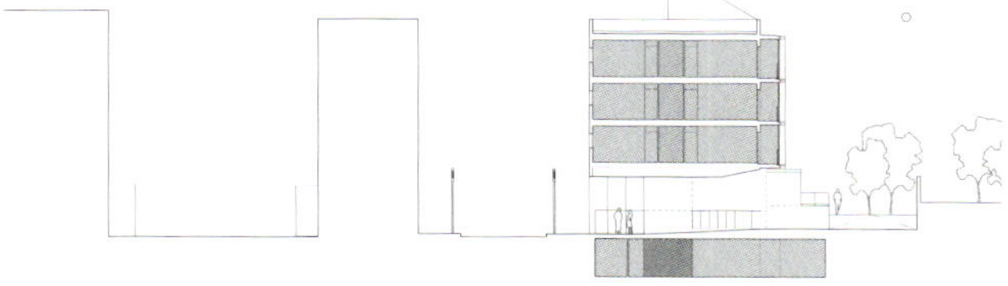

SECTION T1

1/400

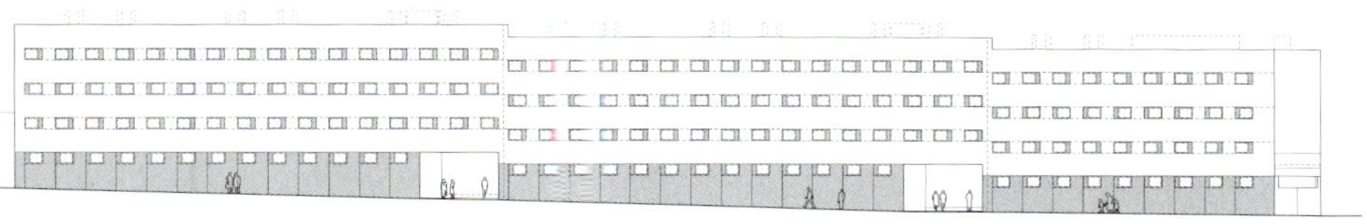

NORTH ELEVATION - BIESCAS STREET

1/400

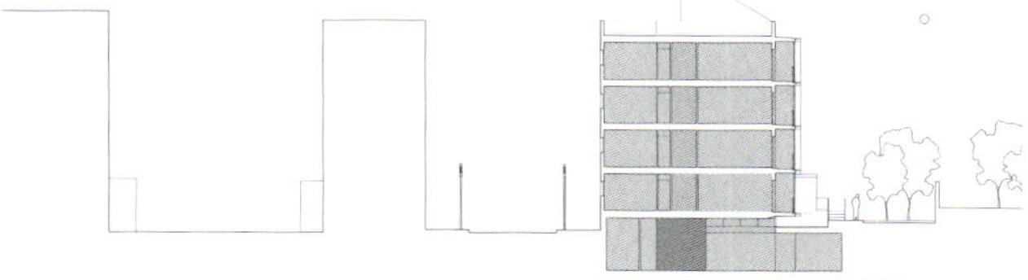

SECTION T2

1/400

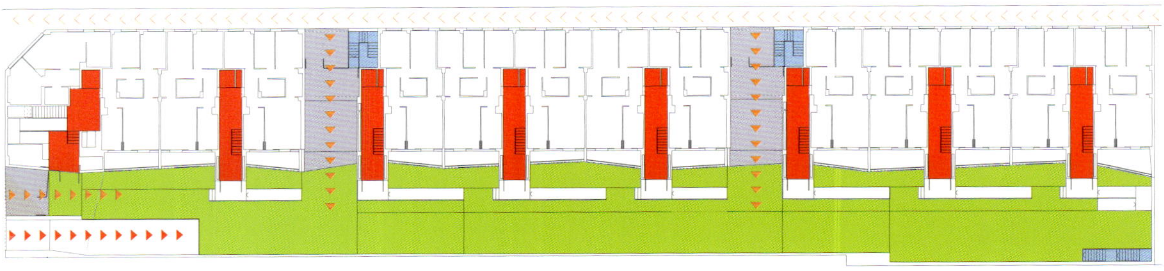

CIRCULATIONS AND ACCESS TO GROUND FLOOR

BIESCAS STREET — PEDESTRIAN ACCESS — PARKING ACCESS — DOORWAYS

SEMI-PUBLIC INNER SPACE — PEDESTRIAN ACCESS TO BASEMENT FLOOR

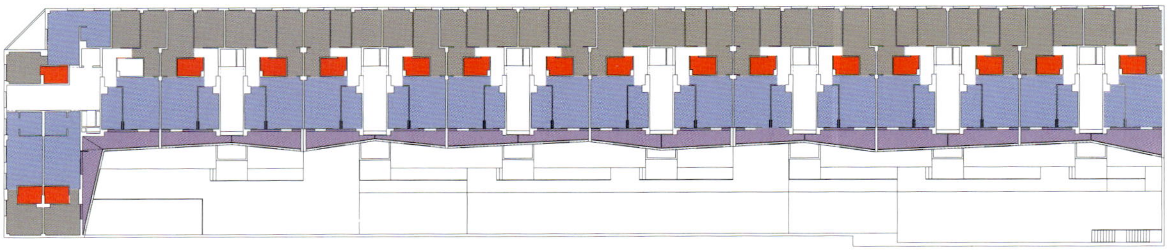

OUTLINE OF USES

LIVING ROOMS + KITCHENS - Daytime Rooms — BEDROOMS - Nightime Rooms — BATHROOMS

TERRACES — CIRCULATIONS

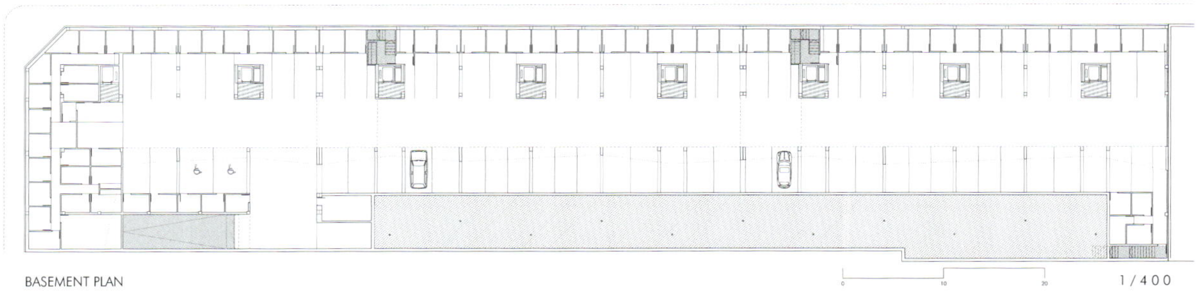

BASEMENT PLAN 1/400

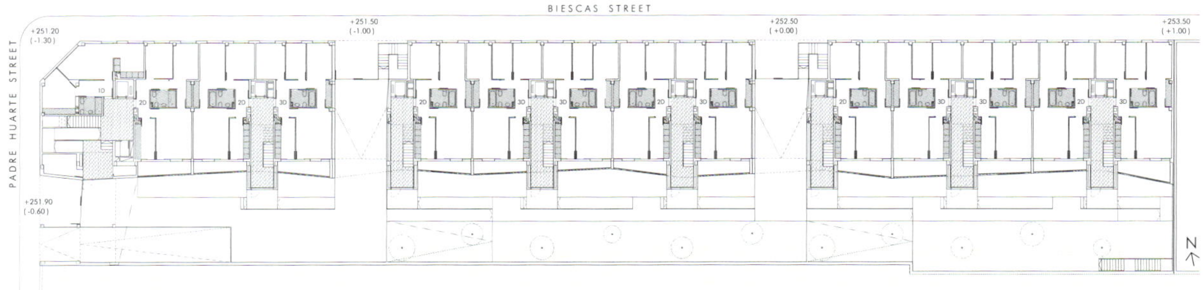

HOUSING TYPES: 01 HOUSING 3D+1B: 34 UNITS 02 HOUSING 2D+1B: 24 UNITS 03 HOUSING 1D+1B: 10 UNITS

GROUND FLOOR PLAN 1/400

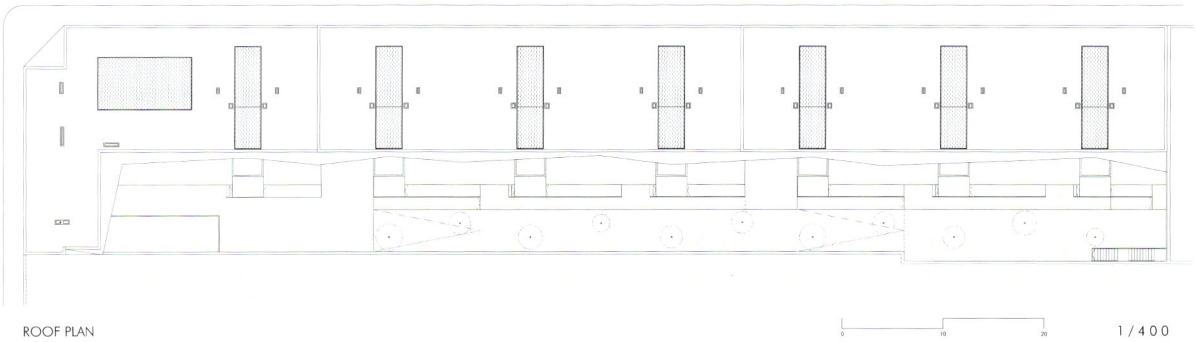

ROOF PLAN 1/400

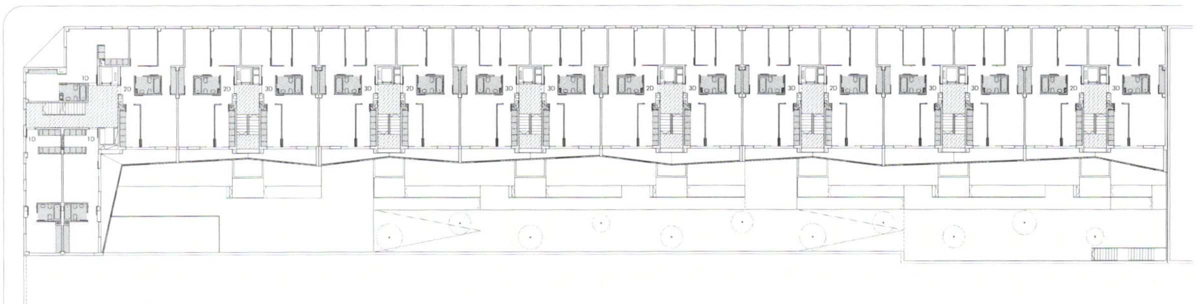

HOUSING TYPES: 01 HOUSING 3D+1B: 34 UNITS 02 HOUSING 2D+1B: 24 UNITS 03 HOUSING 1D+1B: 10 UNITS

TYPICAL FLOOR PLAN 1/400

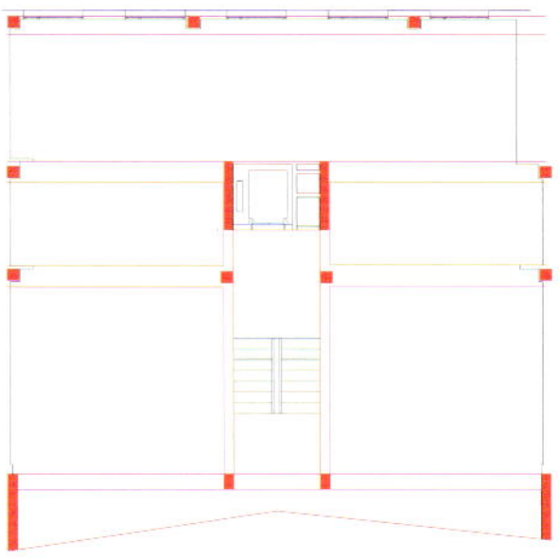

ROUND STRUCTURE

PILLARS ARE LOCATED IN THE PERIMETER OF THE HOUSING; AVOIDING STRUCTURAL INNER ELEMENTS

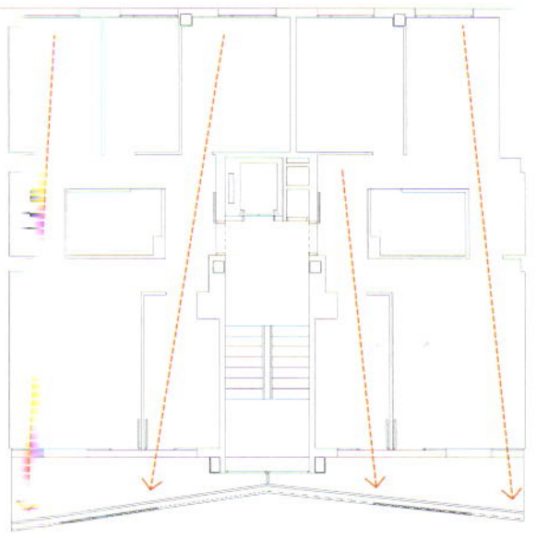

WIDENED LINES OF SIGHT

VISUAL COMFORT - CONTINUITY OF SPACES AND EFFECT OF WIDER LINES OF SIGHT

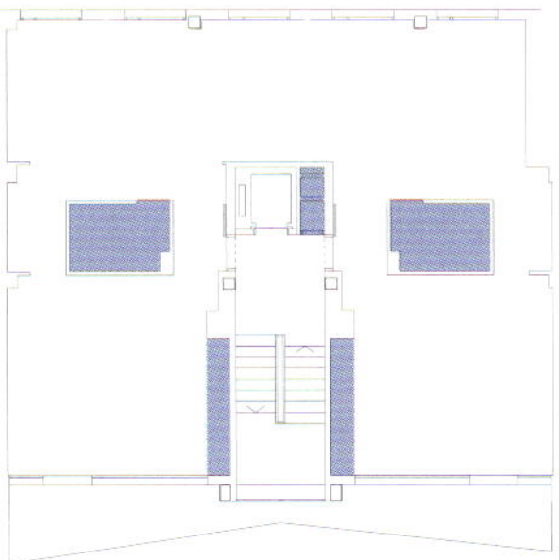

FACILITIES

KITCHEN AS AN EQUIPPED WALL
BATHROOMS IN THE CENTRE, DIVIDING NIGHT AND DAY ZONES

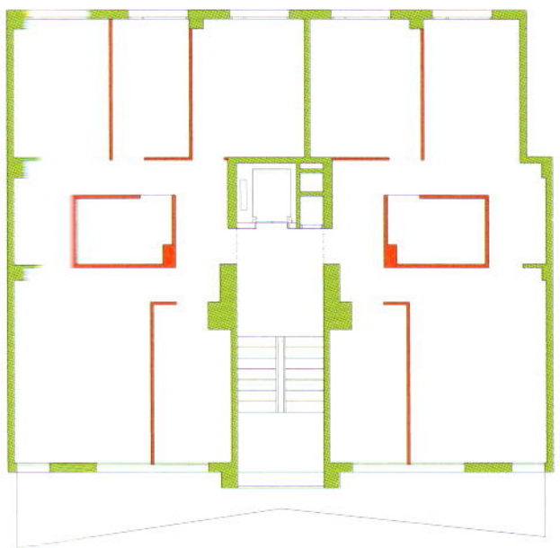

FLEXIBILITY

DISTRIBUTION WITH MINIMAL PASSING AREAS
DIVIDING WALLS OF PLASTERBOARD
POSSIBLE MODIFICATIONS AND CHANGES

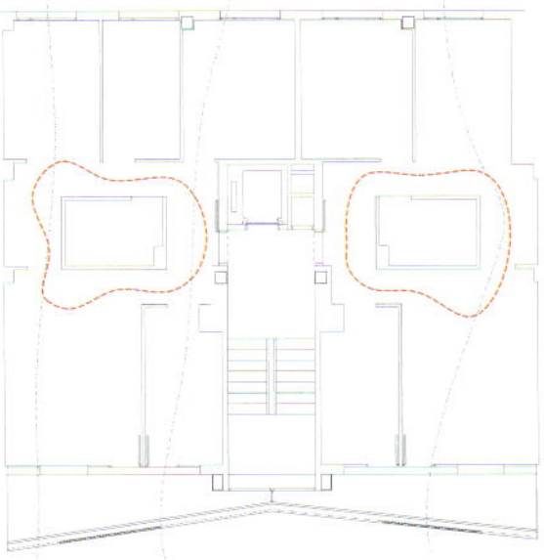

DOUBLE CIRCULATION

POSSIBILITY OF SEVERAL ROUTES

CROSS VENTILATION

DOUBLE ORIENTATION TO NORTH AND SOUTH

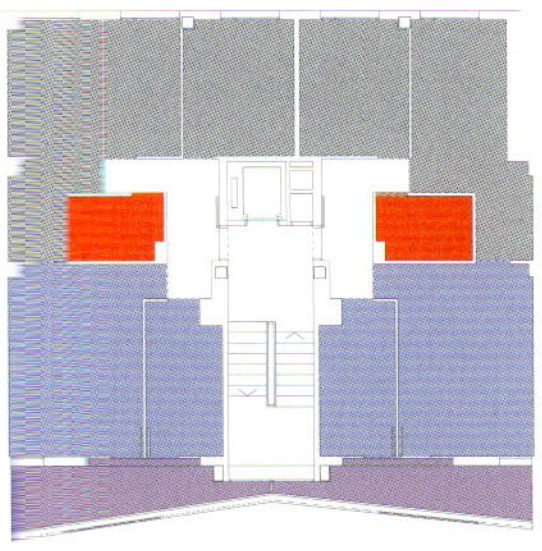

OUTLINE OF USES

- DAYTIME ROOMS
- NIGHTIME ROOMS
- BATHROOMS
- TERRACES

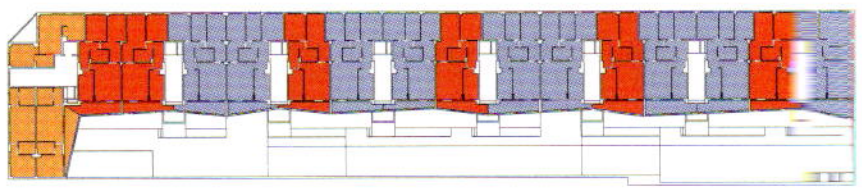

HOUSING UNITS (TOTAL : 68 UD.)

1D (45 m2) - 10 U. 2D (60 m2) - 24 U. 3D (70 m2) - 34 U.

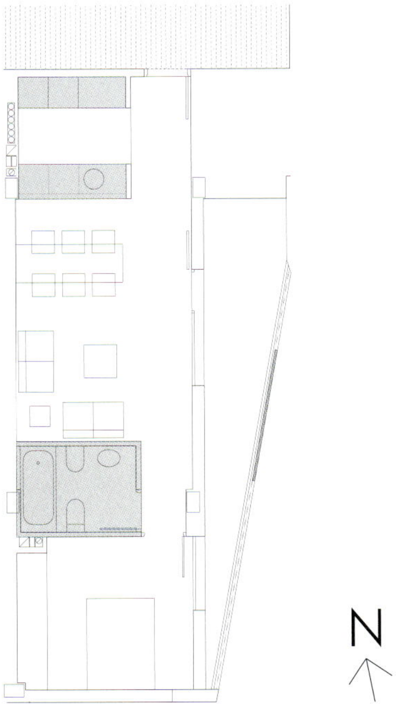

HOUSING 1D+1B: (10 UD.)

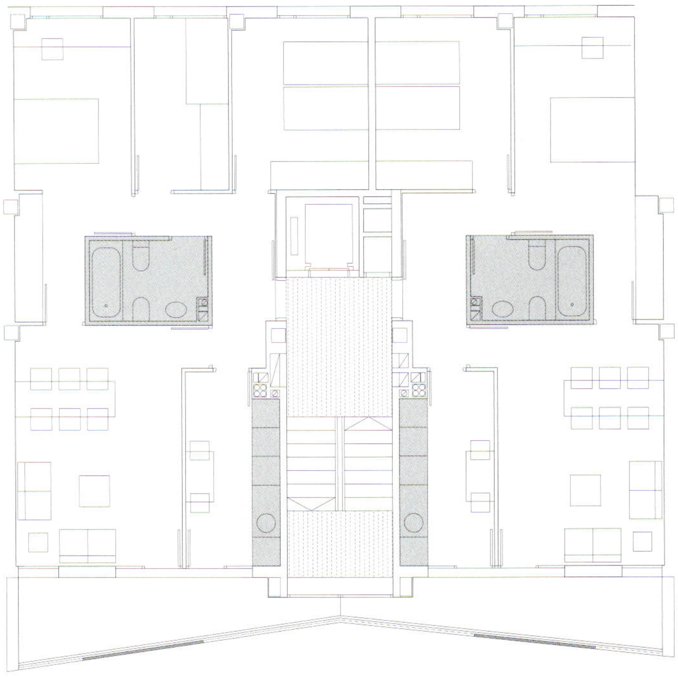

HOUSING 3D+1B (34 UD.)

HOUSING 2D+1B (24 UD.)

HOUSING UNITS

1/100

Architect: CEBRA
Location: Herning, Denmark
Area: $4300m^2$

Fuglsang Cuts (Competition – 1st Prize)

Idea: The residential blocks have been developed based on the municipally determined building body. All apartments are given conceptual cuts to create terraces with optimal sunlight and views. This gives the rigid building volume a varied mode of expression.
The estate consists of five blocks, the design of which is based upon a wish of creating a flexible concept able to contain different accommodation units within the same idea and idiom. The buildings were made so that they can be adapted to each individual resident and to the demand and will con-sequently be suitable for a wide segment of residents. The flats in the five blocks are situated around a centrally placed staircase (the core of the house); on the ground floor there is a parking area (also for bicycles) for the residents, depots and rooms for garbage cans. Around this core it is possible to set up three, four or five flats, which can be arranged horizontally, vertically or as a combination and the arrangement can be different in the other blocks.
The basic idea is to build all five blocks with three to five flats depending on requirements as to size and inner arrangement. Afterwards the intention is to imitate "a stroke with an axe" in order to give the building an aesthetic distinctive character as well as functional advantages. The strokes are placed where they will give the optimum conditions for view and daylight in the depth of the building.
Apart from giving view and sunlight the strokes also create a private terrace for each flat. The inside arrangement of the flats consists of regular rooms which are contrasted by the oblique angles made by the strokes of an axe. Due to the combination of the different arrangement of the three storeys of the blocks, the number of flats and the placement of the strokes you will see five blocks with a different exterior, yet architecturally closely related.

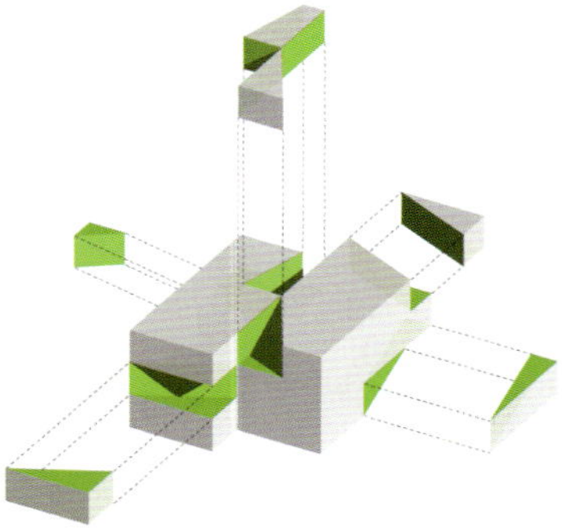

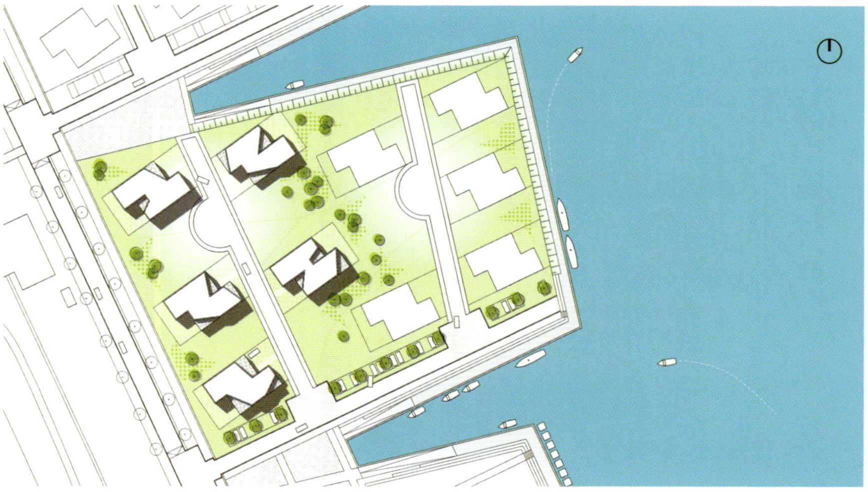

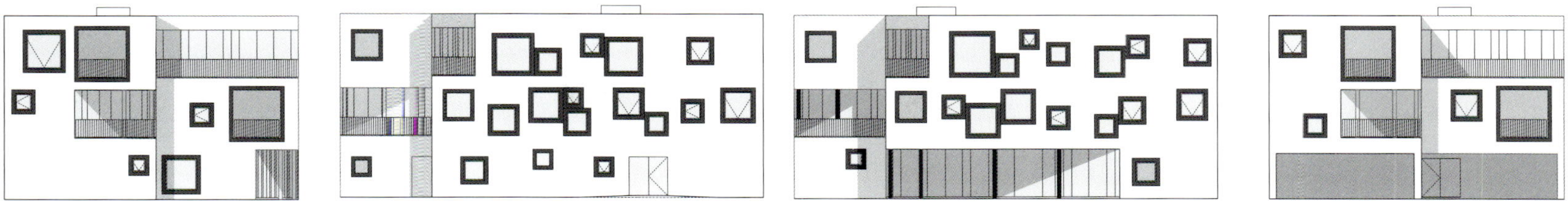

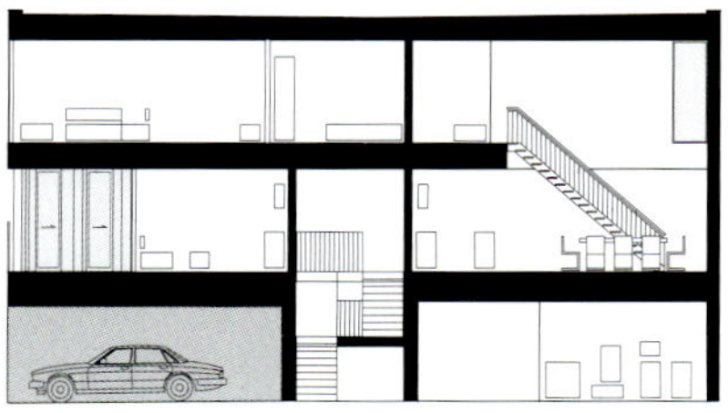

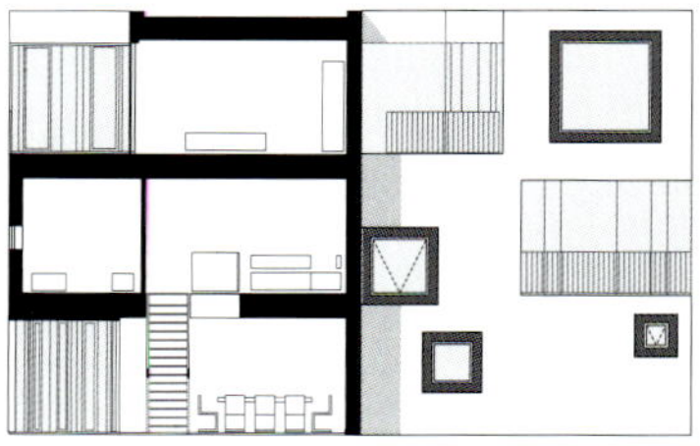

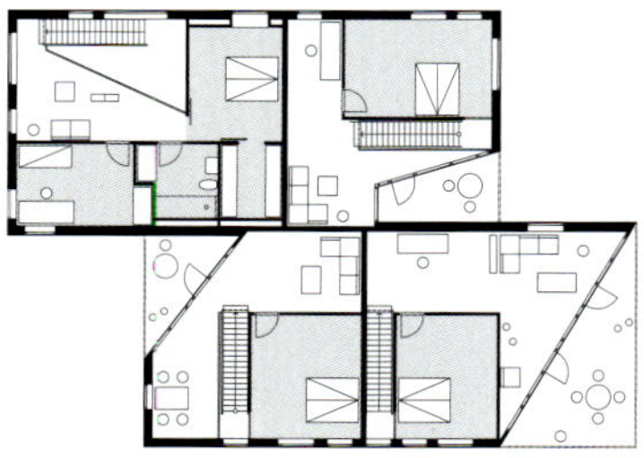

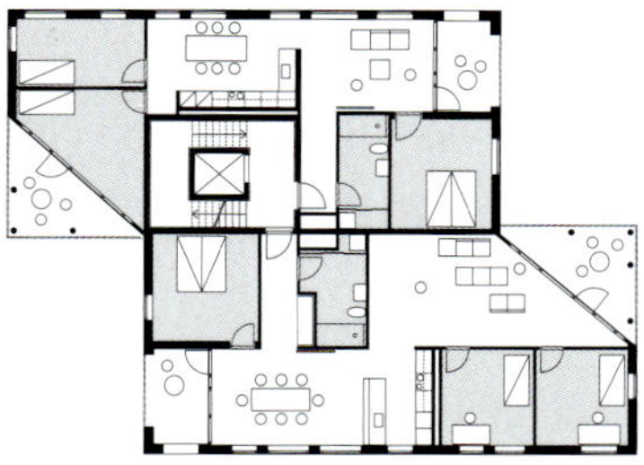

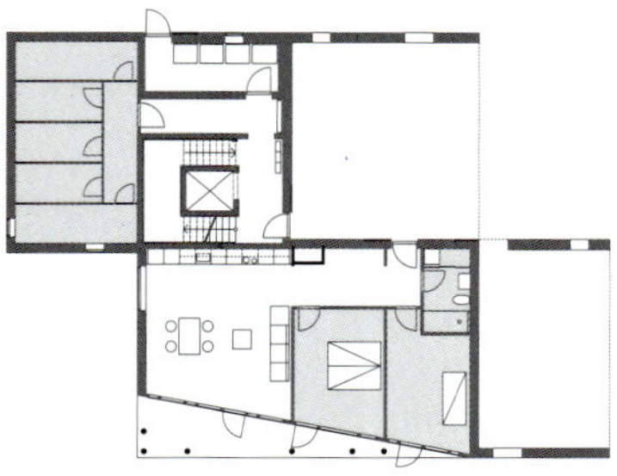

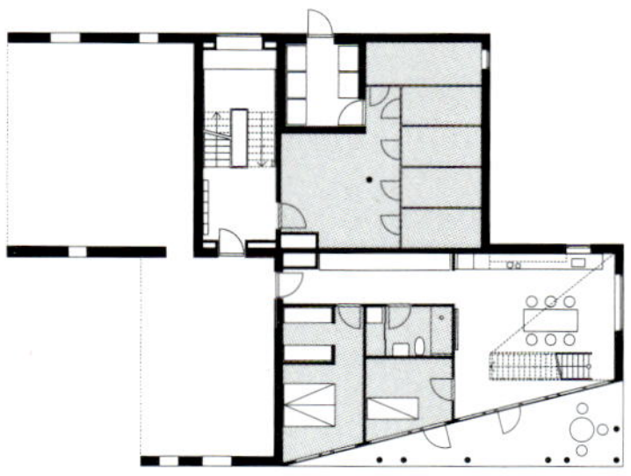

Architect: Ivan Matušík, Sebastian Nagy
Firm: sebastian nagy architects
Location: Spojovacia 30, Nitra, Slovakia
Area: 1000m²
Photographer: Bobo Boska, Ivan Karlik

TRIANGOLO

The apartment house TRIANGOLO was created as a dialog between natural predispositions, town and the construction site locality. The Nitra's upland, Zobor and meander of the river Nitra represented basic inspirational dominants. Triangular parcel and the roof of harmonic curve together with dispositional and material concept eventually formed architecture into its final resemblance. From typological point of view the building is a three-tract, the corridors of which are lit along the entire height by daylight from terminal vertical skylight. In the architectonic expression dominates mono-materialism, two facades of identical appearance and curved section of the roof with domes. The dispositional and also the constructional principle unroll from bearing ferro-concrete cores, which enabled creation of apartments in various size categories. Individual conception is also indicated by the house symbol located by the entrance to the house.

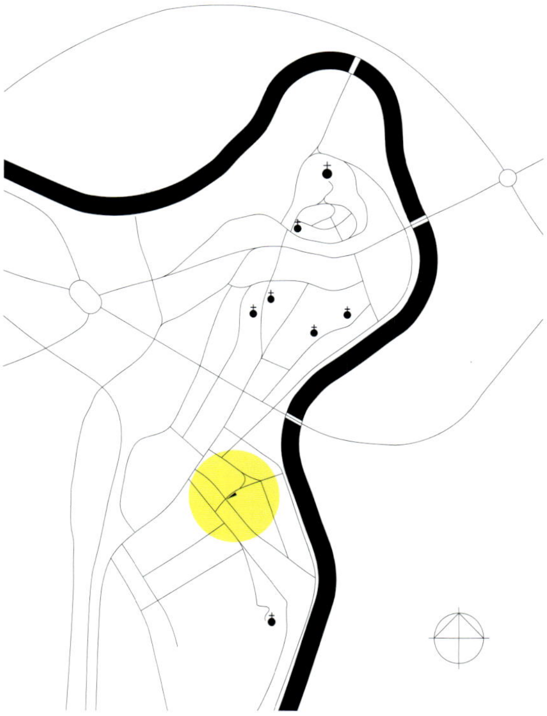

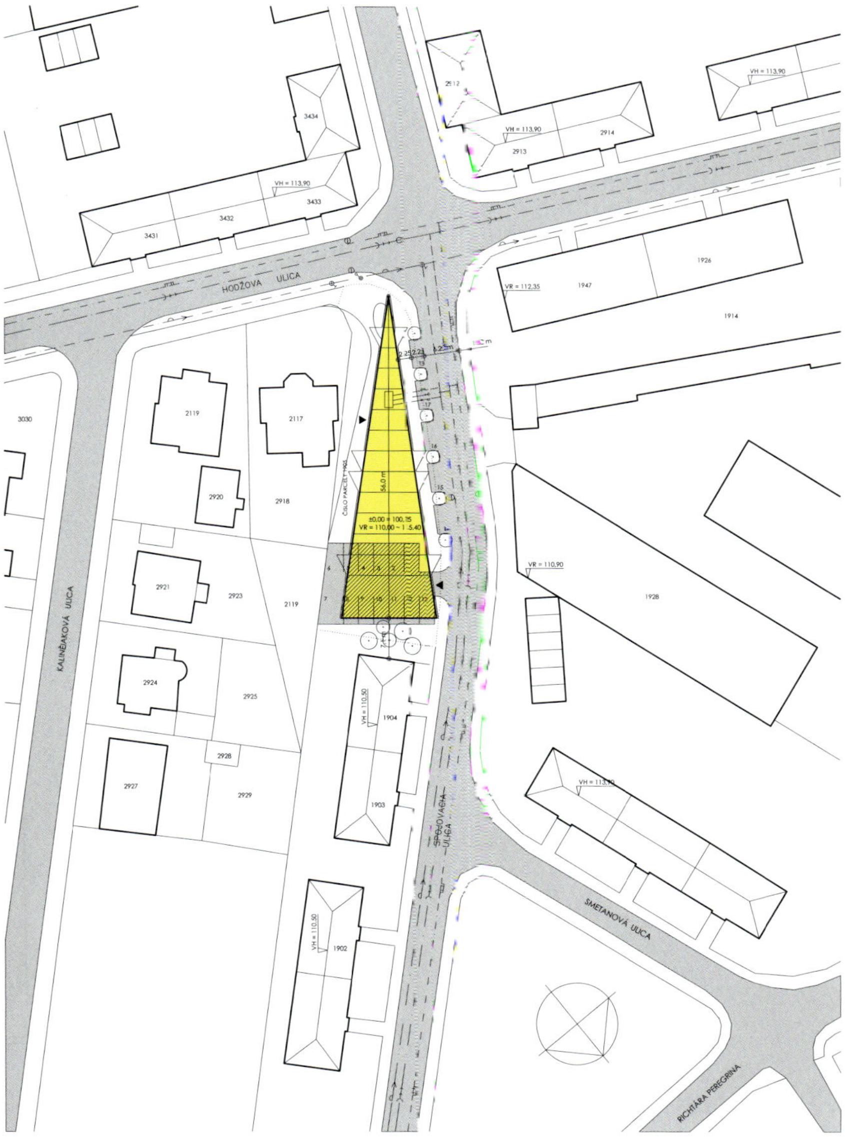
HODŽOVA ULICA
SMETANOVÁ ULICA
RICHTÁRA PEREGRINA

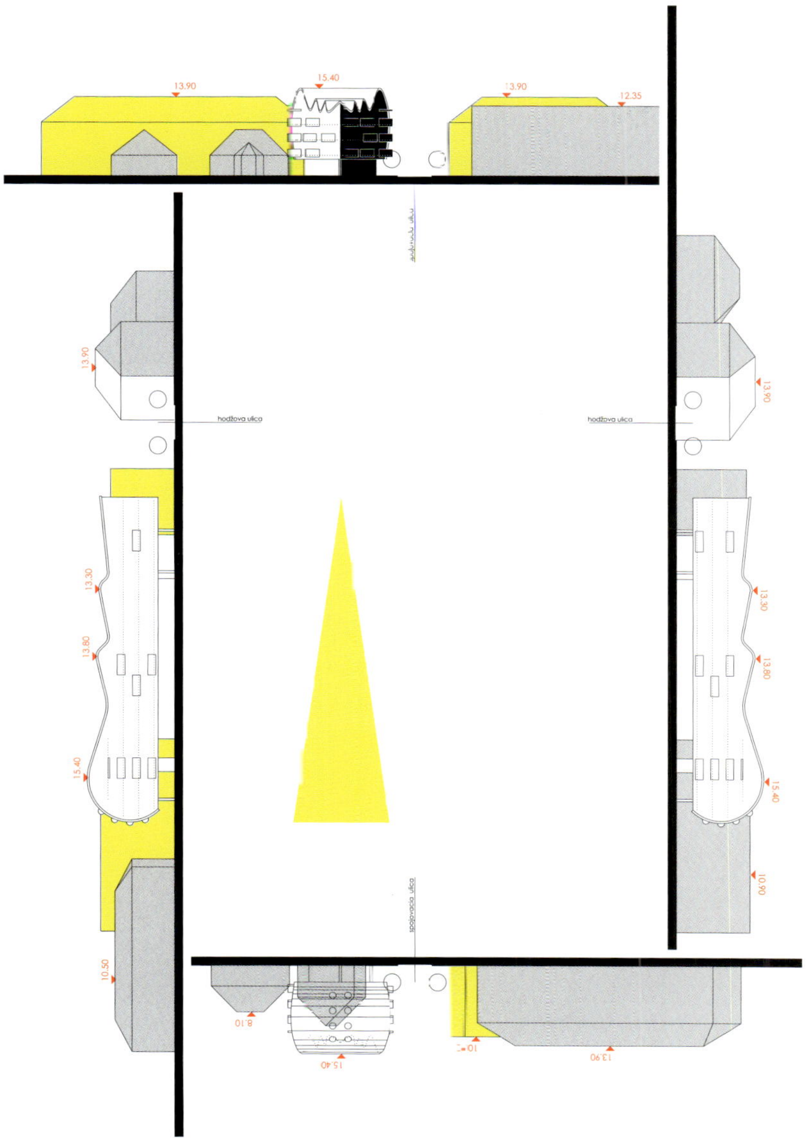

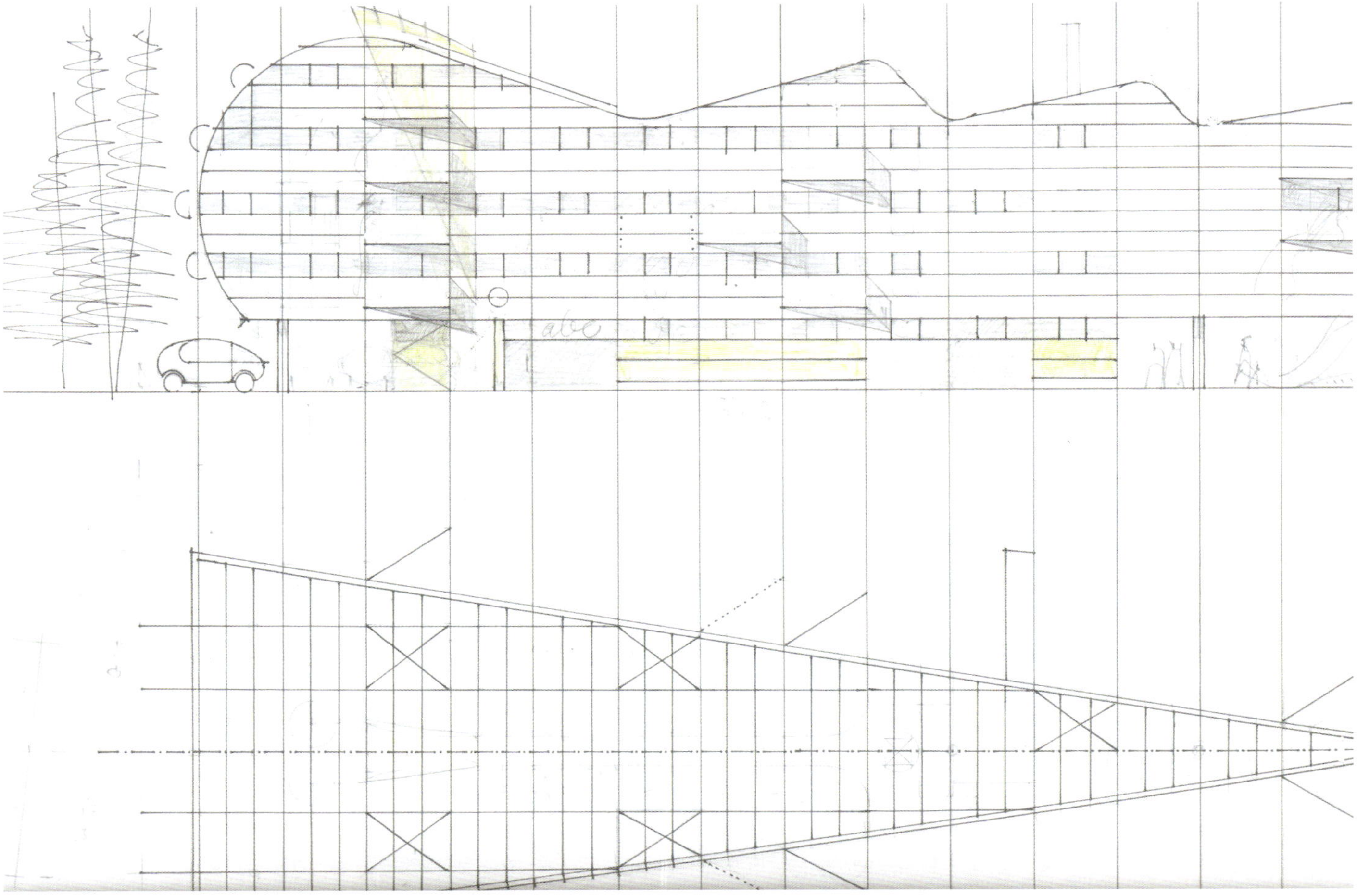

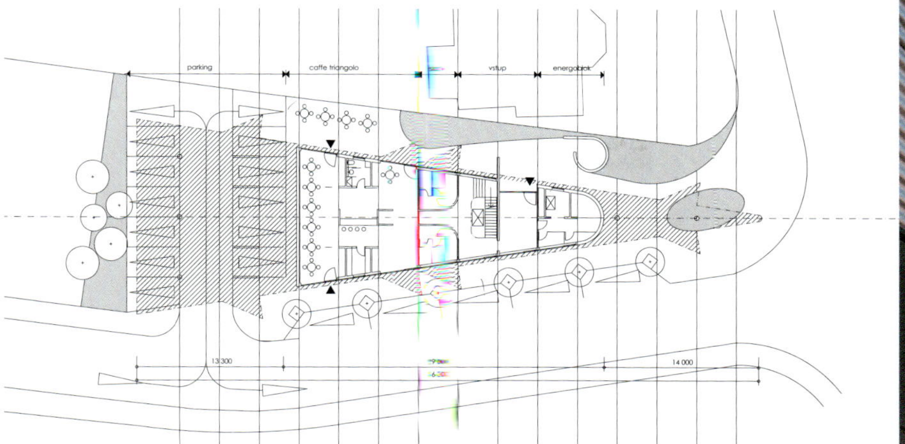
parking
caffe triangolo
vstup
energoblok
13 300
14 000

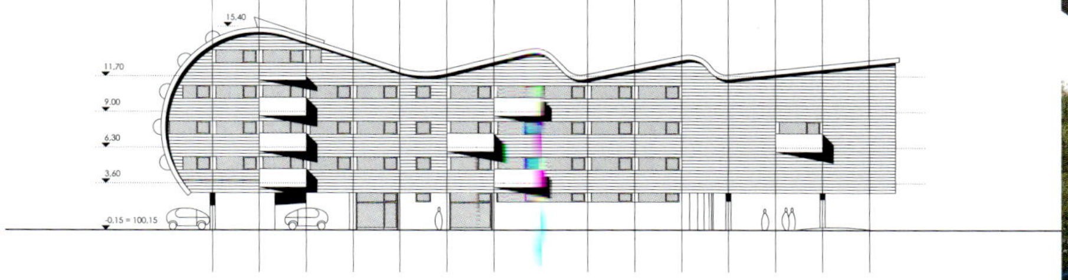
15.40
11.70
9.00
6.30
3.60
-0.15 = 100.15

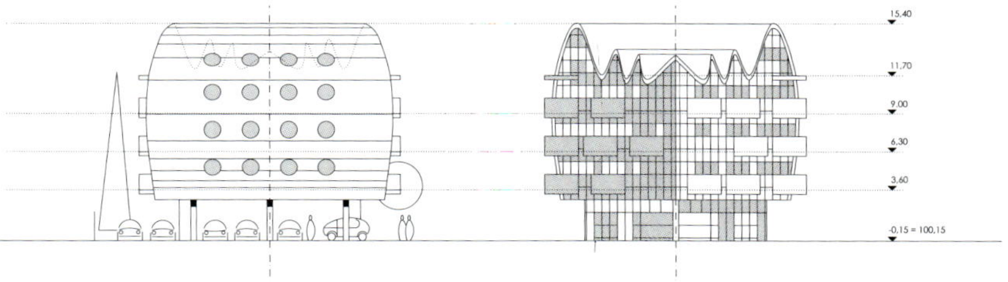
15.40
11.70
9.00
6.30
3.60
-0.15 = 100.15

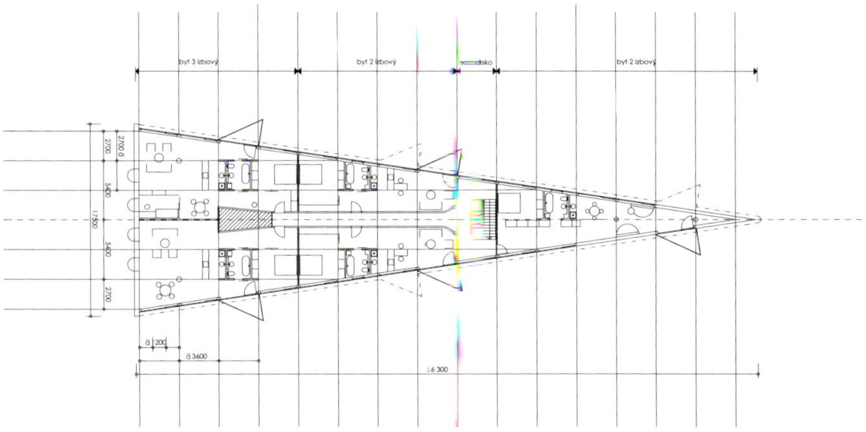
byt 3 izbový
byt 2 izbový
byt 2 izbový

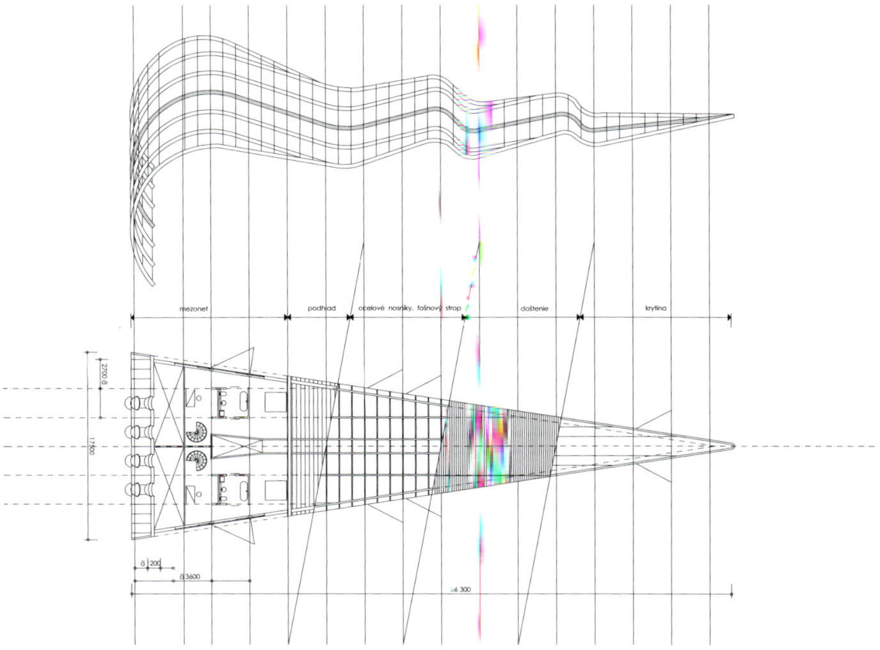
mezonet
podhľad
oceľové nosníky, fošnový strop
doštenie
krytina

TRIANGOLO
2008

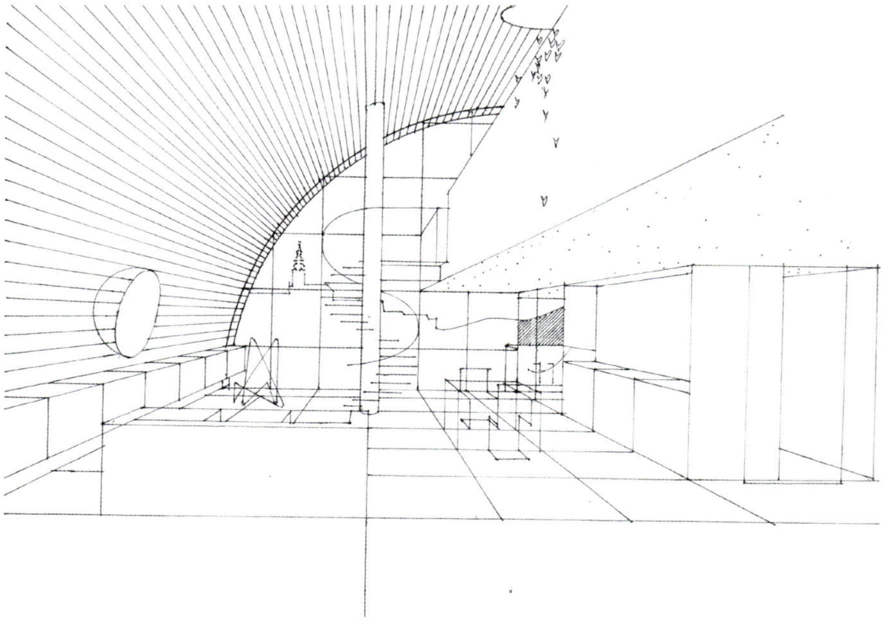

30
SAT
VIII

Architect: José Miguel Roldán Y Mercè Berengué
Firm: Roldán + Berengué, arqts
Collaborator architect: Vicenç Sanz, Zana Bosnic
Structure: Manuel Arguijo
Installations: Manel Comas, Javier Mateos
Technical architect: Joan Rovira i Duran
Construction company: CRC obras y servicios
Location: Europa Square, e.i.o.5, Incasol, Barcelona
Area: 10312.98m^2
Photographer: Jordi Surroca

Social Housing Tower of 75 Units

Grouping together floors from 3 to 3, the image of the tower, perceived in some kind of cinematographic long plan, could approximate to a building of 5 floors of height. Consequently the scale of the windows, according to that law of grouping them, is also transforming in frames of 10 meters height and different thickness. In order to be more accurate, the facade and windows, this last ones with a sequential number and modulated in pieces with similar dimensions to a door (0.8 x 2.10m), succeeding in different planes with a depth that range between 50 cm and 1.2m and emphasises the concept of big inverted tribunes. In each section of the tower the first floor of every serial of 3 is a balcony.

As some of the images in a "short plane" have done for the competition show, this frames avoid vertigo impression because between the interior and the outside of the apartment there is always an intermediate element: balconies, jambs, or the lintels of this big holes.

The election of the elements for the facade's assembly has been done paying attention to a comparative study of several constructive solutions which evaluated the origin of already recycled materials, its natural origin and the capacity for being recycled at the end of its useful life. Likewise, architects also valuated the expense of fabrication energy and the recovered politics of the surplus energy generated during the production process.

Materially, the facade is constructed with a 8 mm thick HPL pannels hanging of hidden structure of recycled aluminium perfiles. Black frames are made of 4mm thick aluminium composite panels which brings equal resistance with a lower weight per m^2 to any other material with the same features.

Synthetically, all materials used in the tower's construction are 100% recyclable and specifically the ones used in the facade come from a 65% and 100% already recycled materials.

In this Project, the apartment unit's distribution begins in the 14th floor and keeps descending so the compacted residential volume from the superior floor releases a space of 3 levels height with a T form in the access. This piece, generous in volume and 'sober' in measures and finishing materials, in its longest dimension it ends being a street with accesses in the extremes and has unusual dimensions for the building's entrance hall of usual public housing buildings.

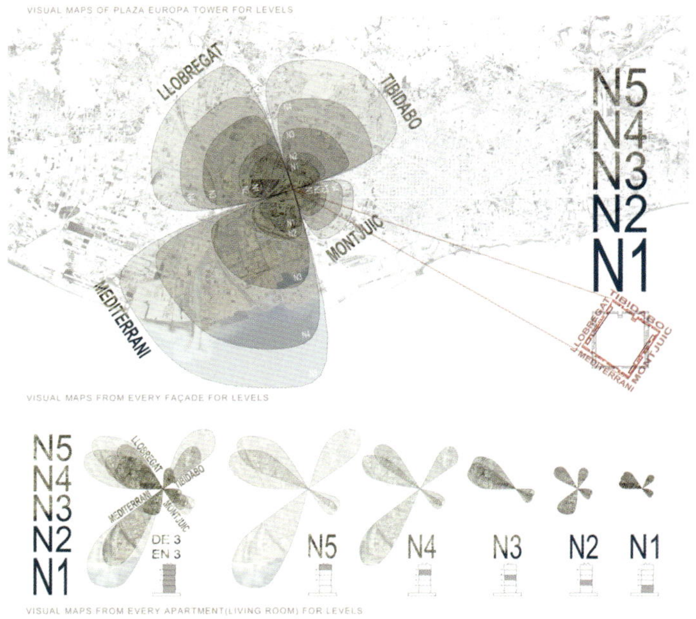

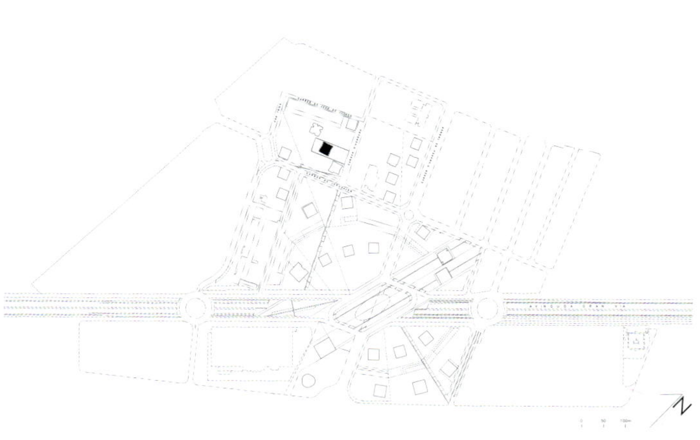

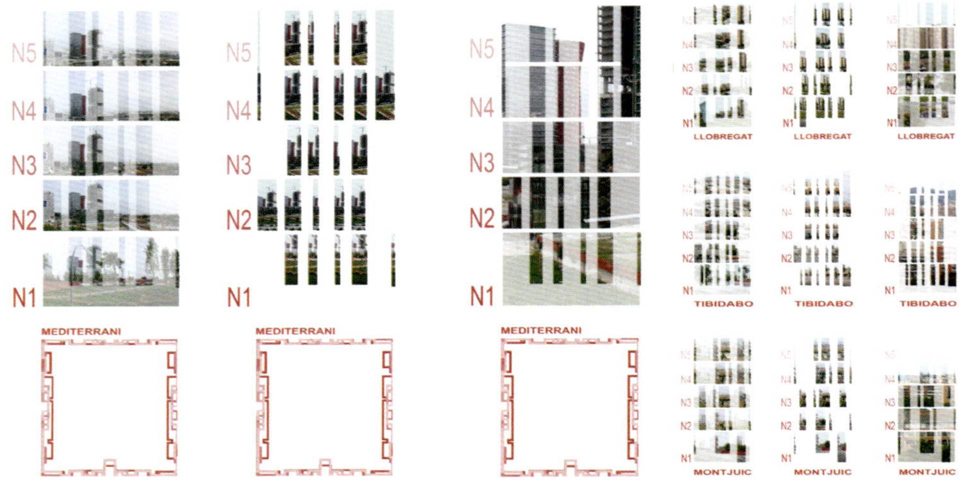
N5
N4
N3
N2
N1
MEDITERRANI
LLOBREGAT
TIBIDABO
MONTJUIC

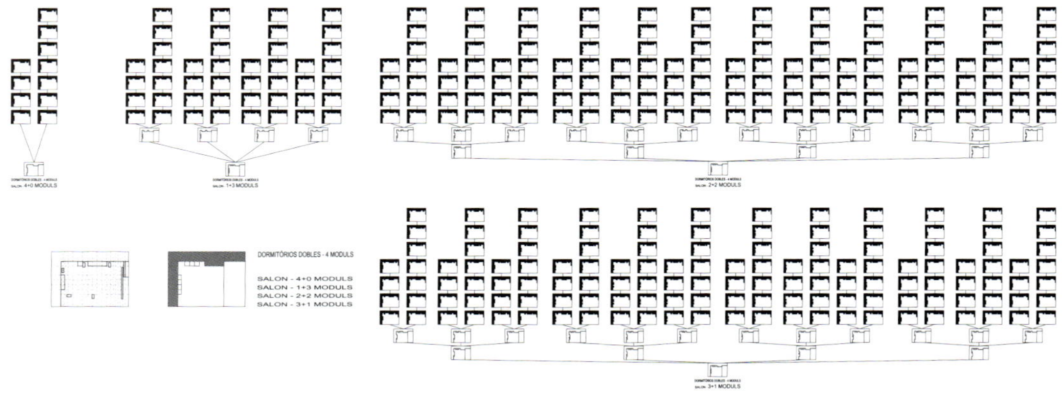
DORMITORIOS DOBLES - 4 MODULS
SALON - 4+0 MODULS
SALON - 1+3 MODULS
SALON - 2+2 MODULS
SALON - 3+1 MODULS

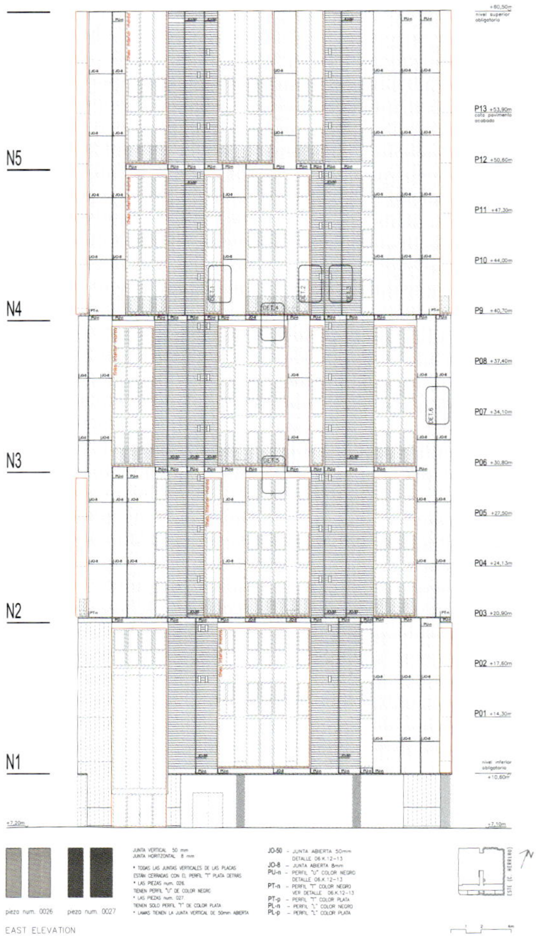
N5
N4
N3
N2
N1
EAST ELEVATION

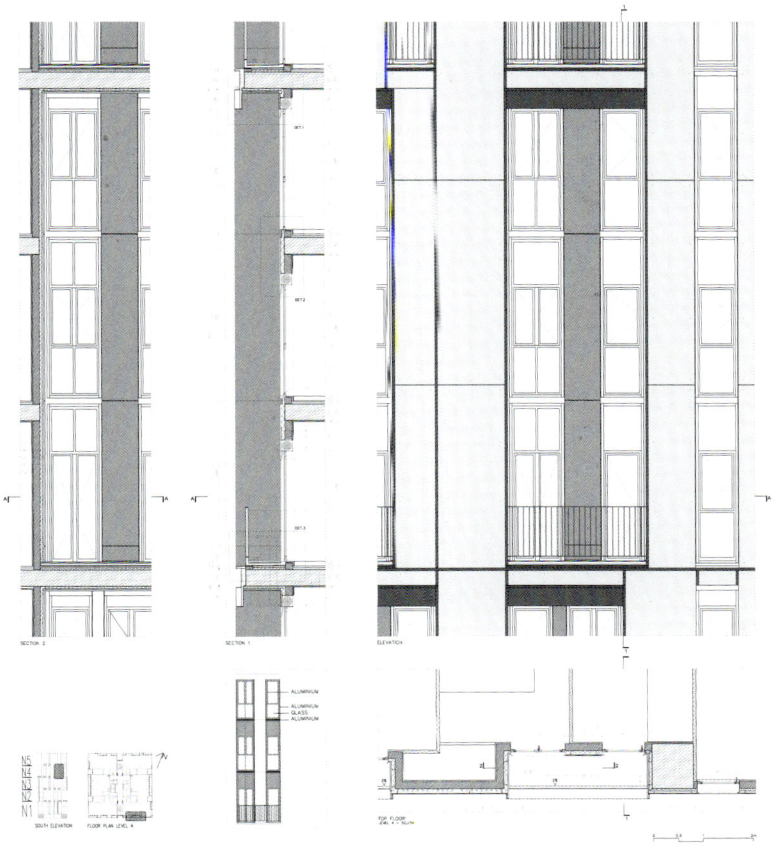

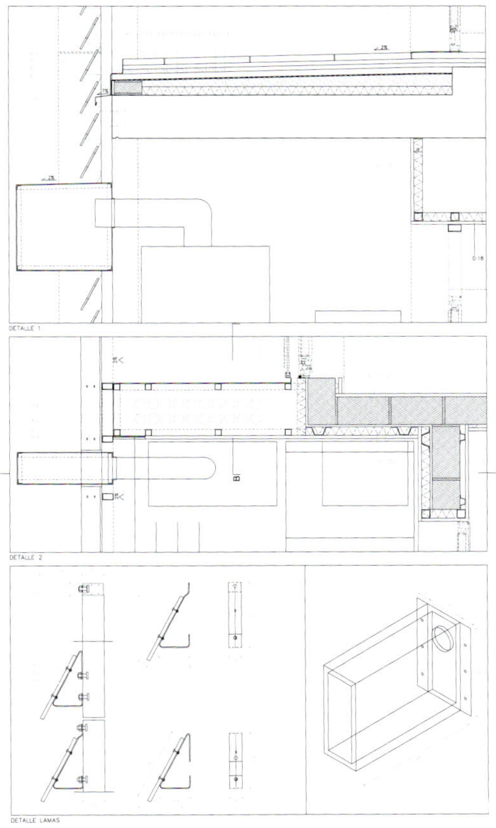

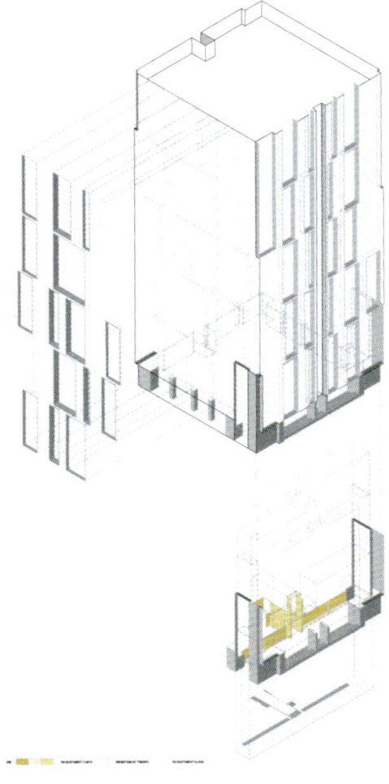

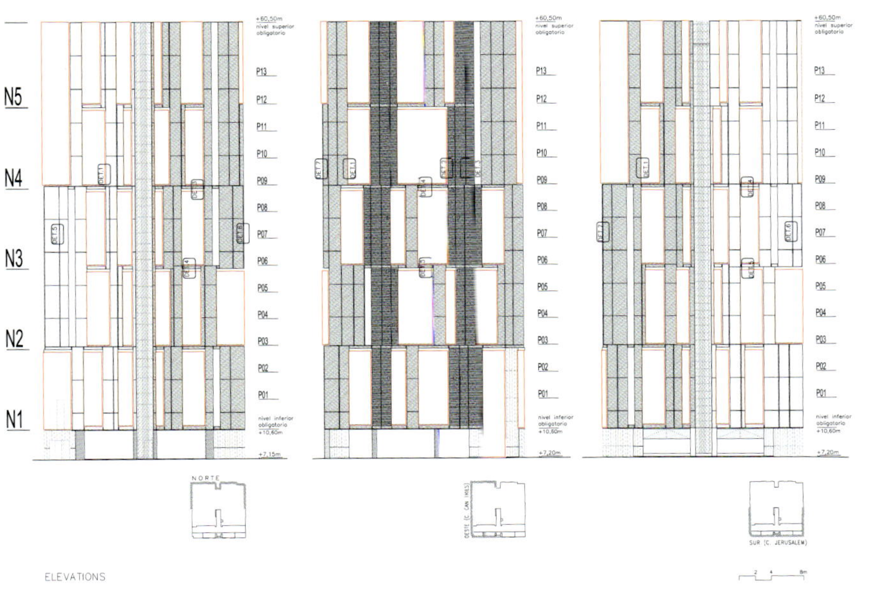

ELEVATIONS

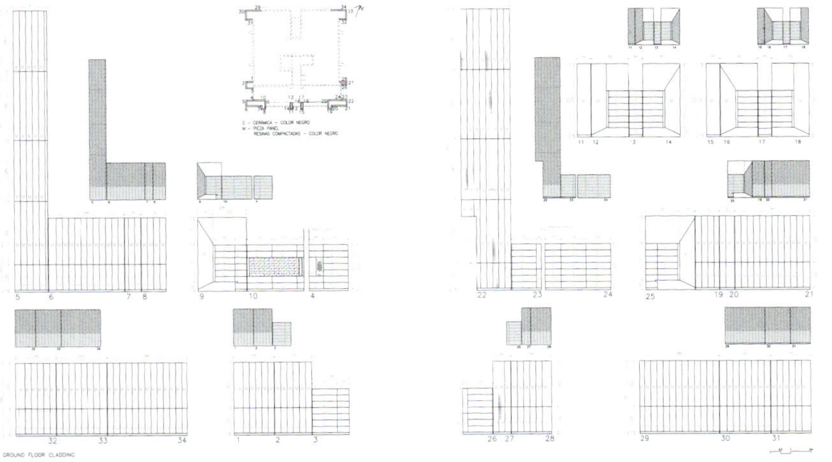

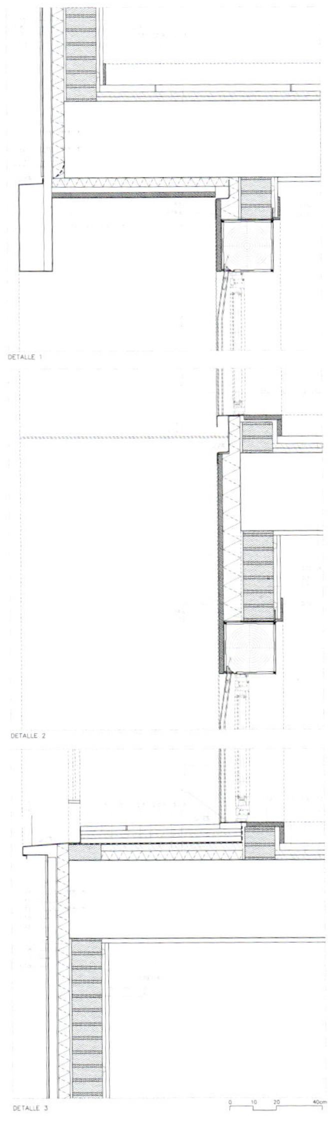

FACADE DETAIL_ BALCONY FRAMES

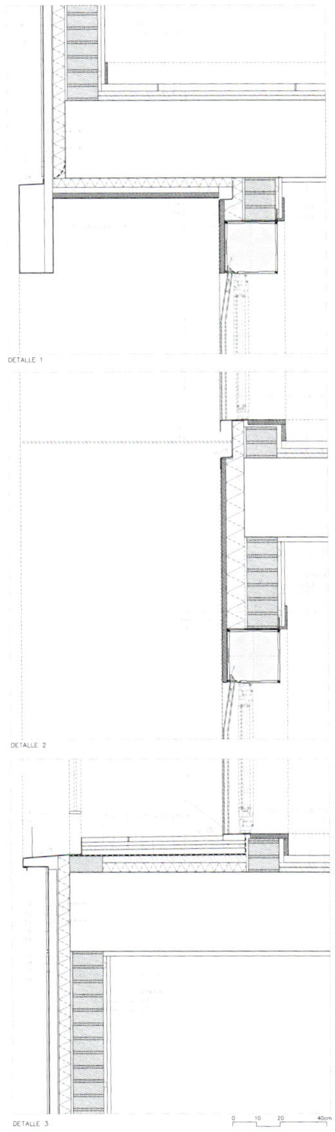

FACADE DETAIL_ BALCONY FRAMES

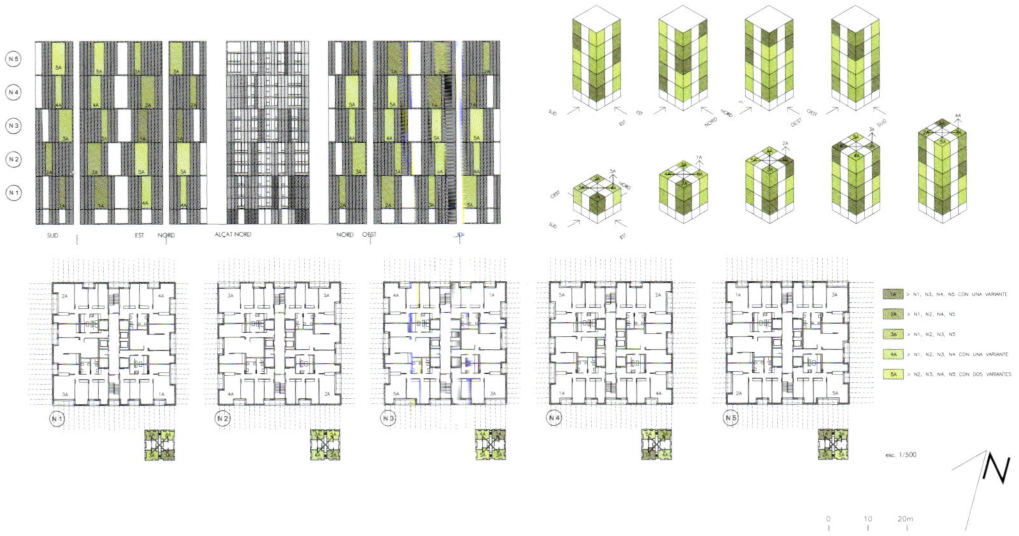

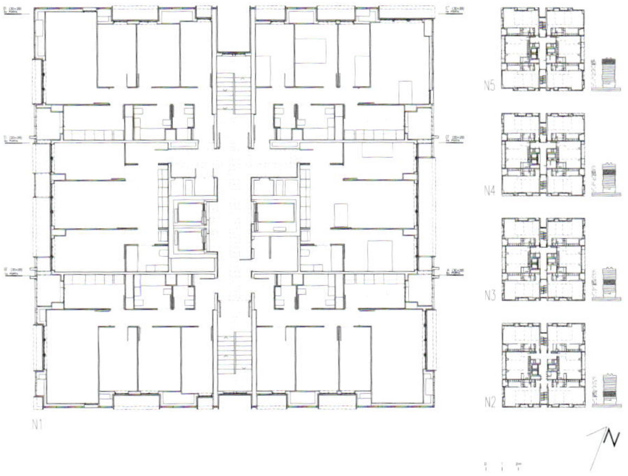

vista SUD-OEST

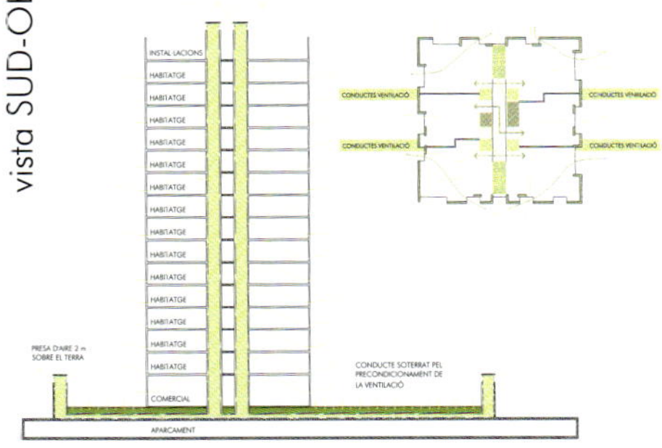

DIAGRAMA COLUMNES VENTILACIÓ

DIAGRAMA ORIENTACIONS

PLANTES 1-3 esc. 1/200

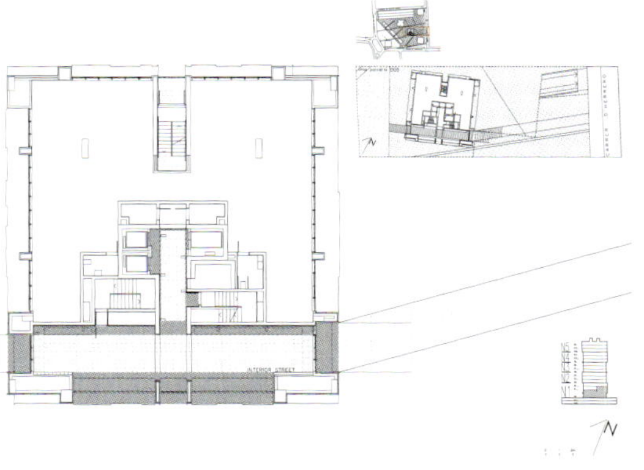

Architect: Kavellaris Urban Design
Firm: Kavellaris Urban Design
Location: Brunswick
Photographer: Peter Bennetts

Jewel

The areas of the designers' investigations were, the entry into the building, constructing key and strategic views and the manner in which the building addresses the prominent corners within its context. The designers incorporated these design strategies to construct a dialogue with the various interfaces which informed the master plan and design process.

The designers incorporated landscaped streets and laneways within the subject site to provide spatial separation between dwellings and enable every habitable space to receive natural light and ventilation. In addition, these separations allow the spaces to be naturally cross ventilated thereby minimising the need for air-conditioning. Low e mission lighting complimented with solar panels further reduced the carbon footprint of the complex.

The rhythmic boxed formed geometries are referential variations of the same architectural language resulting in a sculptured building that evokes a sense of movement. As the building is experienced from the various vantage points, the building form takes on different readings to people in trains, cars, on bicycles or walking pedestrians.

The asymmetrical composition of slipping corners that wrap around the northern and southern edges of the building blurs the boundaries from the converging facades into a continuous architectural expression. The designers studied the column motifs of Robert Venturi's design for the Sainsbury Wing and incorporated the manner in which the columns negotiated the corner intersections into the design. The southern elevation has been specifically design with a constructed view of the building typical of the classical order when viewed from Watson Street. Vertical strip windows capture the opposing perspective providing a visual connection and a conclusion to the journey for occupants that have entered the building from Watson Street. The accented double height entry void that interrupts the rhythmic articulation of the composition emphasizes and clearly announces the entry into the building as seen from Watson Street.

The designers used three species of timber to accent elements of the composition in order to establish an immediate dialogue with the juxtaposed buildings.

East Elevation

North Elevation

South Elevation

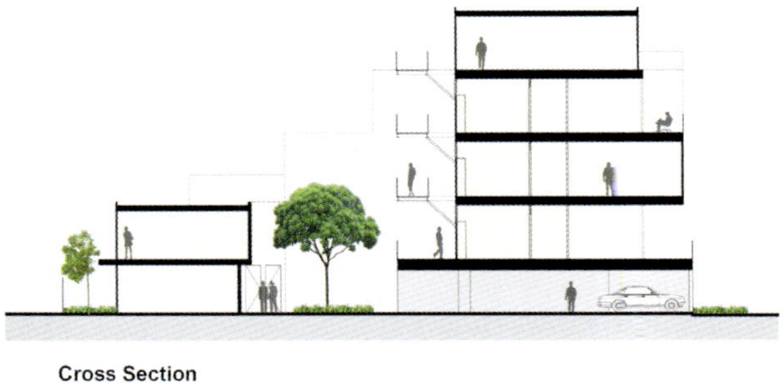

Cross Section

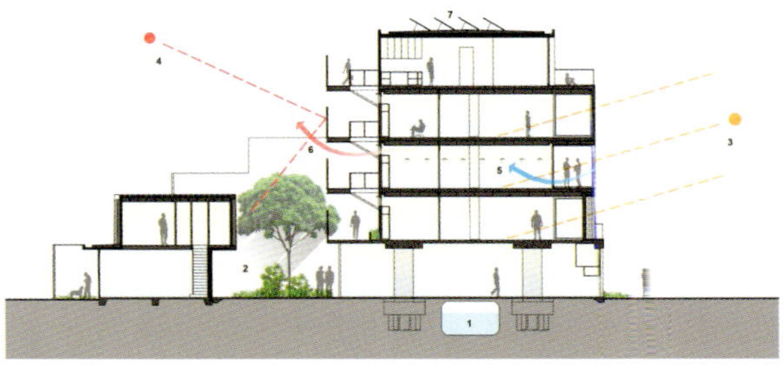

1. water tank to collect and store stormwater for toilet flushing
2. courtyard garden to provide spatial separation
3. morning sun
4. bridge awning to protect building from west sun
5. natural cross ventilation
6. highlight window to expel hot air through cross ventilation
7. solar panels for all communal lighting

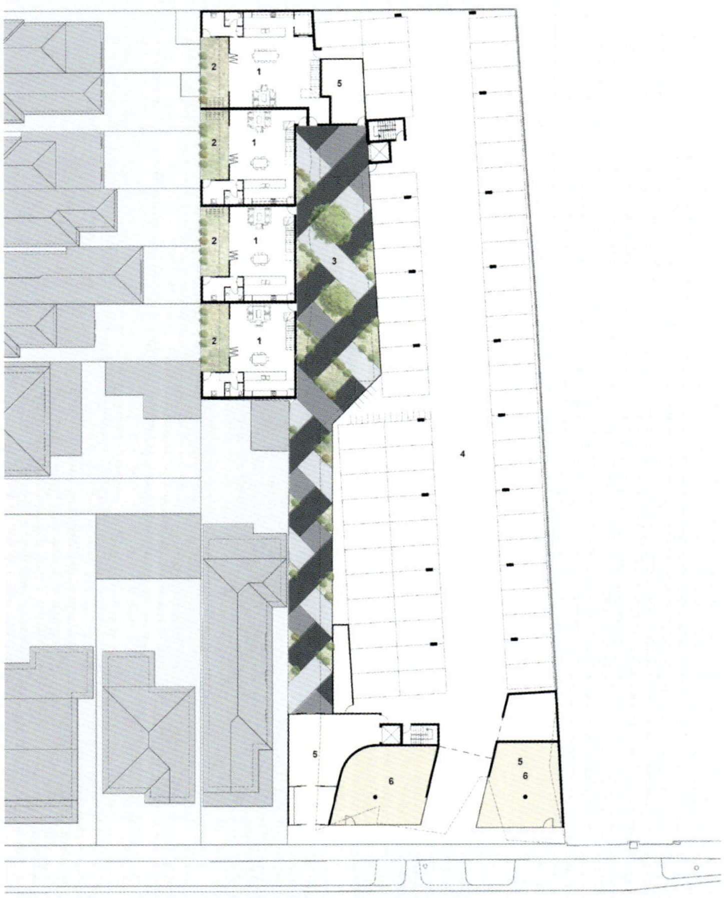

Ground FLoor Plan

Legend

1. townhouse unit
2. townhouse backyard
3. communal courtyard
4. carpark
5. public area
6. commercial space

First Floor Plan

Legend

1. townhouse unit
2. townhouse backyard
3. communal courtyard
4. carpark
5. public area
6. commercial space
7. internalised street corridor
8. apartment unit

Second Floor Plan

Legend

1. townhouse unit
2. townhouse backyard
3. communal courtyard
4. carpark
5. public area
6. commercial space
7. internalised street corridor
8. apartment unit

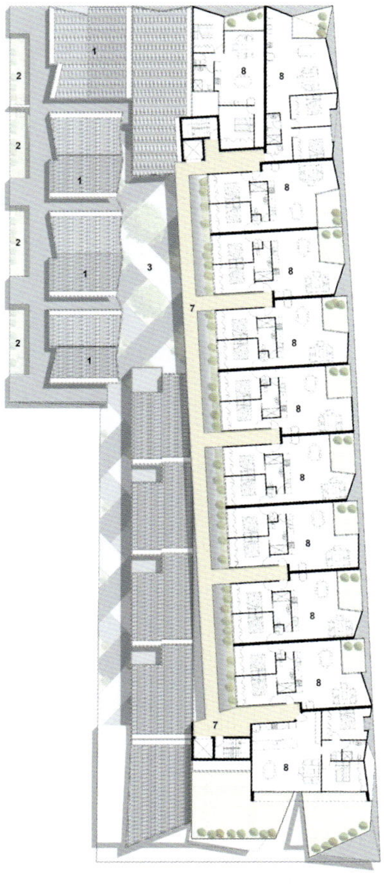

Third Floor Plan

Legend

1. townhouse unit
2. townhouse backyard
3. communal courtyard
4. carpark
5. public area
6. commercial space
7. internalised street corridor
8. apartment unit

Fourth Floor Plan

Legend

1. townhouse unit
2. townhouse backyard
3. communal courtyard
4. carpark
5. public area
6. commercial space
7. internalised street corridor
8. apartment unit

Architect: María José Pizarro, Óscar Rueda.
Firm: Rueda Pizarro Arquitectos
Partners: Alberto Galindo, Ramiro Losada, Laura Montero, Jesús Muñoz
Engineering: PROYTEC, DPI ingenieros
Location: Tonw of Parla, Madrid
Area: 18043.07m²
Photographer: Miguel de Guzmán

156 Transitory Dwellings in Parla

In the middle of 2005 the City Council of Parla, summons a new contest in its exposition: 826 housing were distributed in 6 parcels of public equipment. They also adjudge the right of surface to private companies that pay for the aid. Three years later, in front of other aids never made by unworkability in the management, the housings have dealt. This project was one of the six winners and the designers had to include exactly 156 housings which were those that corresponded to them in the plan.

In the aid, the designers left from the situationists postulates of years 1950s. In its texts they taught to understand the city from the individual experience. The space of the city acquires sense with the use that their inhabitants do of it, of the multiplicity of relations that in the city can be given: to experience the city like a territory where "to construct situations". This way, the designers would like to understand their proposal of houses for young people like a playful reappropiation of the urban space.

The designers propose that the ground is freed to the maximum, creating a field of leisure activities that vary with the subject and the environment. In each one, the houses are organized around a central, illuminated and ventilated space, that allows to establish relations between the users in vertical and horizontal sense, and guarantees the ventilation crossed in all the houses. A vertical nucleus of communications serves five houses as a dormitory, and one of two, by plant. In its interior, all the space is opened to an extra outer emptiness, that introduces light, ventilation and energy to the house and are simultaneously a multipurpose space. Faced with the standard program-Bedroom-Hall Kitchen, we propose a configurable space, without program, multipurpose in its use. The only piece fixes are the fog rooms, on which halls and dormitories pivot, introducing certain randomness in the composition of the facade. A closing of you lick of hard anodizing aluminum of 18 cm of width, surrounds the facade, dressing the construction, and obtaining a contemporary image in the frame of a limited budget that did not surpass the 700 Euros square meter on grazing.

MAPA DE ESPACIOS LIBRES Y ZONAS DE INFLUENCIA. PARLA, 2005

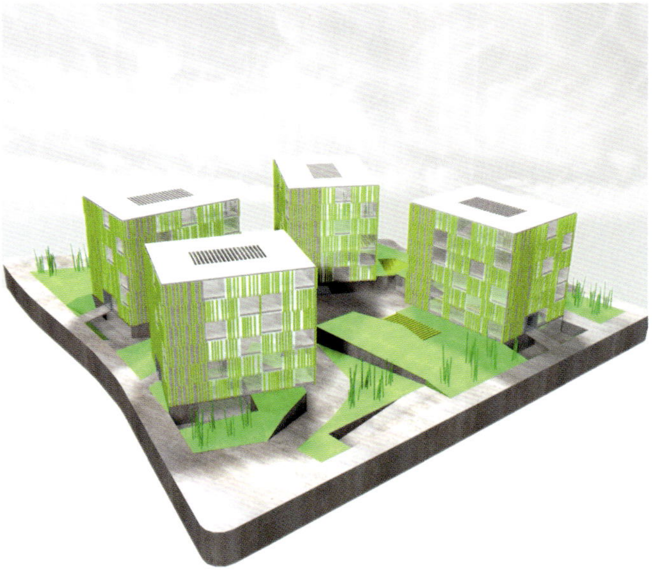

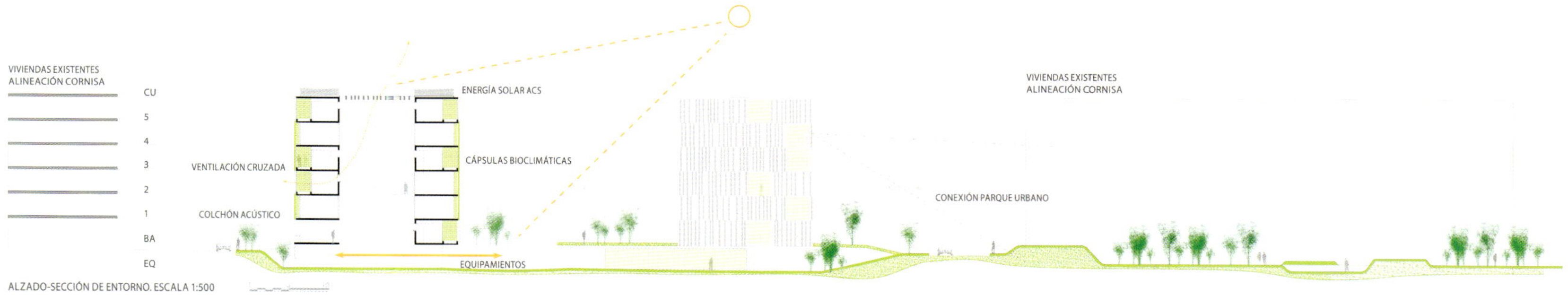

ALZADO-SECCIÓN DE ENTORNO. ESCALA 1:500

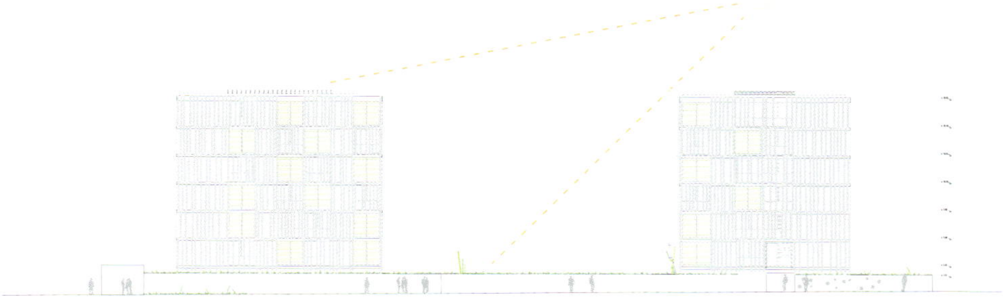

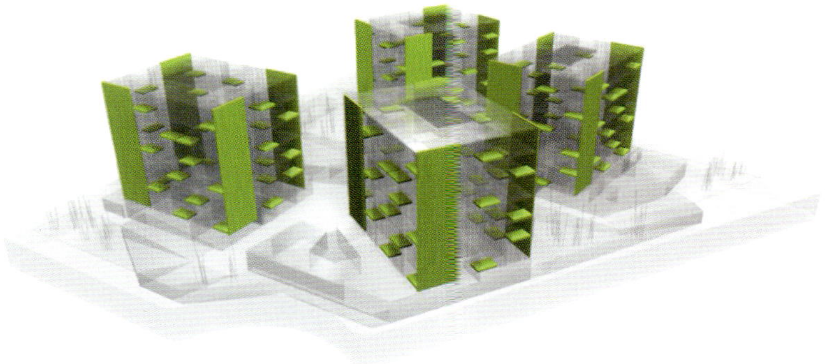

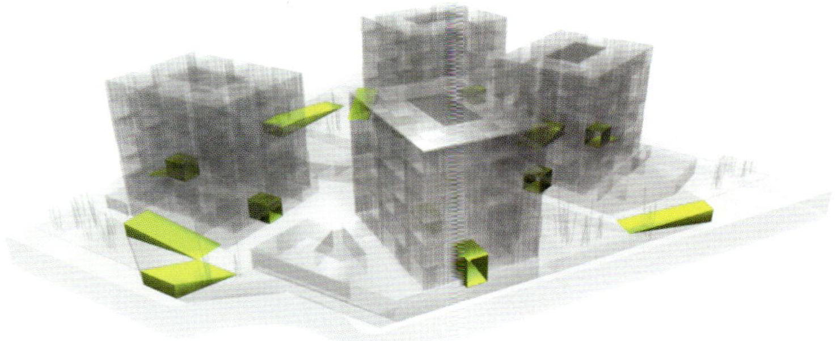

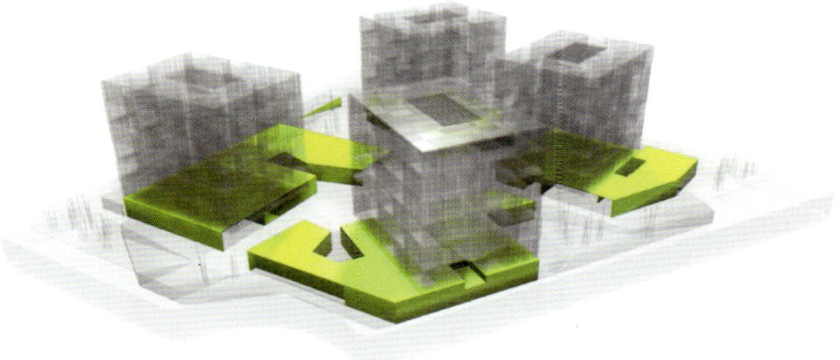

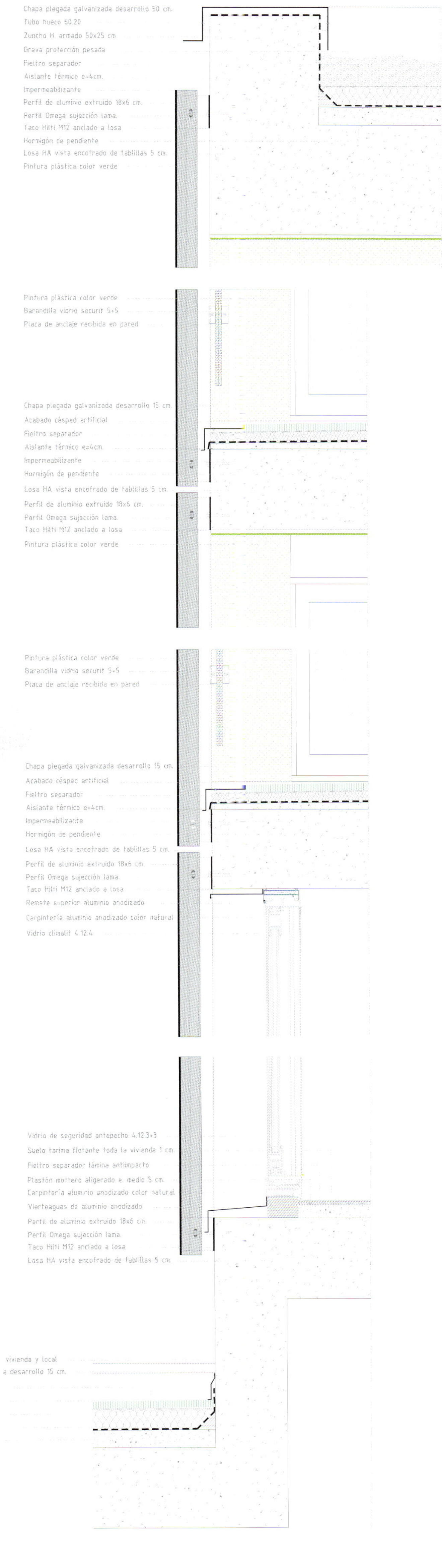
Chapa plegada galvanizada desarrollo 50 cm.
Tubo hueco 60.20
Zuncho H. armado 50x25 cm
Grava protección pesada
Fieltro separador
Aislante térmico e=4cm.
Impermeabilizante
Perfil de aluminio extruido 18x6 cm.
Perfil Omega sujección lama.
Taco Hilti M12 anclado a losa
Hormigón de pendiente
Losa HA vista encofrado de tablillas 5 cm.
Pintura plástica color verde
Pintura plástica color verde
Barandilla vidrio securit 5+5
Placa de anclaje recibida en pared
Chapa plegada galvanizada desarrollo 15 cm.
Acabado césped artificial
Fieltro separador
Aislante térmico e=4cm.
Impermeabilizante
Hormigón de pendiente
Losa HA vista encofrado de tablillas 5 cm.
Perfil de aluminio extruido 18x6 cm.
Perfil Omega sujección lama.
Taco Hilti M12 anclado a losa
Pintura plástica color verde
Pintura plástica color verde
Barandilla vidrio securit 5+5
Placa de anclaje recibida en pared
Chapa plegada galvanizada desarrollo 15 cm.
Acabado césped artificial
Fieltro separador
Aislante térmico e=4cm.
Impermeabilizante
Hormigón de pendiente
Losa HA vista encofrado de tablillas 5 cm.
Perfil de aluminio extruido 18x6 cm.
Perfil Omega sujección lama.
Taco Hilti M12 anclado a losa
Remate superior aluminio anodizado
Carpintería aluminio anodizado color natural
Vidrio climalit 4.12.4
Vidrio de seguridad antepecho 4.12.3+3
Suelo tarima flotante toda la vivienda 1 cm.
Fieltro separador lámina antiimpacto
Plastón mortero aligerado e. medio 5 cm.
Carpintería aluminio anodizado color natural
Vierteaguas de aluminio anodizado
Perfil de aluminio extruido 18x6 cm.
Perfil Omega sujección lama.
Taco Hilti M12 anclado a losa
Losa HA vista encofrado de tablillas 5 cm.
vivienda y local
a desarrollo 15 cm.

PLANTA DE SITUACIÓN ESC 1:1000

3.00 M
2.40 M
3.00 M
3.00 M
5.60 M
5.60 M
3.00 M
8.40 M
1
5
10M

Architect: Bjarke Ingels, Thomas Christoffersen
Firm: Big-bjarke Ingels Group
Project leader: Ole Elkjaer-Larsen, Henrick Poulsen
Project manager: Finn Nørkjær, Henrik Lund
Collaborators: Hopfner Partners, Moe & Brodsgaard, Klar
Location: Copenhagen, DK
Area: 61000m²

8 House

8 HOUSE is located in Southern Ørestad on the edge of the Copenhagen Canal and with a view of the open spaces of Kalvebod Fælled. It is a big house in the literal sense of the word. A house offers homes in all its bearings for people in all of life's stages: the young and the old, nuclear families and singles, families that grow and families that become smaller. The bow-shaped building creates two distinct spaces, separated by the centre of the bow which hosts the communal facilities of 500 m². At the very same spot, the building is penetrated by a 9 meter wide passage that connects the two surrounding city spaces: the park area to the west and the channel area to the east. Instead of dividing the different functions of the building, for both habitation and trades, into separate blocks, the various functions have been spread out horizontally. The apartments are placed at the top while the commercial program unfolds at the base of the building. As a result, the different horizontal layers have achieved a quality of their own: the apartments benefit from the view, sunlight and fresh air, while the office leases merge with life on the street.

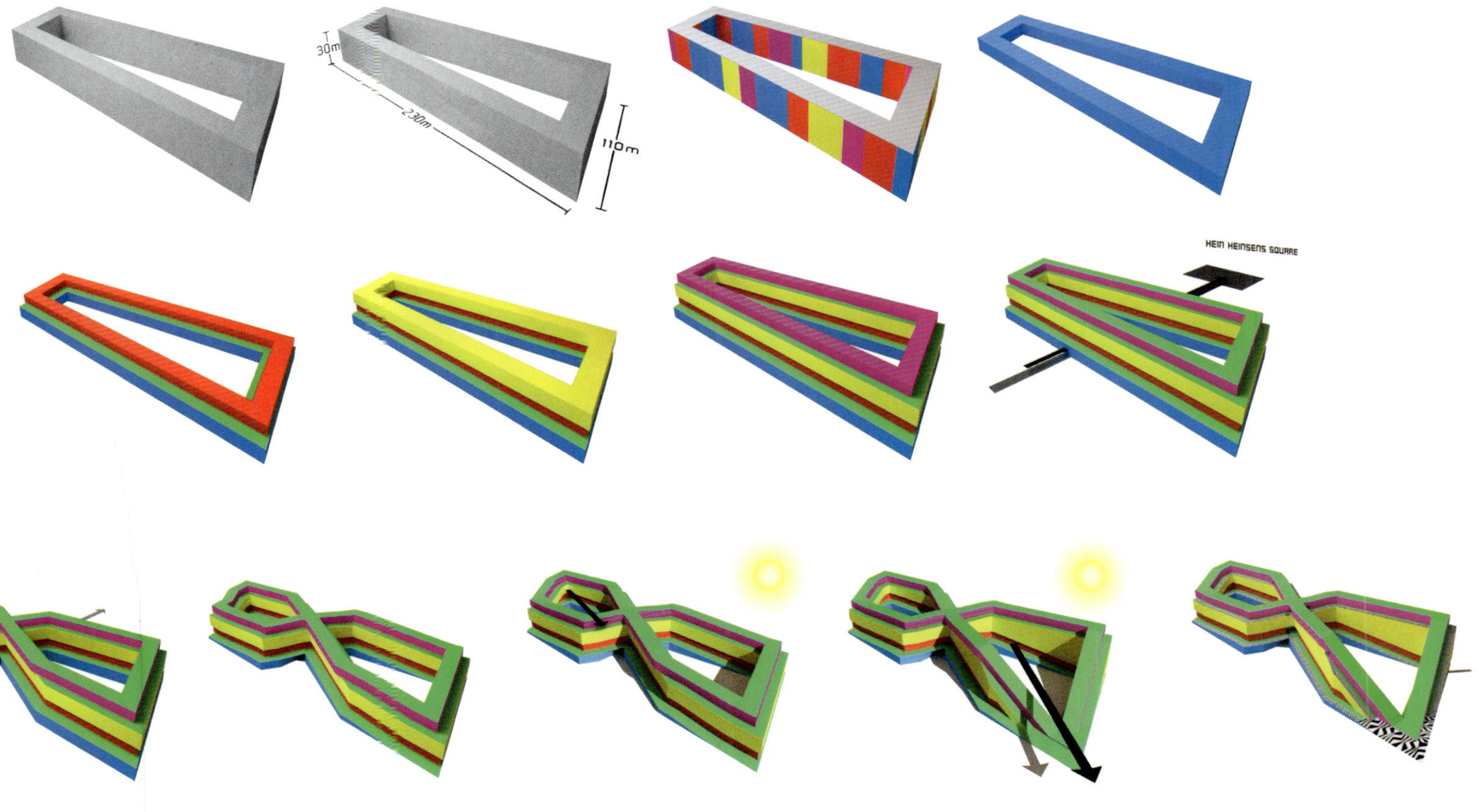
30m
230m
110m
HEIN HEINSENS SQUARE

82.st.T2

Architect: María José Pizarro, Óscar Rueda
Firm: Rueda Pizarro Arquitectos
Partners: Alberto Galindo, Ramiro Losada, Gonzalo del Val and Laura Montero
Engineering: Jofemar
Location: Ville of Vallecas, Madrid
Area: 8232.64m²
Photographer: Miguel de Guzmán

Memoria 64 Vpp Emvs Vallecas, Madrid

To build in environments as the ville of Vallecas means discovering new codes, dialogue with new monuments of the "city entropic" portrayed by Robert Smithson in the late 1960's in an article in Artforum.

Decisions are made with parameters closest to the laws of thermodynamics and field theoretical models of physics than traditional scopes: entropic landscape or thermodynamics are some new codes to which the designers turn to the development of this building.

With these premises, the designers intend to establish a compromise between the strict enforcement of the rules of the urban game (a volume absolutely preconfigured) and in-depth study of it to find the key that lend identity to the architectural volume. The designers establish a series of operations in the given solid by the rules that they let to achieve a more comfortable housing climate, temperature and light:

1. The volume is emptied in the northeast facade to get the largest possible facade with the better orientation (south) and adjust the excess build.
2. Lower the height on the south side and create a slope to get more sunlight on the walls less favorable.
3. The ground floor is 2.85 meters set back to save the corner and create a landscaped area prior that allows more privacy in the houses located on ground floor.
4. From the sixth to the eighth floor leans the facade, forming a continuous envelope that solves the problem of recessed penthouses and ground floor corner the same way.

The volume is completed with a system of energy saving and climate protection of the building system: 1. Design of windows heat sensors to south and windows shields to north, reinforced the status of enclosures in front to increase the isolation. 2. Exterior insulation system, which wraps around the building as a 'quilt', and maintains the mass inside the enclosure to increase the thermal inertia and, therefore, the temperature stability. 3. Boost for water and heating with solar panels on deck.

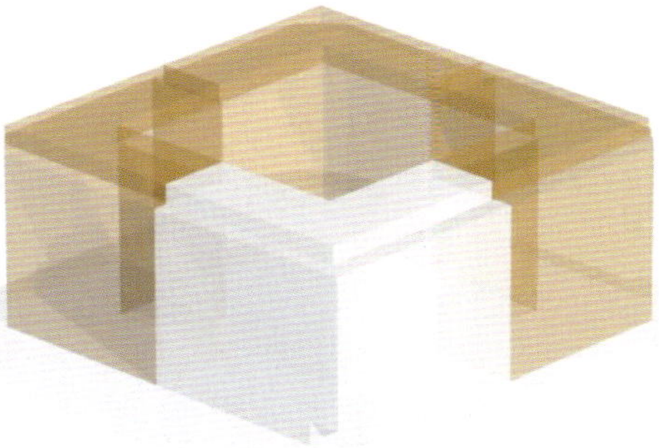

1. PLANEAMIENTO: SÓLIDO CAPAZ

2. ESTRATEGIA DE DESOCUPACIÓN: MAS SUPERFICIE FACHADA SUR

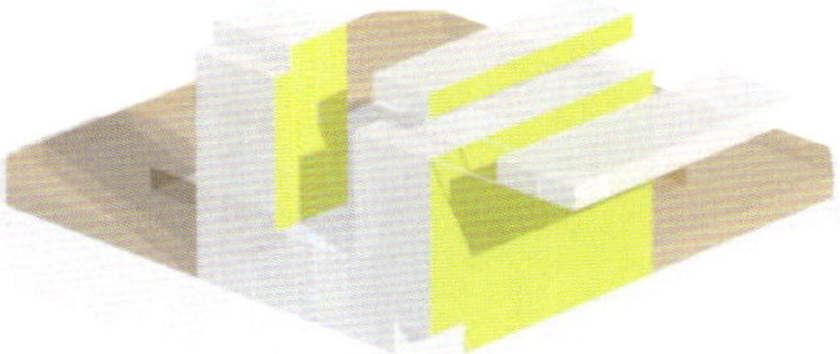

3. ESTRATEGIA DE DESOCUPACIÓN 2: DISMINUCIÓN DE UNA ALTURA

4. MODELADO DEL SÓLIDO: ÁTICO + P. BAJA

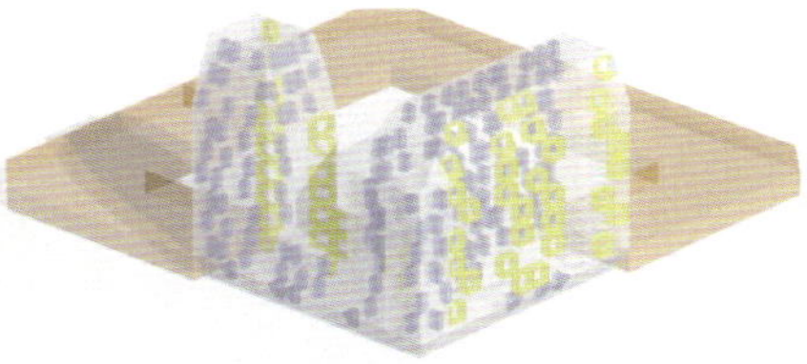

5. APERTURA DE HUECOS: CAPTADORES AL SUR + PROTECTORES AL NORTE

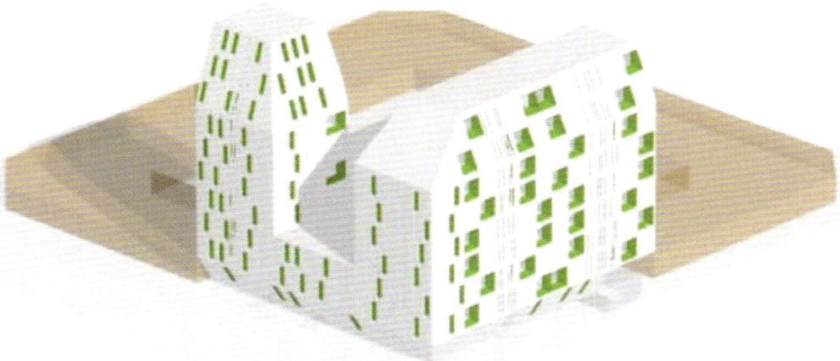

6. PROPUESTA DEFINITIVA

DESARROLLO VOLUMÉTRICO DE LA PROPUESTA

12
37.5
12
37.5

2D
2D
2D
2D
3D
3D
3D

ALZADO NOROESTE

ALZADO NORESTE

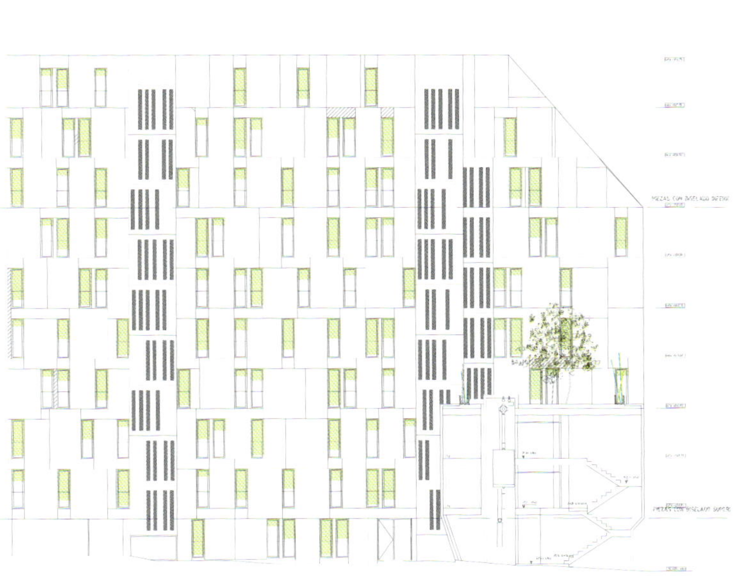

ALZADO SUROESTE

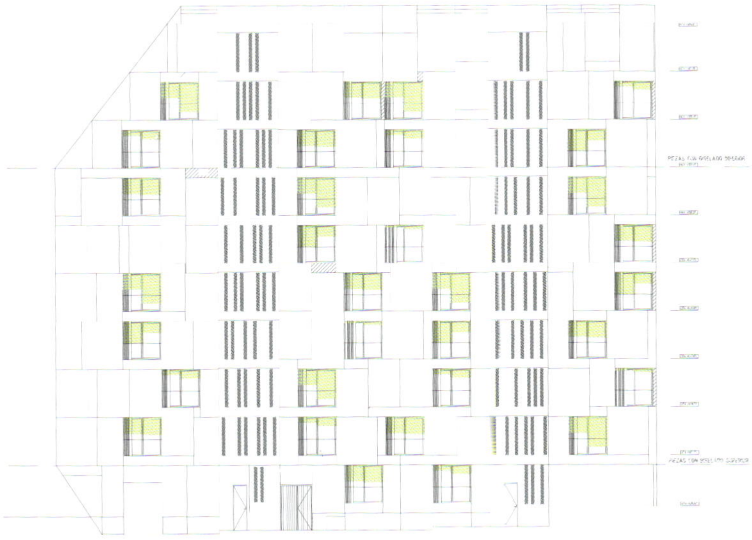

ALZADO SURESTE

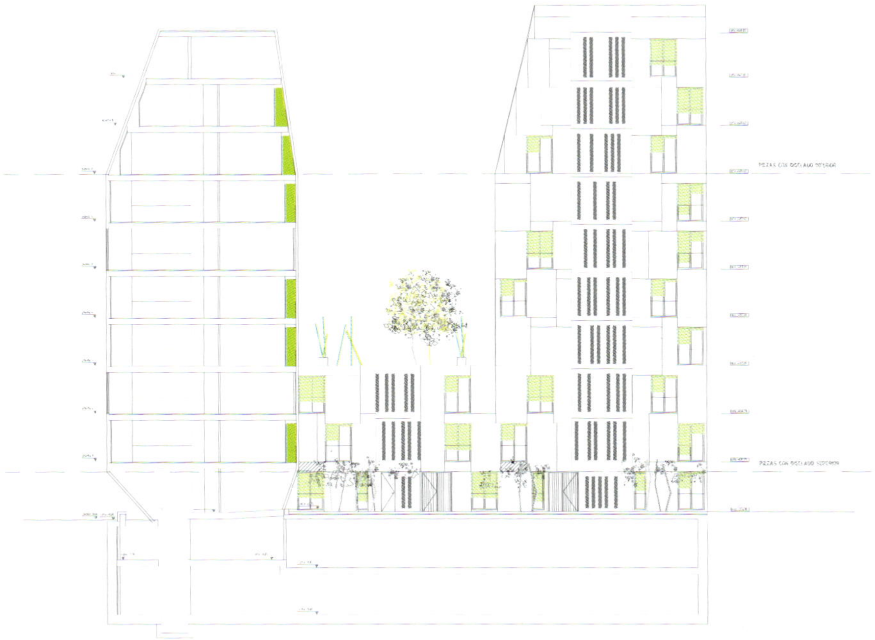

CUBIERTA PERIMETRAL DEL SOTANO

1.- GRAVA DE PROTECCIÓN, e= 3 cm
2.- LÁMINA GEOTEXTIL DE PROTECCIÓN
3.- PLANCHA DE POLIESTIRENO EXTRUSIONADO e= 4 cm, D= 35 Kg/m3
4.- DOBLE LÁMINA ARMADA DE IMPERMEABILIZACIÓN BITUMINOSA DE 4 Kg/m2
5.- HORMIGÓN CELULAR DE PENDIENTES ESPESOR MEDIO= 6 cm
6.- LÁMINA DE POLIVINILO DE BARRERA DE VAPOR
7.- FORJADO RETICULAR DE BOVEDILLA CERÁMICA e= 30+5 cm
8.- GUARNECIDO YESO NEGRO + ENLUCIDO YESO BLANCO e= 1 cm

CUBIERTA PATIO INTERIOR Y TERRAZA

9.- BALDOSA Y RODAPIE DE HORMIGON 5cm
10.- MORTERO DE AGARRE e= 9 cm
11.- DOBLE LÁMINA ARMADA DE IMPERMEABILIZACIÓN BITUMINOSA DE 4 Kg/m2
12.- LÁMINA ANTI-IMPACTO DE POLIETILENO e= 3 mm
13.- HORMIGÓN CELULAR DE PENDIENTES ESPESOR MEDIO= 6 cm
14.- PLANCHA DE POLIESTIRENO EXTRUSIONADO e= 4 cm, D= 35 Kg/m3
15.- FORJADO RETICULAR DE BOVEDILLA CERÁMICA e= 30+5 cm
16.- GUARNECIDO YESO NEGRO + ENLUCIDO YESO BLANCO e= 1 cm
17.- FORRADO DE CHAPA DE ACERO GALVANIZADO ESTIRADA EN FRIO
18.- PLETINA DE ACERO GLAVANIZADO 50.10 ANCLADA A CUBIERTA
19.- LOSA DE HORMIGON A. e=15cm. ACABADO FRATASADO

CERRAMIENTO FACHADA EXTERIOR Y SEPARACION ENTRE VIVIENDAS

20.- MORTERO COTETERM ACABADO EN BLACO COLOR A ELECCIÓN DE LA DF
20A.- MORTERO
21.- EPS DE DENSIDAD 20 KG/M3
22.- TABLERO HIDRÓFUGO (e=19mm)
23A.- ANCLAJE DE BASTIDOR CON TACOS HILTI A FORJADO MEDIANTE PERFILES L 47x7
23.- CHAPA GRECADA (e=35mm)
24.- TRASDOSADO DE CARTON-YESO
25.- SOLADO DE TARIMA FLOTANTE
26.- LAMINA ANTIMPACTO
27.- MORTERO DE NIVELACION 5cm
28.- ESTRUCTURA AUXILIAR COMPUESTA CON PARES U 80x40 Y PERFILES OMEGA TRANSVERSALES 50.40.1,5
29.- BARANDILLA DE VIDRIO STADIP 3+3 Y PERFILES METALICOS 60.20 ANCLADOS A LA FÁBRICA
30.- TRASDOSADO DE PLADUR CON AISLAMENTO ACÚSTICO
31.- TRASDOSADO LADRILLO HUECO DOBLE+ENLUCIDO YESO
32.- HORMIGON DE PENDIENTE
33.- DOBLE LAMINA DE IMPERMEABILIZACION
34.- ACABADO DE CESPED ARTIFICIAL
35.- PINTURA DE RESINAS COLOR A ELEGIR POR LA DF
36.- FALSO TECHO PLADUR

HUECO DE VENTANA

37.- VIERTEAGUAS DE ALUMINIO ANODIZADO
38.- CAJÓN DE PERSIANA DE PVC COLOR A ELECCIÓN DE LA DF. CARRILES PARA PERSIANA Y REMATE SUPERIOR CON AISLAMIENTO T
39.- CARPINTERIA DE ALUMINIO ANODIZADO CORREDERA
40.- ACRISTALAMIENTO DOBLE TIPO CLIMALIT 4/6/4

SEPARACIÓN HORIZONTAL ENTRE VIVIENDAS P. BAJA Y GARAJE

41.- SOLADO DE TARIMA FLOTATE
42.- LAMINA ANTIMPACTO
43.- MORTERO DE NIVELACION 5cm Y AISLAMIENTO TERMICO POLIESTIRENO e=4cm, D=35 Kg/m3
44.- FORJADO RETICULAR DE BOVEDILLA CERÁMICA e= 30+5 cm
45.- GUARNECIDO YESO NEGRO + ENLUCIDO YESO BLANCO e= 1 cm

SOLADO DE PATIO-PLAZA SOBRE GARAJE

46.- ADOQUÍN DE HORMIGÓN SOBRE CAMA DE ARENA SOBRE FORJADO RETICULAR DE BOVEDILLA CERÁMICA e= 30+5 cm CON DOBLE LÁMINA ARMADA DE IMPERMEABILIZACIÓN BITUMINOSA DE 4 Kg/m2

SOLADO DE PATIO-PLAZA SOBRE TERRENO NATURAL

47.- ADOQUÍN DE HORMIGÓN SOBRE CAMA DE ARENA SOBRE SOLERA DE HORMIGÓN CON MALLAZO SOBRE TERRENO NATURAL COMPACTADO

SEPARACIÓN HORIZONTAL ENTRE PLANTAS DE GARAJE

48.- PAVIMENTO CONTINUO IN SITU DE HORMIGÓN PULIDO ENDURECIDO AL CUARZO
49.- FORJADO RETICULAR DE BOVEDILLA CERÁMICA e= 30+5 cm
50.- GUARNECIDO YESO NEGRO + ENLUCIDO YESO BLANCO e= 1 cm

SOLERA PLANTA SÓTANO

51.- PAVIMENTO CONTINUO IN SITU DE HORMIGÓN PULIDO ENDURECIDO AL CUARZO
52.- LOSA DE HORMIGON DE CIMENTACION
53.- LAMINA IMPERMEABILIZANTE EN "POLIVINILO"
55.- TERRENO NATURAL VACIADO

BANCO PERIMETRAL EXTERIOR VENTILACION GARAJE

56.- MURO DE 1/2 PIE DE LADRILLO TOSCO Y TRATAMIENTO ANTIGRAFITTI AL EXTERIOR. ENFOSCADO Y PINTURA AL SILICATO AL INTERIOR CON LOSA INFERIOR DE HORMIGON ARMADO e=15cm

MURO CONTENCIÓN A CALLE EXTERIOR

57.- CAPA REGULARIZADORA DE MORTERO DE CEMENTO
58.- VIGA CORONACIÓN ATADO CABEZA PILOTES
59.- ENFOSCADO DE MORTERO HIDROFUGO SOBRE MALLAZO METÁLICO DE REFUERZO #10x10x#4
60.- FÁBRICA DE TABICÓN DE LADRILLO HUECO DOBLE
61.- CAMARA BUFA
62.- LAMINA IMPERMEABILIZANTE TIPO "DELTA DRAIN"
63.- PANTALLA DE PILOTES CON GUNITADO DE HORMIGÓN SOBRE MALLA METÁLICA EN ESPACIO ENTRE PILOTES
64.- CUÑA DE MORTERO
65.- APOYO DE FORJADO
66.- REJILLA DE VENTILACION DE ALUMINIO 20x20 cm
67.- REGISTRO-REJILLA DE ALUMINIO 40x40 cm
68.- LOSA ARMADA PARA APOYO FÁBRICA 1/2 PIE LADRILLO
69.- CANALETA DE MORTERO DE CEMENTO
70.- BANDA LONGITUDINAL DE POLIESTIRENO EXPANDIDO e= 2 cm
71.- TUBO P.V.C. Ø90 CONECTADO A RED HORIZONTAL DE SUMIDEROS DE GARAJE

MURO CONTENCIÓN A PATIO-PLAZA

73.- MURO DE HORMIGÓN ARMADO 30 cm CON PINTURA AL SILICATO AL INTERIOR
74.- TUBO DE DRENAJE EXTERIOR SOBRE CAMA DE HORMIGÓN
75.- TUBO P.V.C. Ø90 CONECTADO A RED HORIZONTAL DE SUMIDEROS DE GARAJE
76.- LAMINA IMPERMEABILIZANTE TIPO "DELTA DRAIN"
77.- LOSA ARMADA DE HORMIGÓN
78.- TERRENO NATURAL COMPACTADO

RECOGIDA DE AGUAS

79.- CANALON ACERO GALVANIZADO 2 mm
80.- REJILLA DESAGUE EN TERRAZA
81.- GOTERON EN PANEL PREFABRICADO

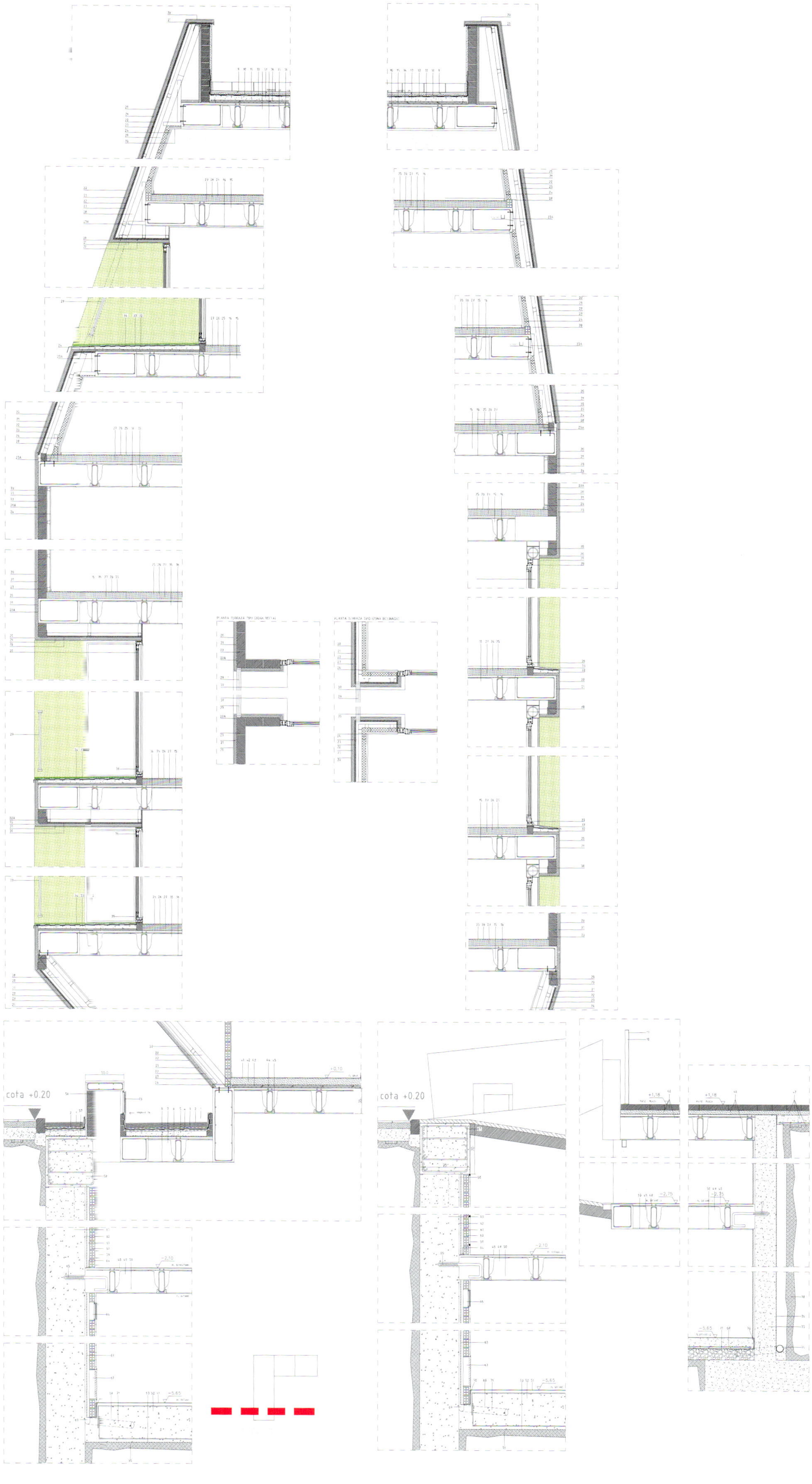

2D
2D

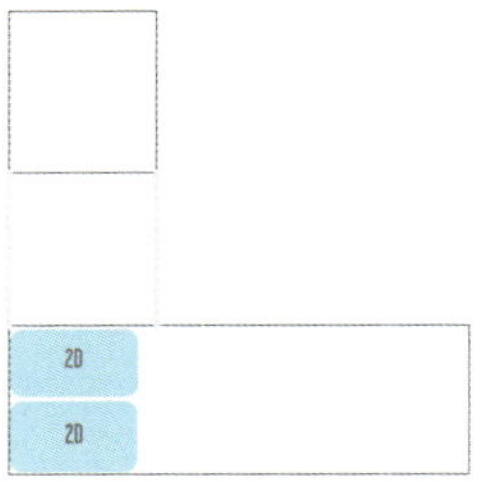
2D
2D

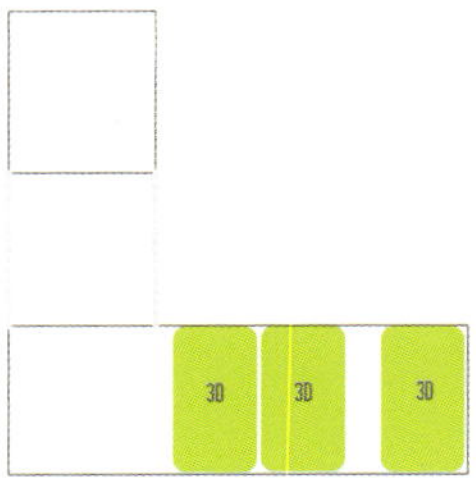
3D
3D
3D

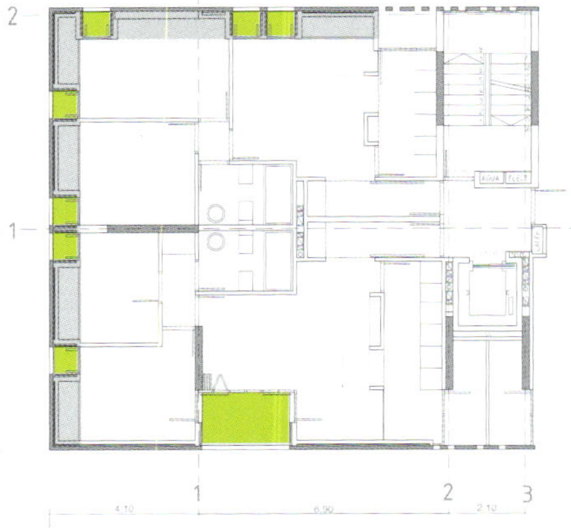

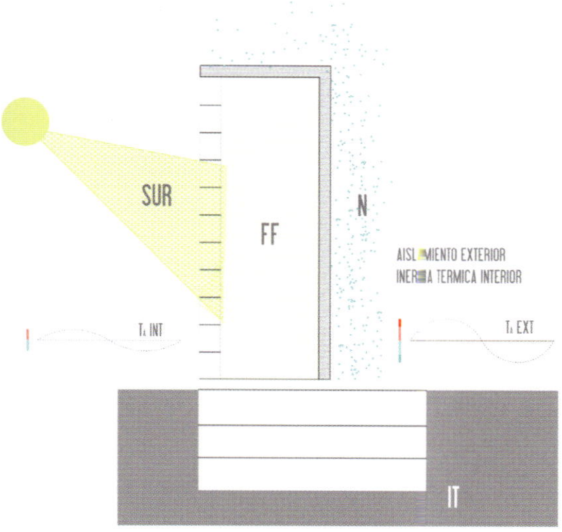
SUR
FF
N
AISL MIENTO EXTERIOR
INER A TERMICA INTERIOR
IT

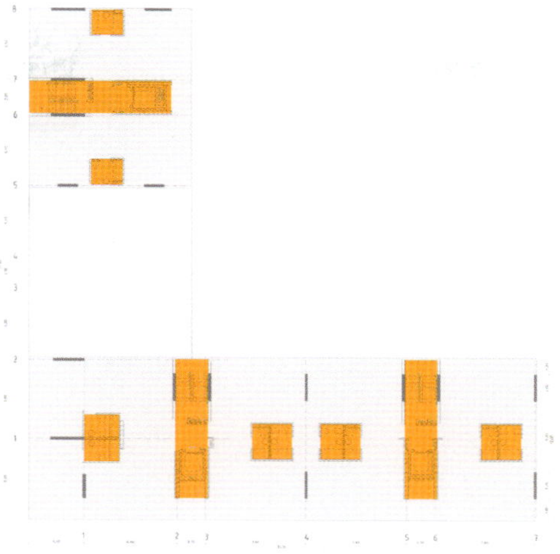

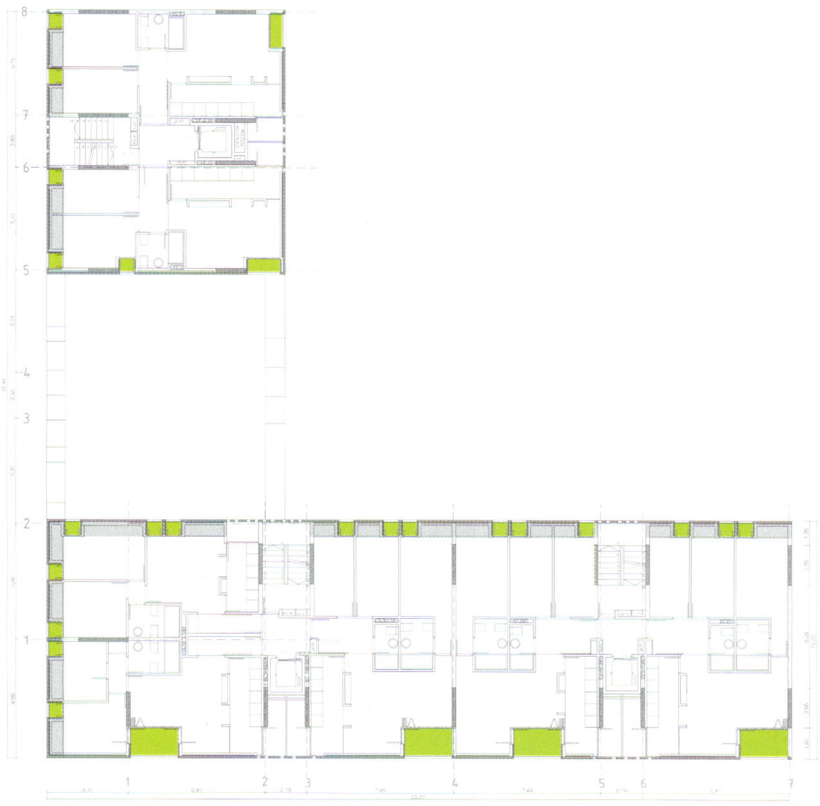

PLANTA OCTAVA_cota 658.40

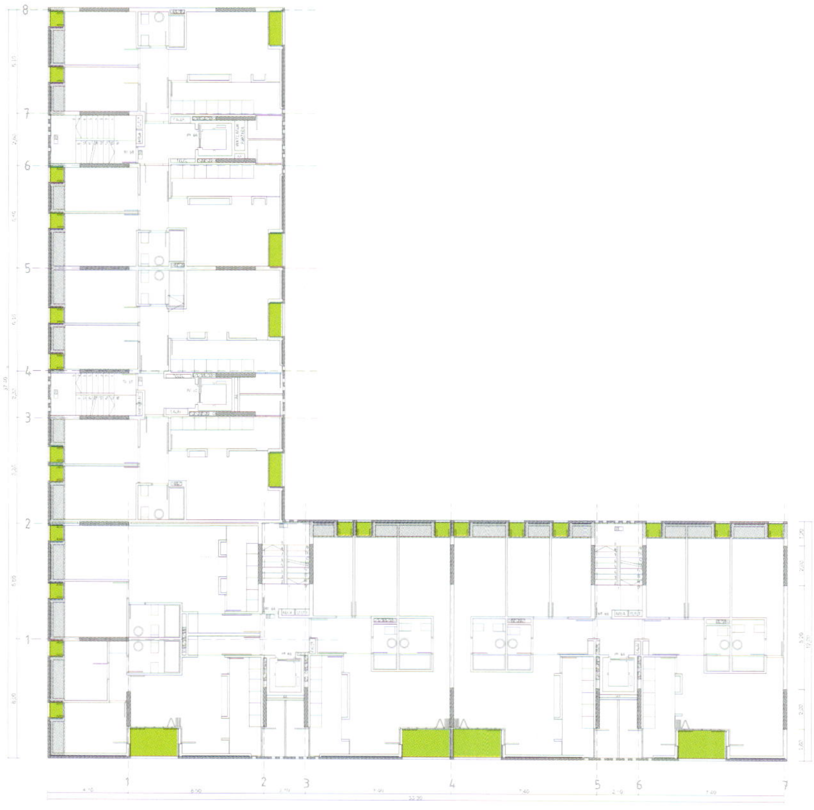

Architect: Peter Kis, Csaba Valkai, Ivett Tarr, János Sarusi Kis, Tamás Ükös, Gábor Balázsa, Gábor Divinyi, Barbora Tuckova
Firm: Kis Péter Építészműterme
Strucure: FRT Raszter Mérnökiroda – Attila Farsang, Orsolya Dudinszky
Garden design: s73 – Sándor Mohácsi, Péter Balogh
Static: mTm Mérnökiroda – Péter Markovits, Róbert Kazsimérszky
Mechanist: Art- tax Szolgáltató Szövetkezet – András Oltvai, Zsolt Lengyel
Electronics engineer: Hochplan Kft. – Dezső Sax, Karola Kárpáti
Public services: László Kádár
Lift: Schindler Kft. – Gábor Balogh
Transport: Közlekedés Kft. János – Szegő, Angéla Kovács
Fire protection: Tűz-Ért-Ő Bt. – Sándor Venczel
Environment: Kotschy Bt. – András Kotschy
Akustic: Elekné Karsai Dr.
Area: 1218m²
Materials: concrete
Photographer: Zsolt Batár, Daniel Németh, Sándor Szabó

Práter Street

Rise of value

The firm will generate the development of the certain site, the sorrounding blocks and its wider environment. The integrated building form generates an inner garden, that visually connects the street to the inner space of the new block. Besides giving the green experience to the space, this solution lets natural sunlight in, and give natural light to the inner street and apartments. The designers connected public spaces to this inner garden, that adds a certain value, and completes the simple function of the apartments. Using the advantage of the location, the designers could create two roof terraces, that give extraordinary panoramic view to the city. The vegetation and its design elements create a continuous, L-shaped garden and public space, which at the same time separate and connect the two block of the new building.

Organization of inner spaces

A clear and uncomplicated longitudinal zone connects all the functions (apartments, parking, public spaces). This zone and the developed structural system allow the fexibility that appears in the organization of the functions, particularly in the layout of the apartments.

Sustainability and efficiency

Besides using contemporary building methods, the designers focused on the sustainability of the building by using durable and persistent materials. The building suggests elegance and balance by its proportion, scale and inner spaces, while it is functionally organized, frugal. The cost estimation and the consturction plane was continuously controlled during the design procedure.

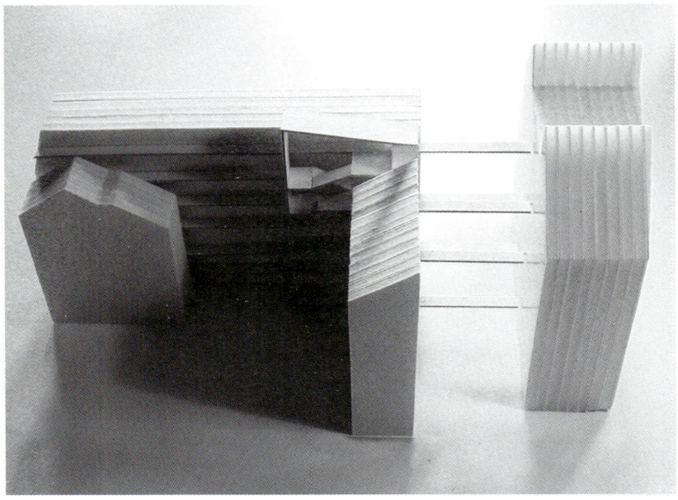

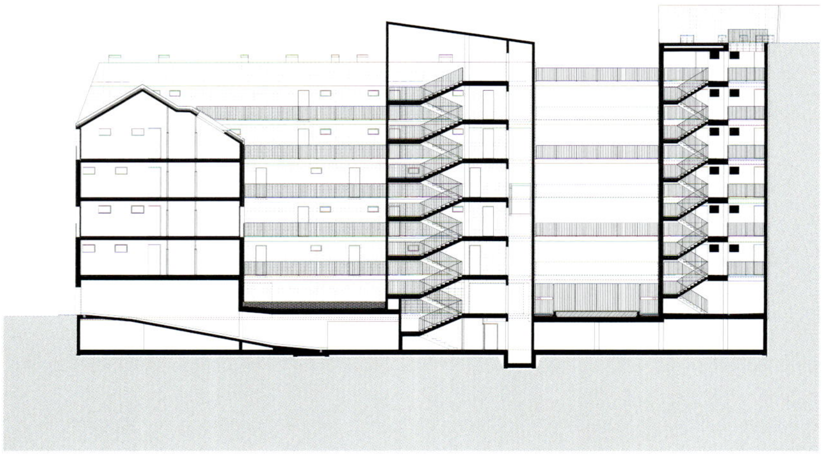

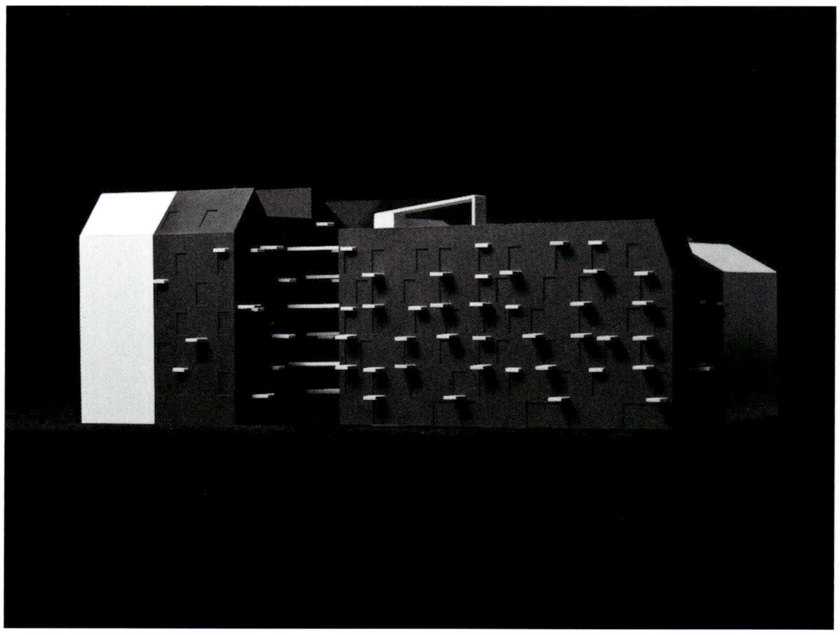
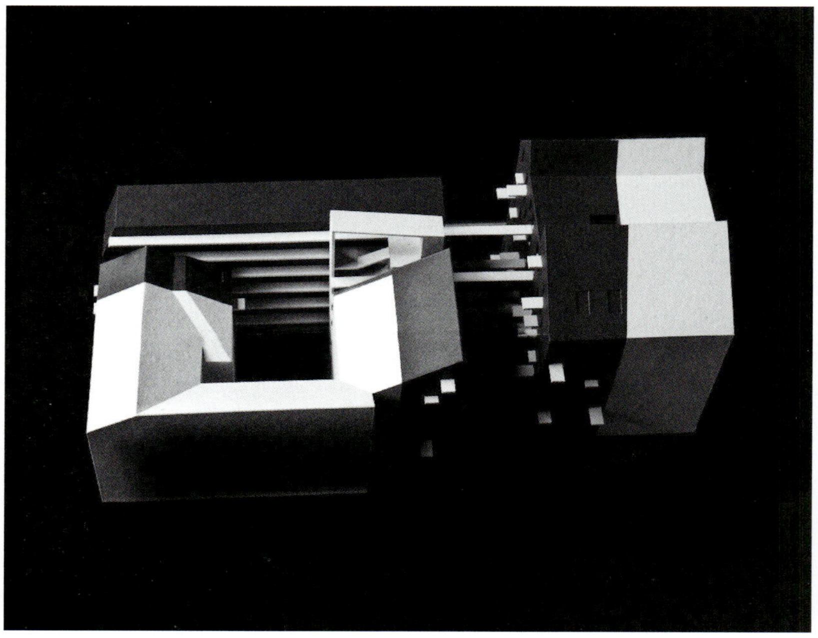

SZAKK

basement

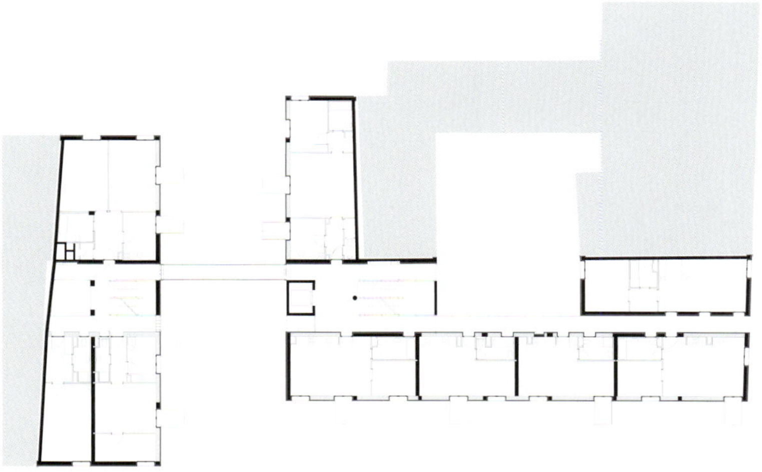

first floor

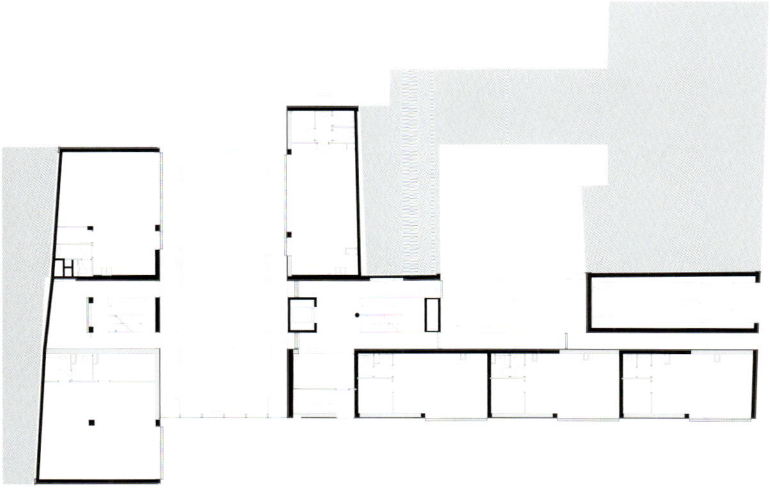

ground floor

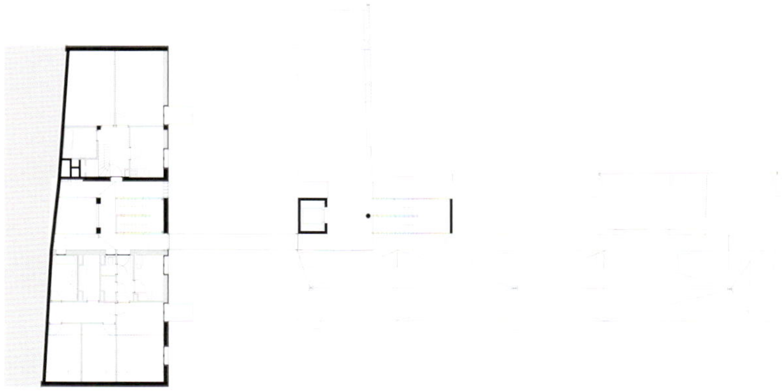

6nd floor

Architect: Marlies Rohmer
Firm: Architectural office Marlies Rohmer
Location: IJburg Amsterdam NL
Area: 10625m²
Photographer: Marcel van der Burg, Luuk Kramer, Roos Aldershoff Fotografie

Floating Houses

The Netherlands has a history of living close to water and of coping with its caprices. That means living on land protected by dykes, on mounds, on shore or floating. Only recently have floating homes been eligible as a significant solution to Holland's modern housing needs. Canals with houseboats are of course a familiar sight in Dutch cities and one may find the occasional floating hotel or restaurant. But these are always individual units and bear more resemblance to boats than to houses. In recent years, however, there has been an increase in the number of water-based housing developments that share more characteristics with land-based housing. These floating dwellings form part of an urban design. They are financially classified as immovable properties, and compete with land-based accommodation in their interior volume and comfort level. The new water-based developments can incorporate several forms of living with the water. Besides floating homes, they may include amphibious homes and homes that stand free of the water on mounds, dykes or other waterside situations. The IJburg district of Amsterdam is to have complete floating neighbourhoods, with jetties instead of paved footpaths and city plazas.

The growing enthusiasm for living beside or on the water has two pragmatic motives. Firstly, rising sea levels and increased precipitation will make it necessary to dedicate ever larger areas of land to water storage basins and peak overflow zones. Secondly, some hold that there is already a crucial shortage of new building land. Not everyone shares that view, however, the strategic consolidation of existing towns produces a sharper urban/rural contrast, making many greenfield developments superfluous or even detrimental. Living and working on the water is in effect the multiple utilization of space. It is also a way of redeveloping obsolete dockland areas and flooded quarries. Another, more aesthetic, argument in favour of living on the water is that it fosters a feeling of liberty and of closeness to nature.

Haringbuisdijk
Kadegebouw
IJburglaan
10
20
30
40
50 m

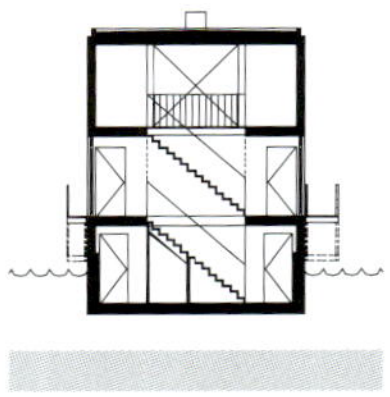

doorsnede A

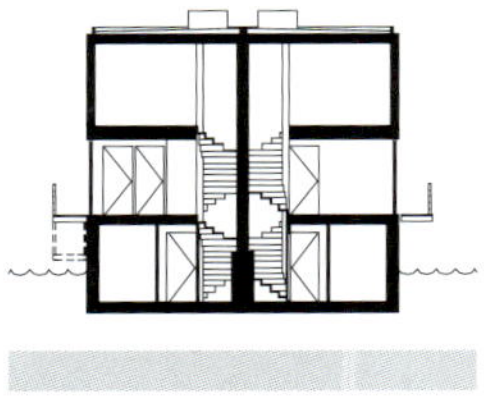

doorsnede A

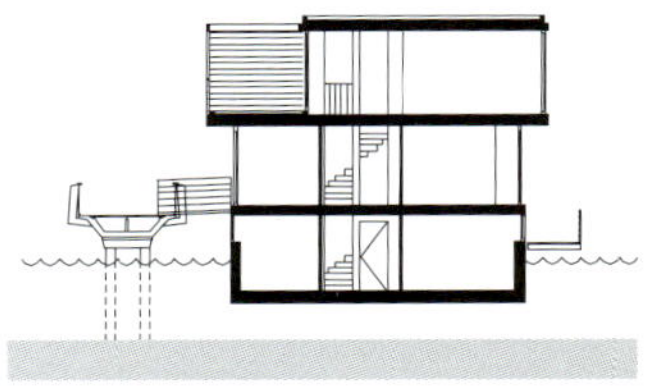

doorsnede B

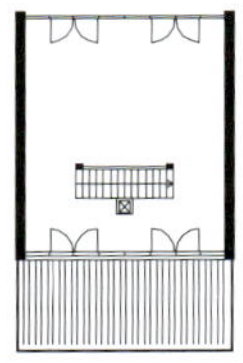

laag 01

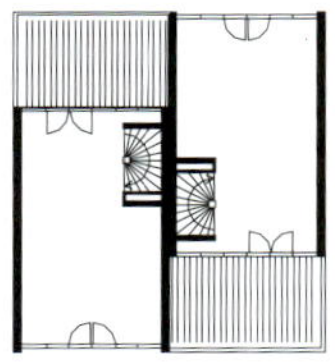

laag 01

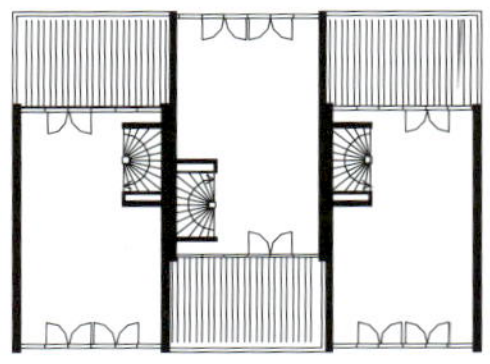

laag 01

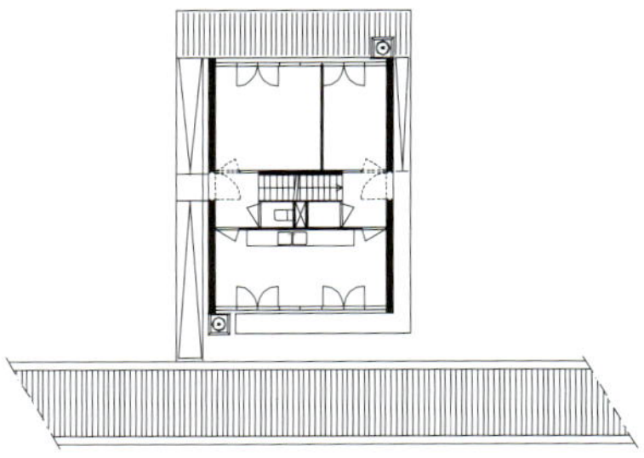

laag 00

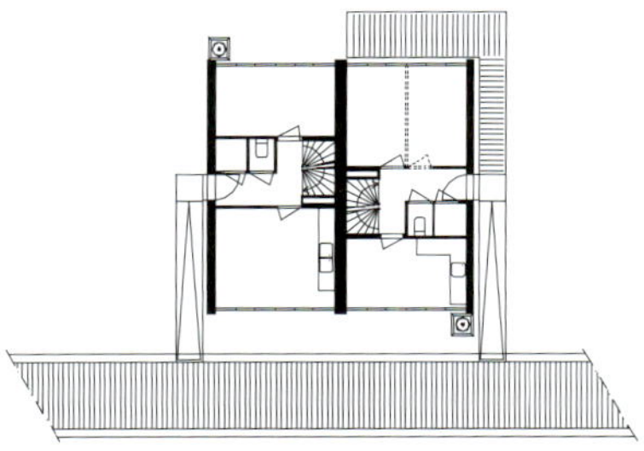

laag 00

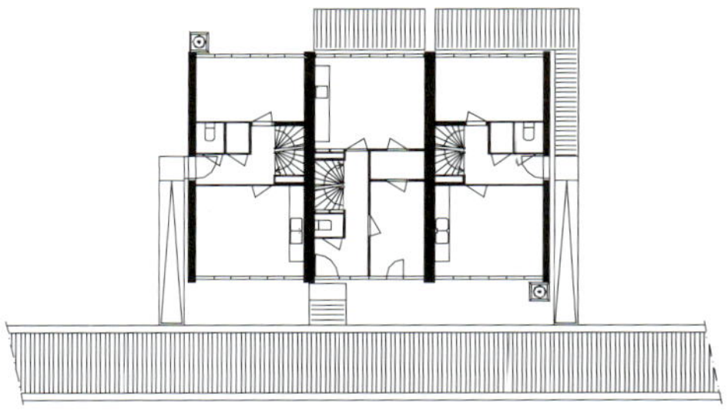

laag 00

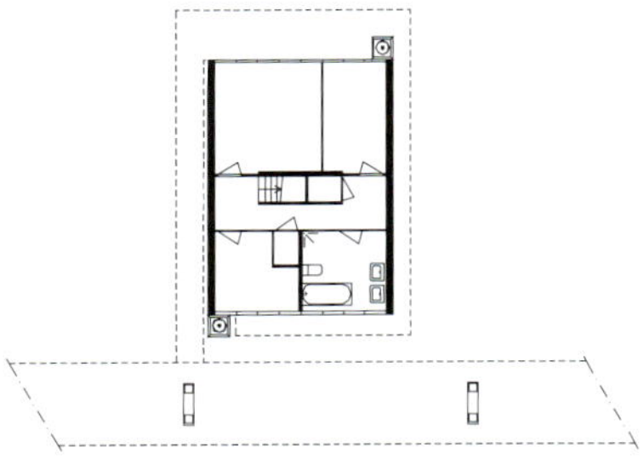

laag -01

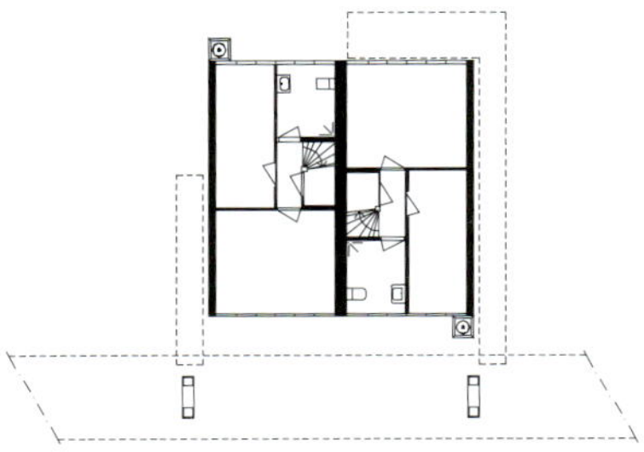

laag -01

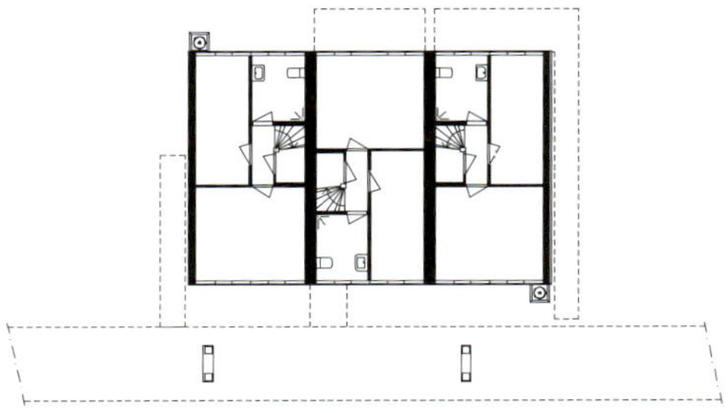

laag -01

PRESENTATIE

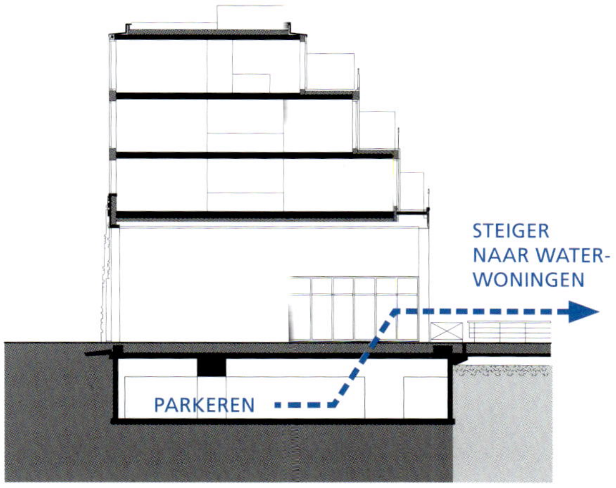

PARKEREN IN KELDER KADEGEBOUW

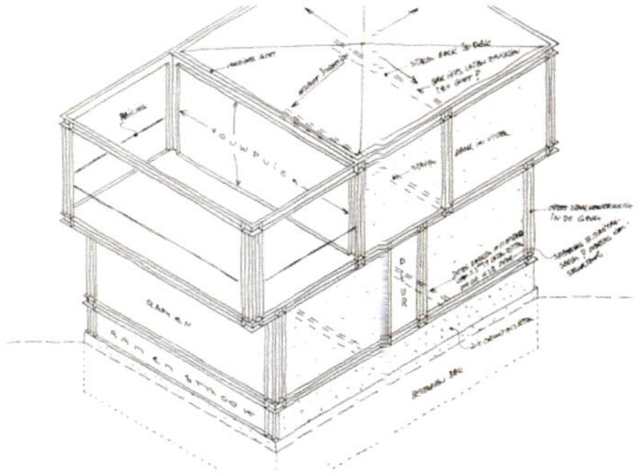

281

Architect: Ute Schneider, Han van den Born, Markus Appenzeller, Irma van Oort, Kees Christiaanse and Ruurd Gietema
Firm: KCAP Architects&Planners
Location: Amsterdam
Area: 15000m²
Photographer: Paulien Borst

Django Building

'Django Building' is an apartment complex in Amsterdam's Zuidas development area. The building occupies lot 8 of Zuidas' Gershwin, cluster III for which KCAP previously completed a mass study. It comprises of 108 rental apartments, commercial spaces and a one-storey underground parking garage by Inbo architects. With its 9 storeys, the Django building is the lowest city block between the highest building in the area, the 29-storey twin tower Amsterdam Symphony designed by the Architecten Cie, and the 20-storey Duke Tower designed by Ateliers Lion. The total program of 15000 m² is distributed over two buildings, a sunken garden and several terraces. The lower floors, with mainly commercial spaces, are defined by a printed glass facade. Above, black brickwork and a repetitive but slightly shifted pattern of anodised window frames and balconies dominate the building's appearance. The all-sidedness of the facade enhances the sculptural character of the stone volumes on the glass plinth. Constructively, the application of a load bearing facade allows for large window openings on all sides. All living areas are oriented in two directions and glased from floor to ceiling. Innovative cold and warmth storage, with a closed source system, is employed to ensure sustainable cooling and heating in the apartments.

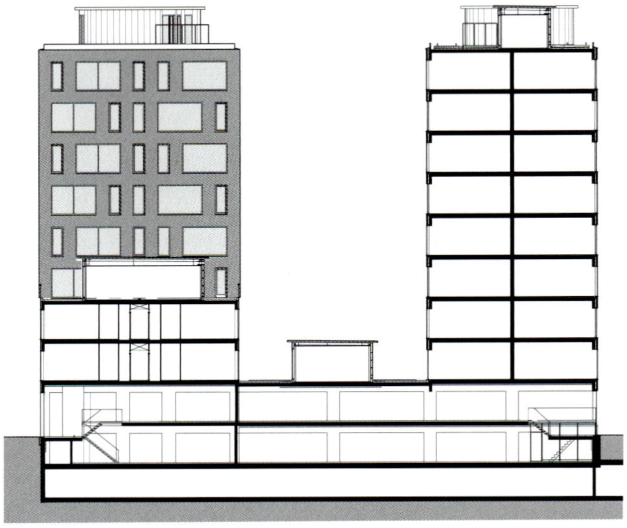

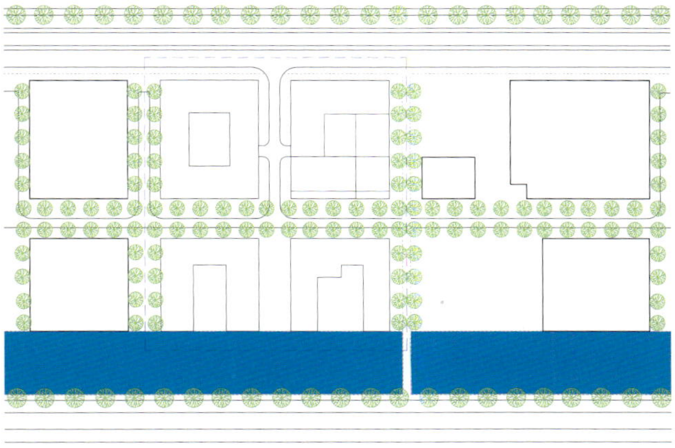

DIANGO.nl

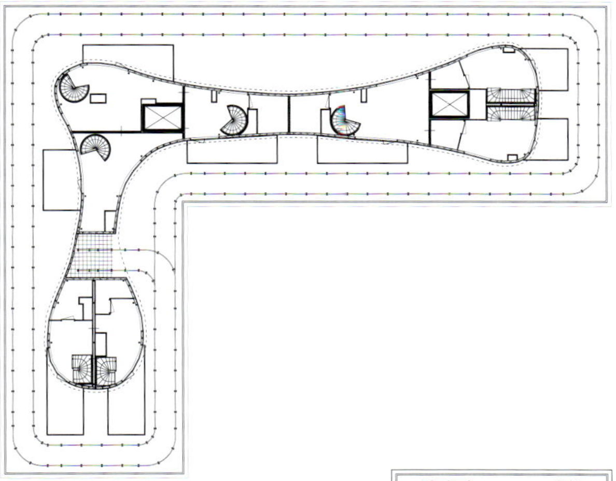
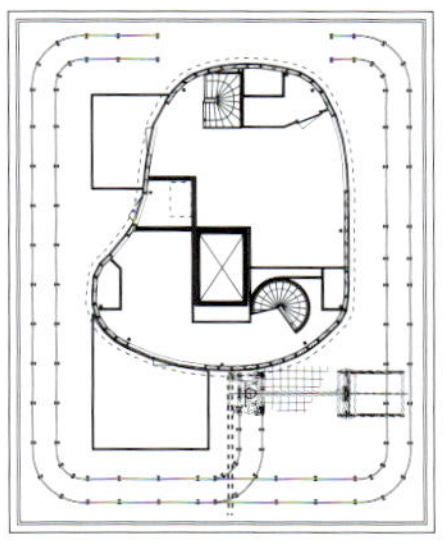
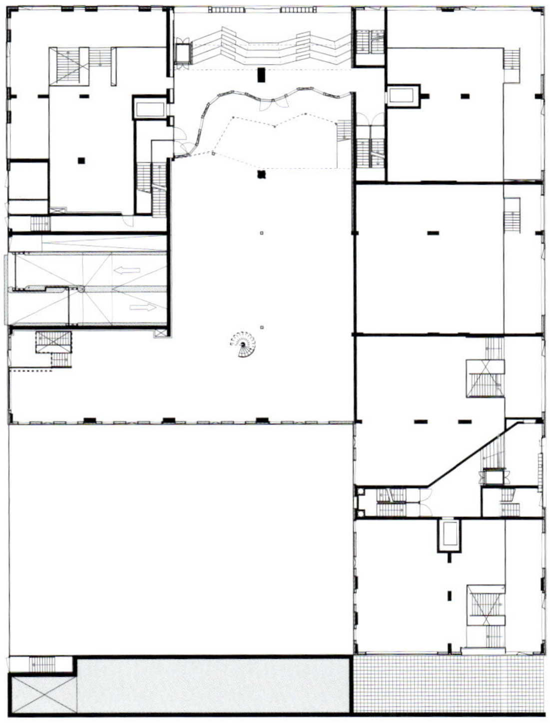
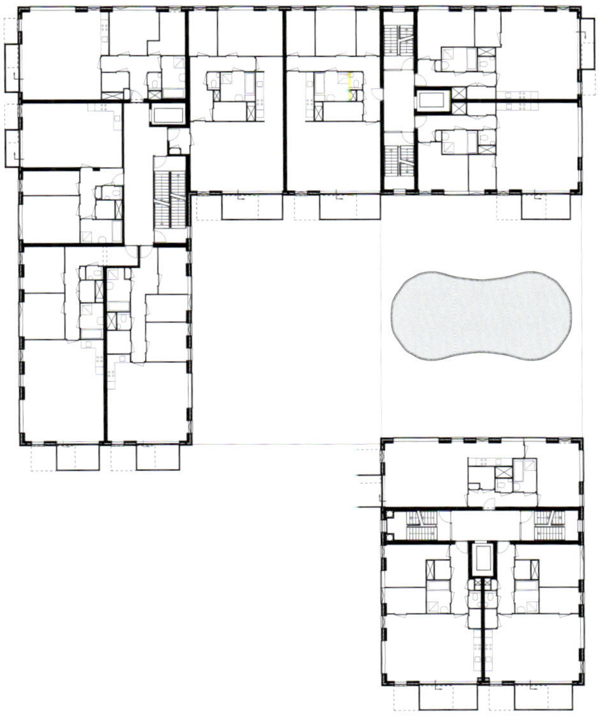
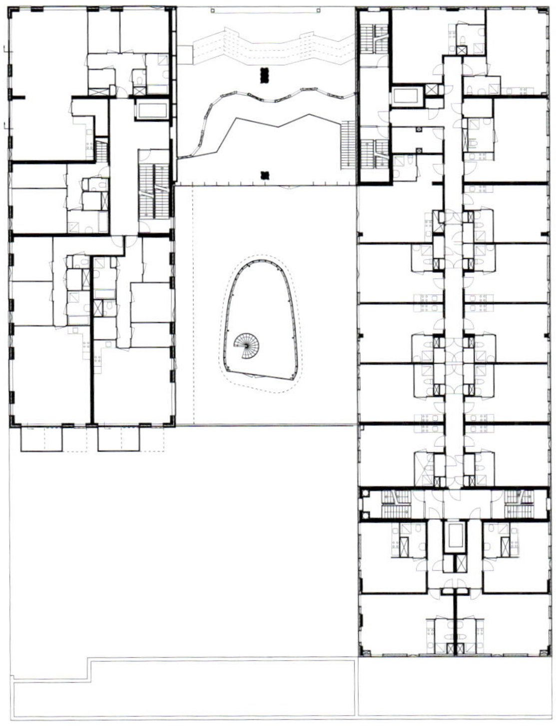

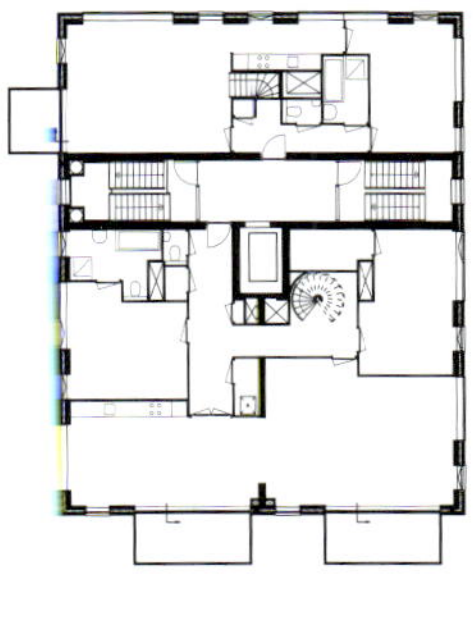

Architect: Fact Architects
Firm: Fact Architects
Design team: Inge Brouwer, Peer Glandorff
Urban design collaborator: Hvdn Architecten
Location: Amsterdam, the Netherlands
Area: 12395m²
Photographer: Luuk Kramer

Zuiderzeeweg

Zuiderzeeweg is a housing project for students.
The building blocks are made from high quality units. The units are prefabricated with balcony, windows, floor heating system and sanitary. After completion the units were transported to the location.
The buildings are sustainable and prepared for future environmental requirements.
The location is temporary, in the future all the buildings will move to a definitive location.
Fact Architects designed a durable and special facade.
The dark wooden façade with the orange texts give the complex a sustainable and robust appearance.
The exterior area is designed with the same clear approach.
The buildings and grass hills enclose a courtyard consisting of lawns and an orange sports field. Along the ring road bicycle parking bays and spaces for bins are situated, separated by with roses overgrown walls.
At the main road concrete blocks are placed that can be used as seating.

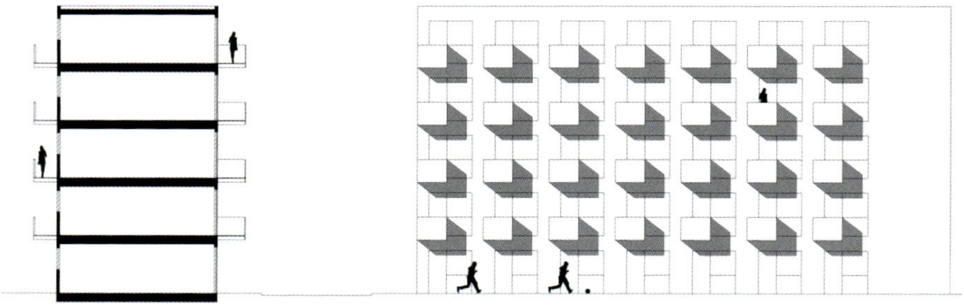

86K
82H
80R
80L

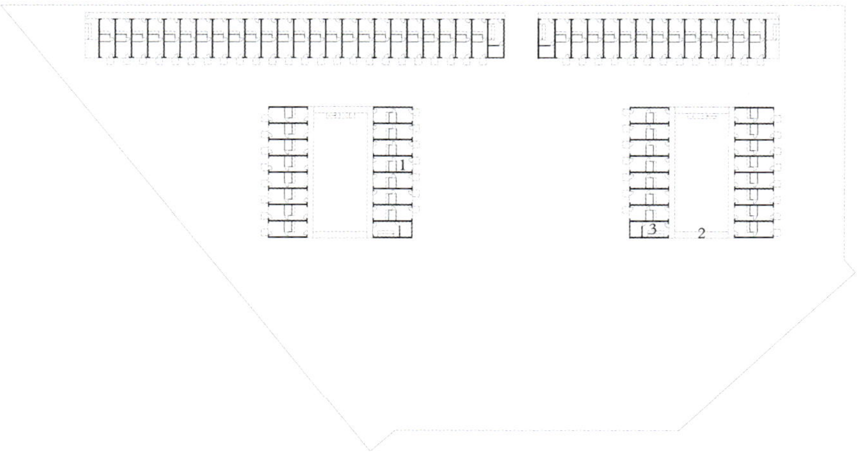

Zuiderzeeweg

Floor 1 - 4

10m.

1. *Apartments*
2. *Corridors*
3. *Staircases*

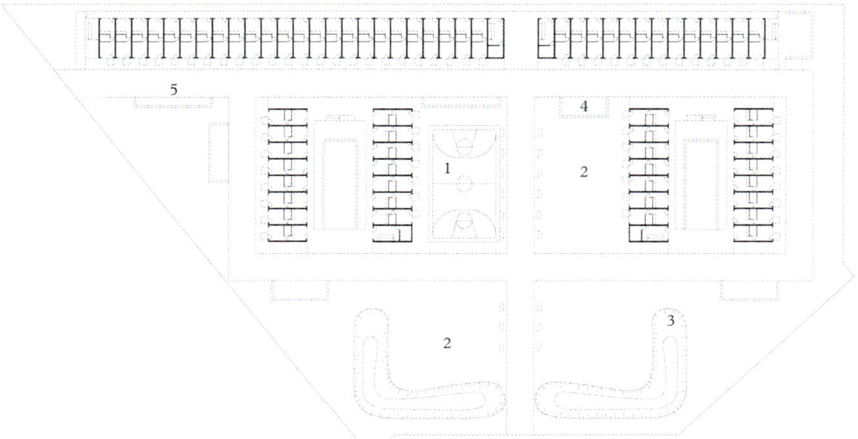

Zuiderzeeweg

Ground floor

10m.

1. *Sports field*
2. *Gras field*
3. *Gras hill*
4. *Bicycle parking bays*
5. *Container bays*

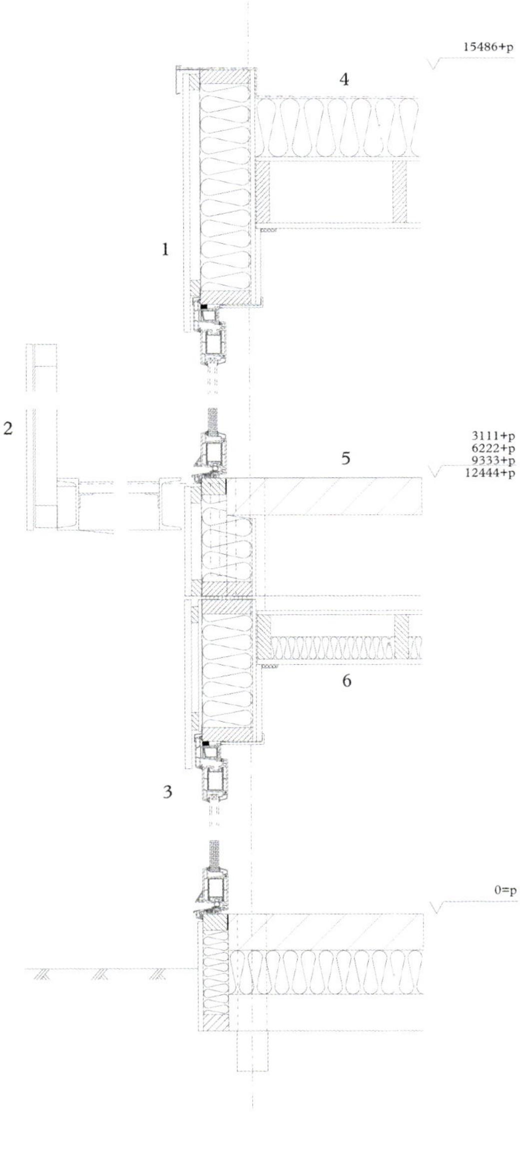

Zuiderzeeweg

Fragment

1. Wooden cladding
Batten
Permeable film
Timber framing
Insulation
Vapor proof film
Plywood
Plasterboard
2. Wooden cladding
Galvanized steel frame
3. Glazing HR++
Syntetic framing

4. Roofing
Insulation
Vapor proof film
Plywood
Timber framing
Plasterboard
5. Precast reinforced concrete floor with floor heating
6. Plywood
Timber framing
Insulation
Plasterboard

Architects: Jorge Martínez , Laura Sánchez
Firm: coco arquitectos
Quantity surveyor: David Gil, Juan Carlos Corona
Structural engineer: Gogayte SL
Mechanical engineer: Geasyt SA
Location: Madrid. Spain
Area: 19000m²
Photographer: Miguel de Guzmán, Ignacio Izquierdo, coco arquitectos

168 Social Housing in Madrid

This project is the result of the winning 1st prize in a competition promoted by the Municipal Housing Agency in Madrid, to build 168 social dwellings. The strong sloped plot was placed in the very edge of the city.
Outlying contexts surrounded by city border highways, fields, malls... require consistent answers. These are places where any action demands the same audacity, for example, a leaning building. The construction expresses with the language of kinetic, fitting its shape parallel to the slope while the ground floor retrieves the level of the plot defining streets.
The designers were required for small dwellings, with just one or two bedrooms, assigned to young people, what allowed the designers to think in new housing concepts.
The building plan is a thin strip, with narrow apartments opened onto two different scenes: one side the city, the other the wide private garden. Every unit has a space which crosses from side to side, opening simultaneously to both of them. The designers seek for dwelling with blurred frontiers between spaces, where inner room shape can be changed, and with the presence of an extra room: a room floating outside the building, attached to the facade, large enough to accommodate any domestic activity.
The designers approach social housing from present society, where customizing your own house is a way to seek for individuality. In this context, the new room — freely set in the façade — expresses this new understanding: a random image with all elevation being unique, every single dwelling different from each other, and only the intervention of the users helping it to acquire significance.

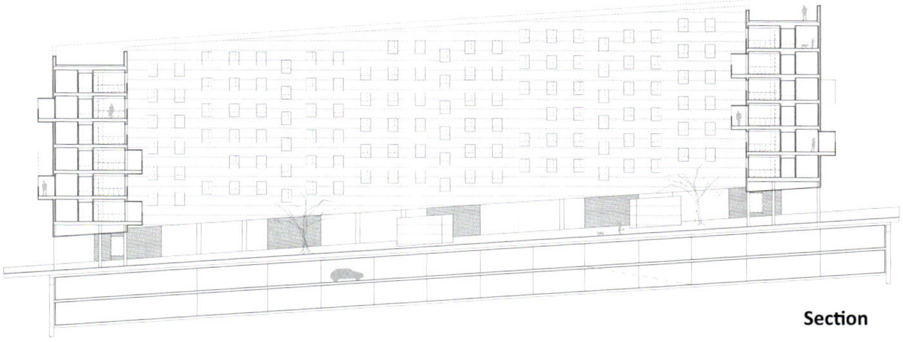

Section

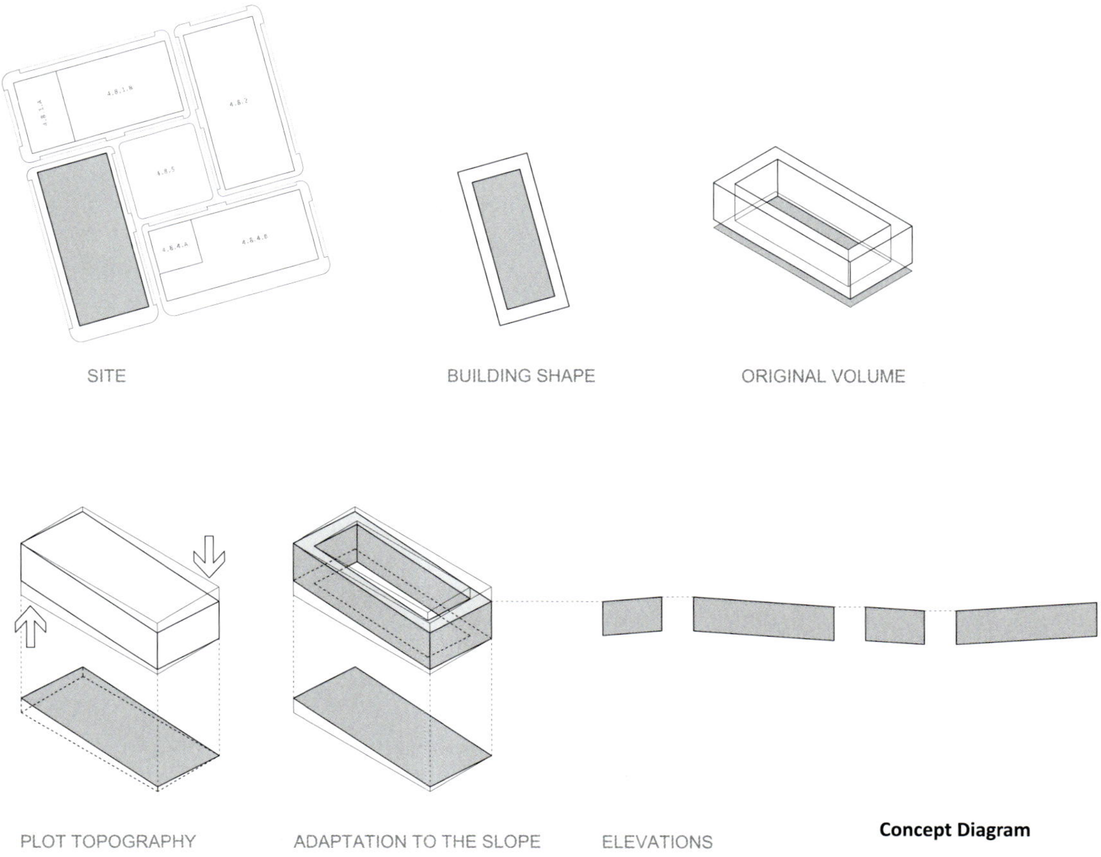

Concept Diagram

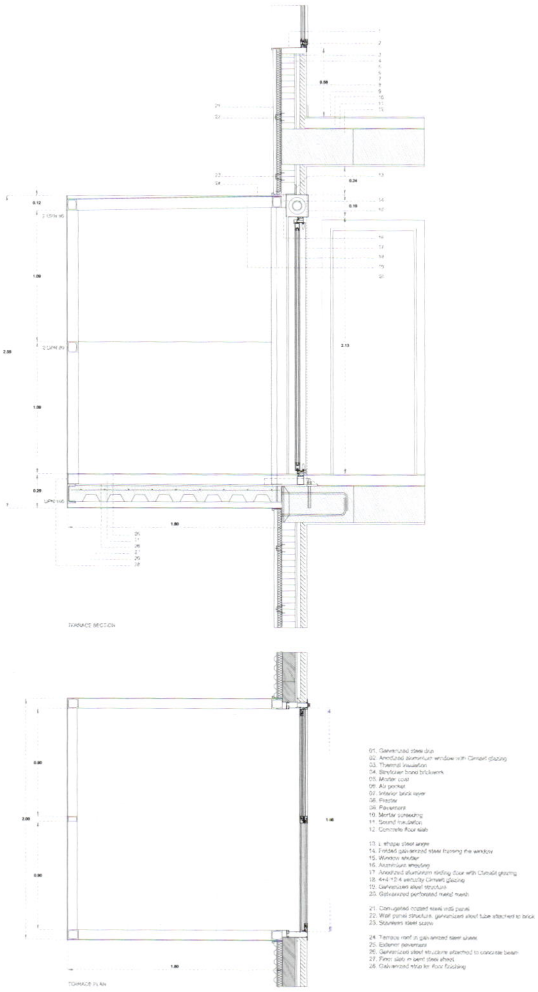

Terrace detail

EXTRA ROOM
BLURRED FRONTIERS
2 FACADES CROSSING SPACE
SERVICES

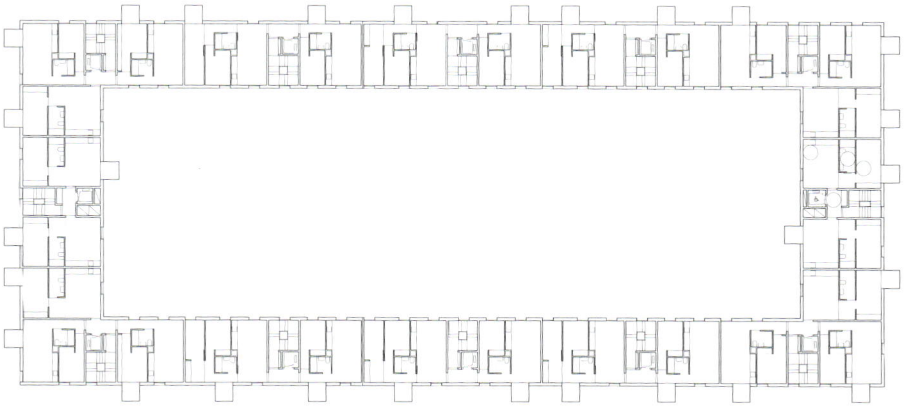

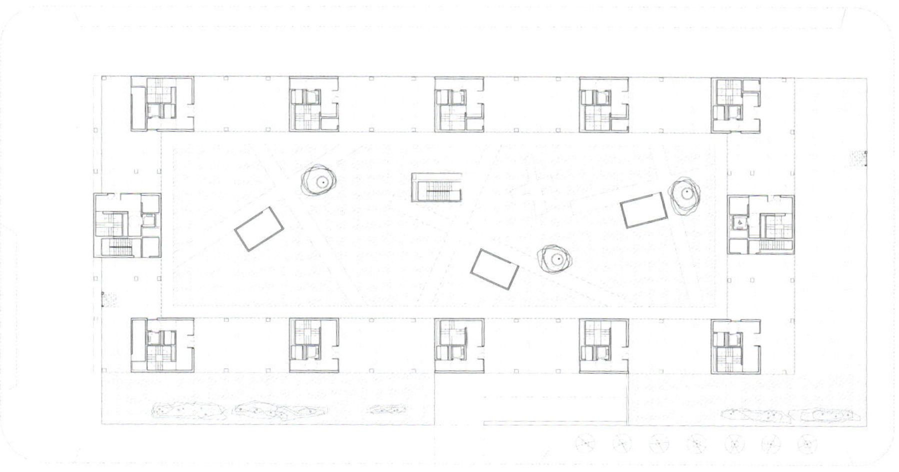

Architect: Thomas Raynaud, MVRDV, Erick Van Egeraat, Manuelle Gautrand, Pierre Gautier
Firm: ECDM, MVRDV, Erick Van Egeraat, Manuelle Gautrand, Pierre Gautier.
Engineering: Kephren, Alto, Peutz Acoustic, Van Santen
Location: Rue Denuzière, Quartier de la Confluence, Lyon (69)
Area: 35060m²
Photographer: Philippe Ruault

The Monolith, Lyon Confluence

Inhabiting Lyon Confluence is, living in the Lyon Centre with its services, its cultural landscape, its animation. Therefore, it's the duty of the project to preserve the qualities and values of the historic city by redefining a new, contemporaneous lifestyle, guided by 2 strong and converging aspirations: a modern environment and a healthy lifestyle, taking into account the requirements for the implementation of sustainable development. Naturally, the purpose is to tend to a more complex city, in a "Tetris attitude" considering combination, to put up scenarios for improbable events, to propose, without mimicry or formal posture, a sedimentation characteristic of the ancient city centres.

The approach is programmatic; the building shelters, contains, adds all together: shops, disabled residence, residential hotel, residential housing, social housing and office.

The project is slice of urbanity, social realism, transversal proposal for diversity; the architecture is with the same quality of facades for all the included programs, the same envelope without any hierarchy or social expressionism, a picture showing a cautious neutrality towards any functional determinism. Maybe some day, same as the traditional workshops in Lyon, the housing will turn into offices and vice versa.

The building is an element of sedimentation, embedded in a complex and rigorous urban composition. The purpose is the inclusion of a building in an urban landscape, a work to the service of the lot, in direct intersection with the surrounding buildings.

The project precisely fits into the templates, alignments, frames, the values that govern the whole of the lot. So the plot consists of 2 parallel constructions, which optimize the characteristics and constraints of the site in terms of view, sunlight and animation disturbances.

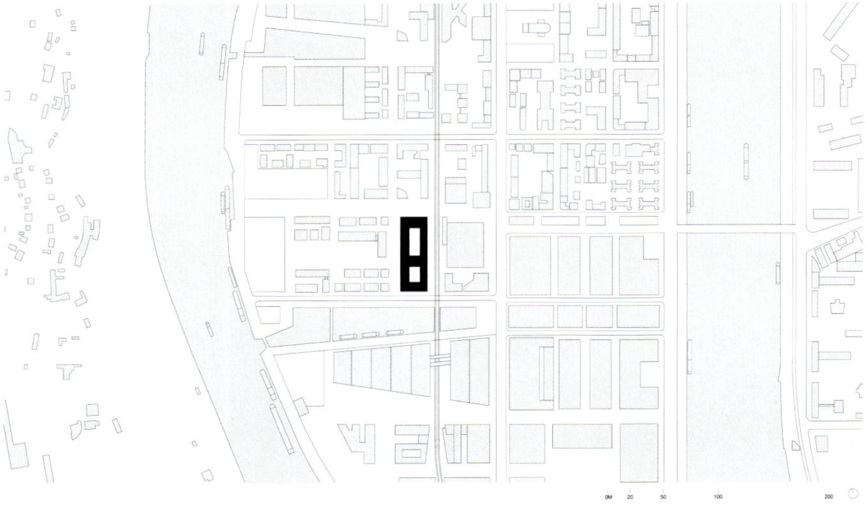

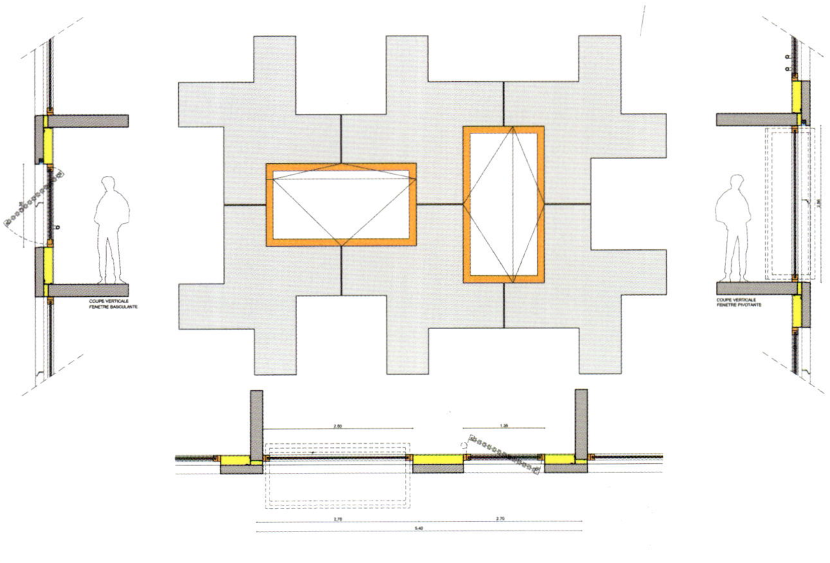

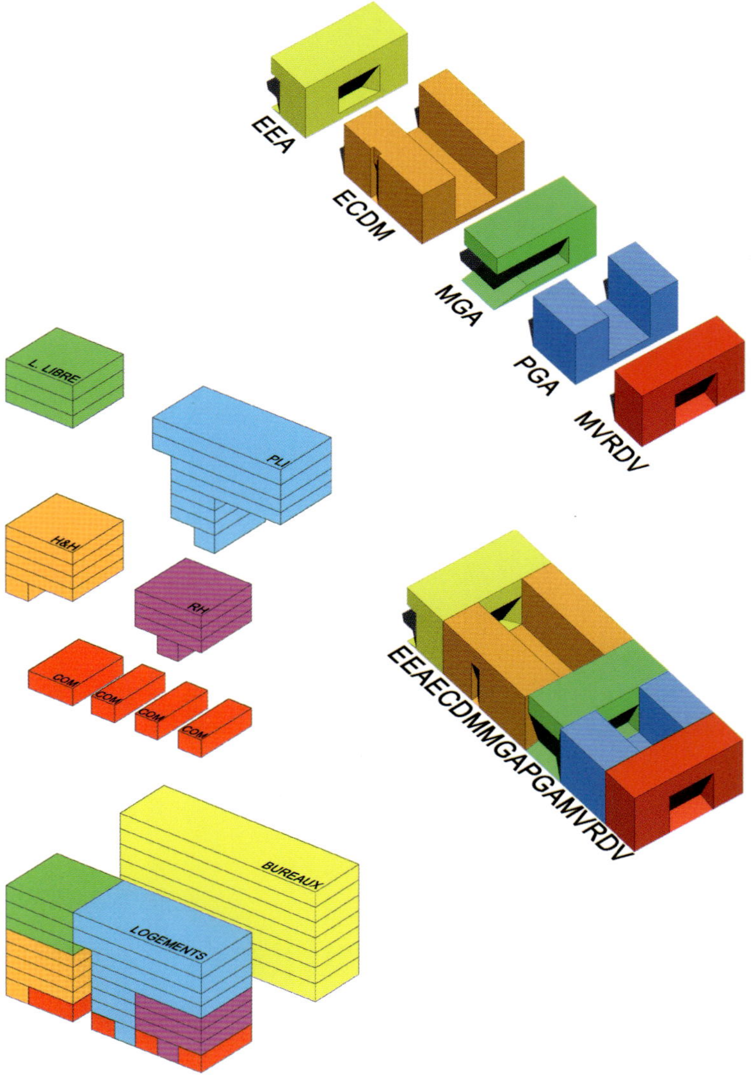
EEA
ECDM
MGA
PGA
MVRDV
L. LIBRE
PLI
H&H
RH
COM
COM
COM
COM
EEAECDMMGAPGAMVRDV
BUREAUX
LOGEMENTS

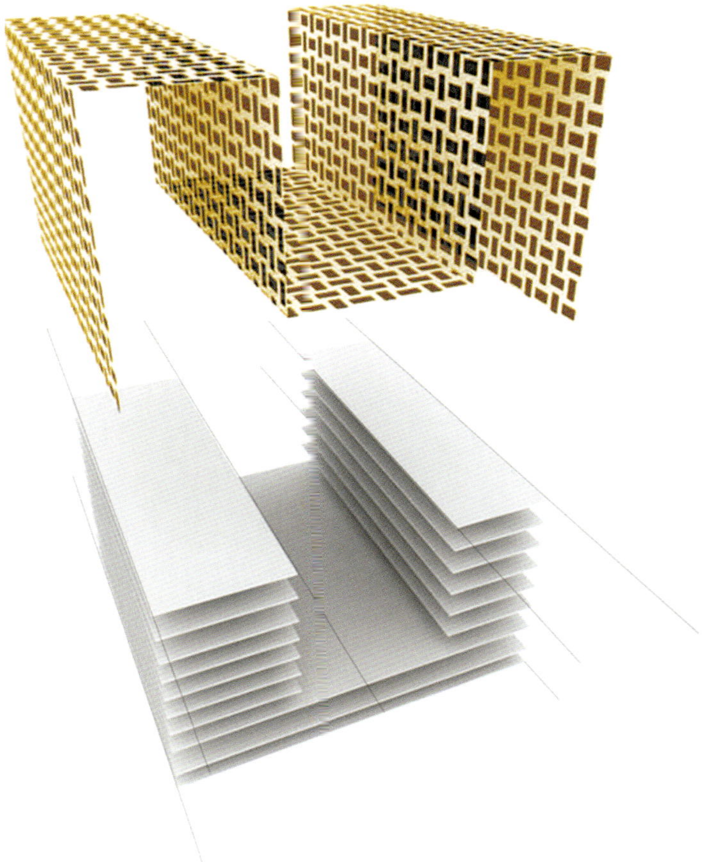

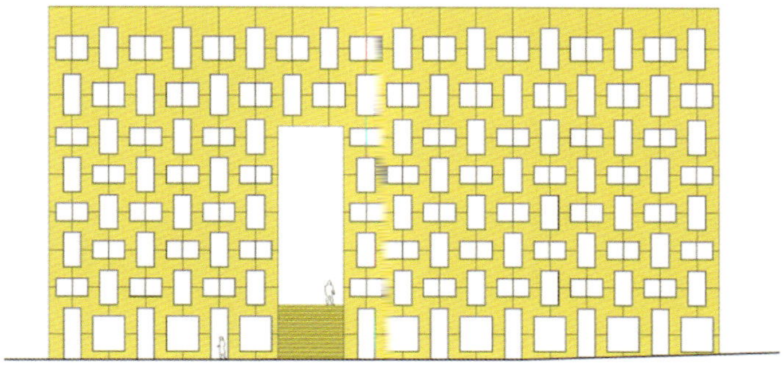

west

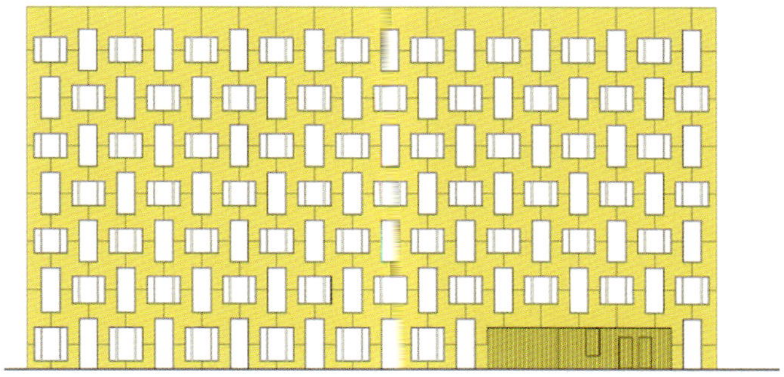

east

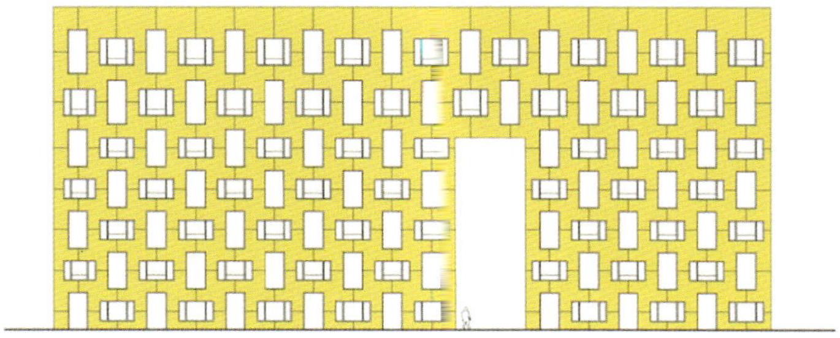

east

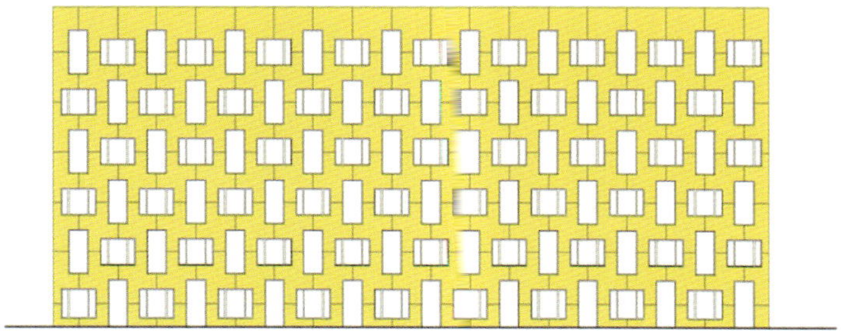

west

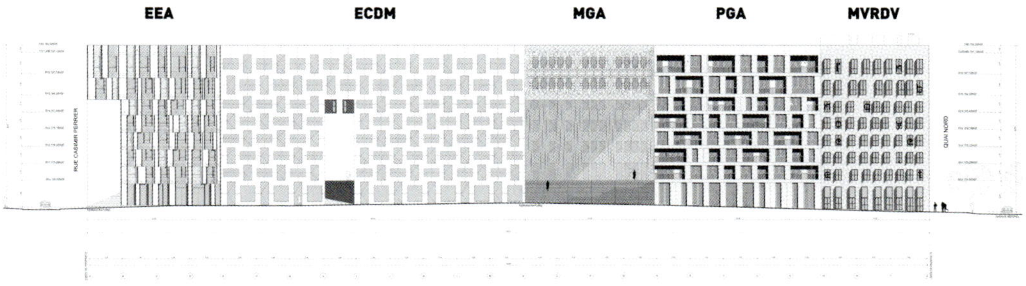

FACADE RUE DENUZIERE

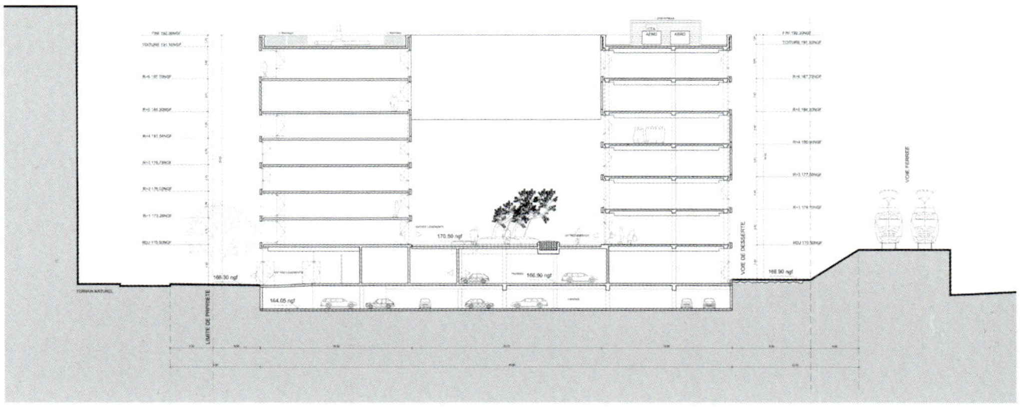

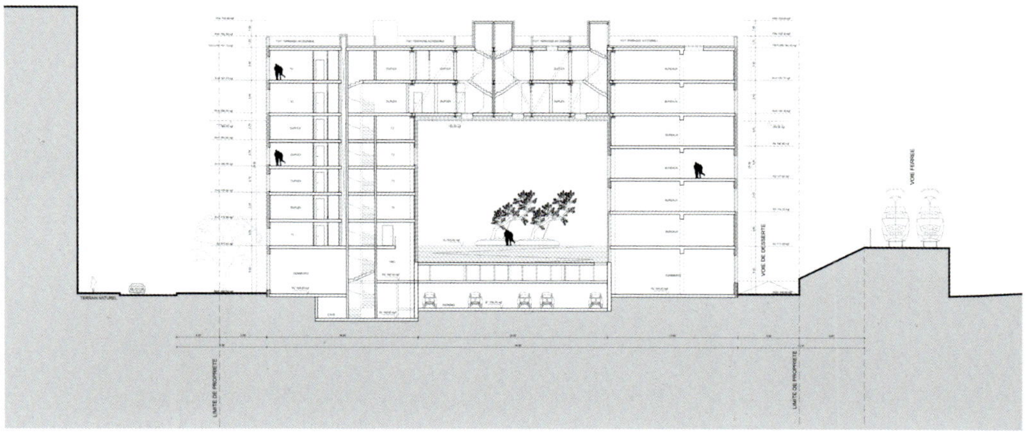

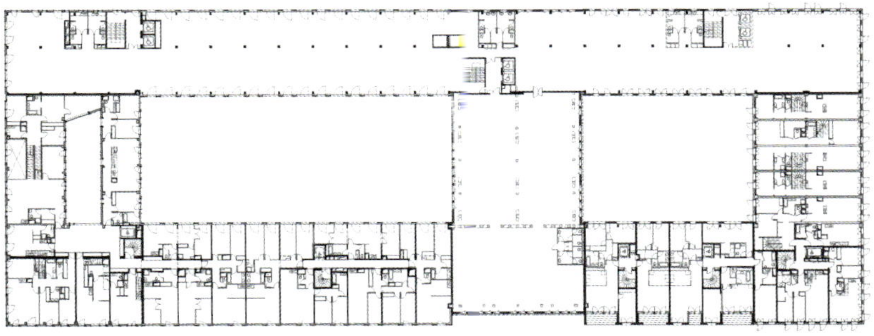

7TH FLOOR PLAN

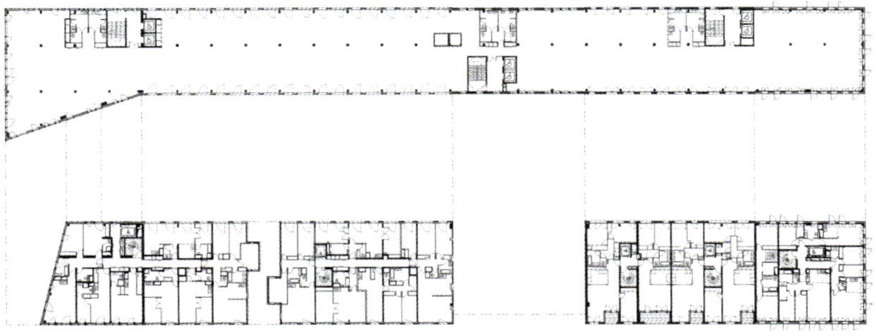

TYPICAL FLOOR PLAN

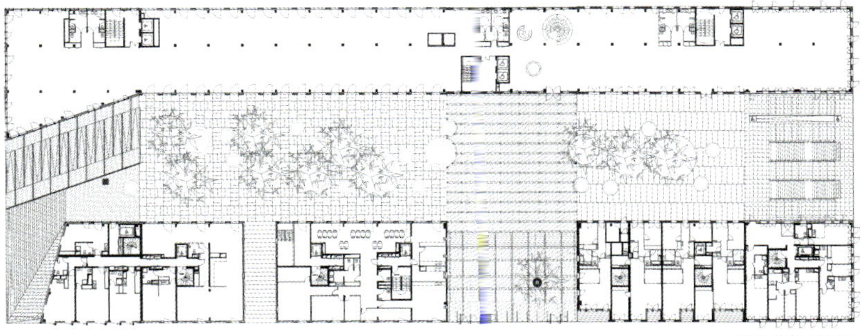

GARDEN FLOOR PLAN

Architect: Marlies Rohmer
Firm: Architectural office Marlies Rohmer
Location: Groningen NL
Area: 9312m^2
Photographer: Rob de Jong, Marcel van der Burg, René de Wit

The Intense City

The former water supply company site on Bloemsingel is part of the City of Groningen's housing construction campaign the Intense City. This building festival covers multiple sites in the city centre, whose infill will enrich the diversity of functions. One theme of the festival is "parochial space", a semi-public space inside an urban block. The location as a whole is typified by the adjacent electricity substation which is listed as an architectural monument, and the ample public green space on the banks of the Ooster Hamrikkanaal (canal).

In this project, sustainability is realized by application of the principle of "richness and abundance". In the past, manpower was cheap and materials were relatively expensive; building were built to last "for ever" and were splendidly decorated and detailed with high craftsmanship, often using very durable materials such as hardwood. Now the reverse is true. Using industrial and less labour-intensive techniques, the designers try to reinstate that rich feeling of the past in a contemporary way. The facade-filling brickwork elements with patterned bonds and relief were therefore prefabricated for this project. This procedure yielded a "rich" facade for a relatively low budget. The industrial prefabrication of custom building components also has advantages for energy efficiency. Prefabrication is less energy-intensive and the transportation of ready assembled components to the building site means less building waste will be produced there. This cuts the energy consumed for the processing and transportation of waste materials.

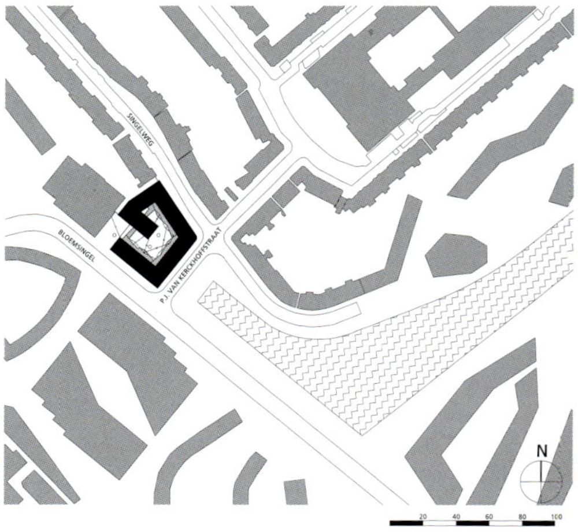

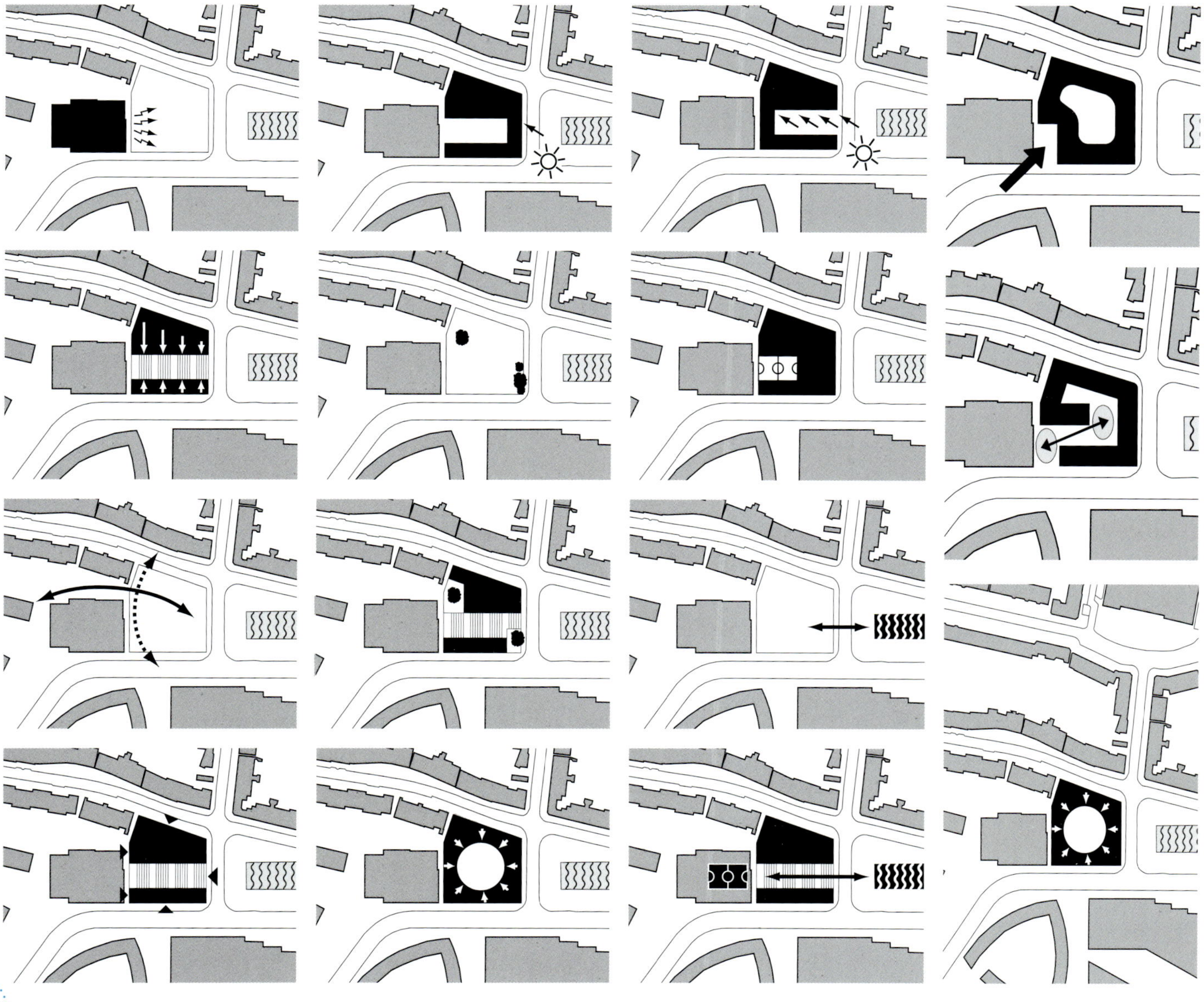

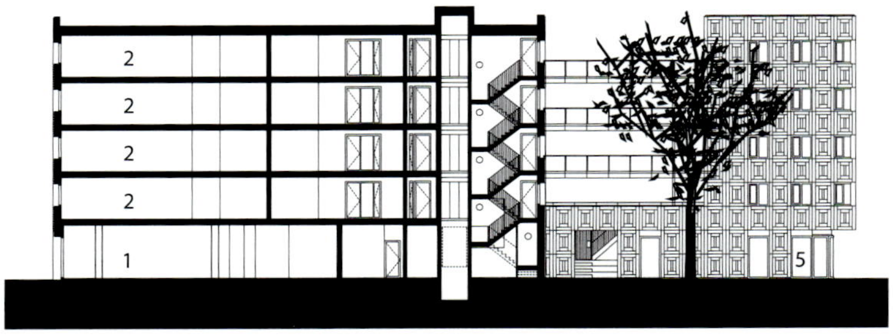

doorsnede AA

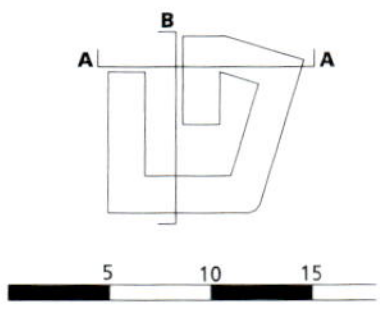

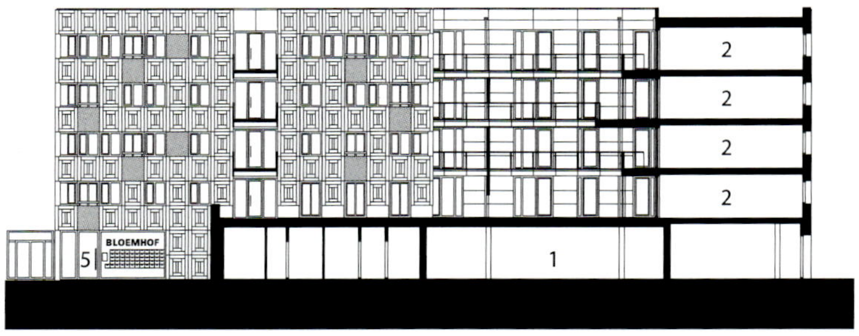

doorsnede BB

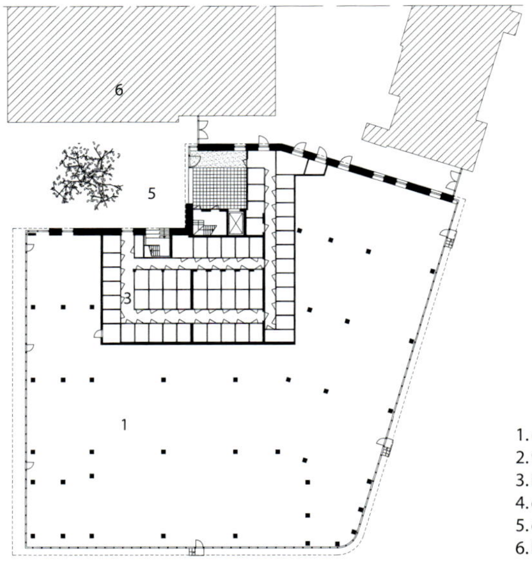

1. commerciële ruimte
2. woning
3. berging
4. collectief binnengebied
5. entreegebied / 'parochiale ruimte'
6. bestaand elektriciteitsgebouw

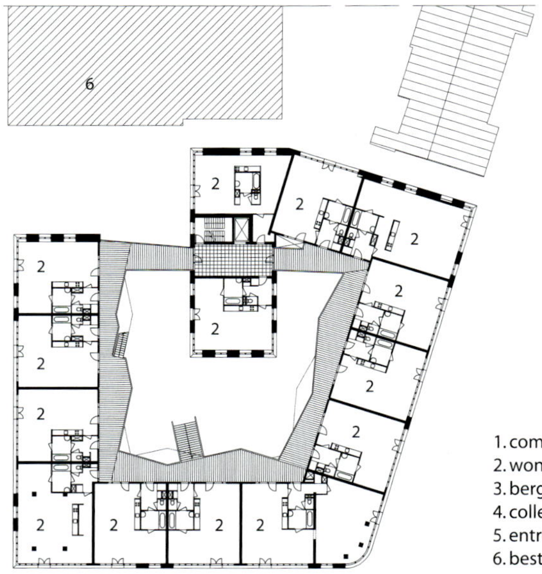

1. commerciële ruimte
2. woning
3. berging
4. collectief binnengebied
5. entreegebied / 'parochiale ruimte'
6. bestaand elektriciteitsgebouw

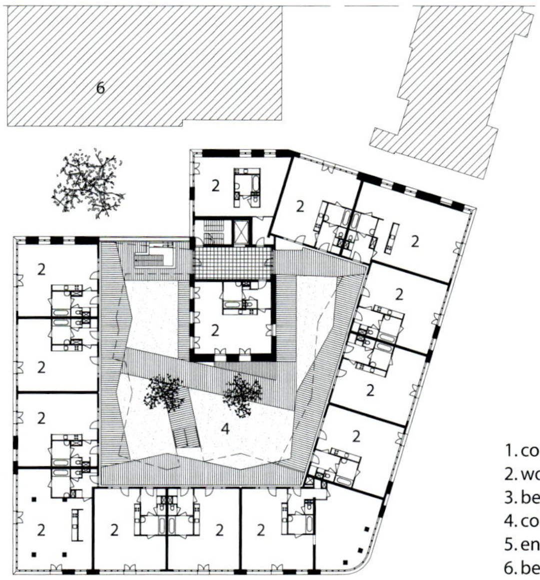

1. commerciële ruimte
2. woning
3. berging
4. collectief binnengebied
5. entreegebied / 'parochiale ruimte'
6. bestaand elektriciteitsgebouw

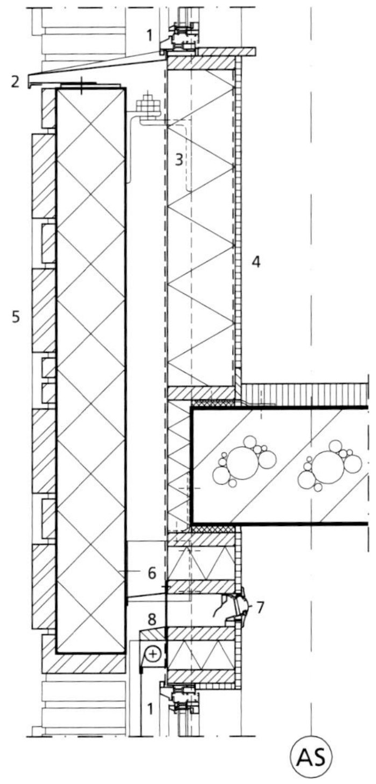

1 aluminium kozijn Reynaers CS38SL
gemoffeld in kleur
2 Compri aluminium waterslag
gemoffeld in kleur
anti-dreun folie onderzijde
3 stalen ophangconstructie tpv betonwar
4 houten binnenspouwblad: Rc ≥ cf EPN
RW-folie, Tyvek o.g.
Verdiepingshoog HSB element,
Minerale wol
Dampremmende folie
Fermacell 15mm o.g.
5 prefab betonnen element vv baksteen:
Kooy 146.P formaat 210 x 98 x 49 mm
6 stalen console tbv. afhangen prefab
gevelelement cf opgave aannemer
7 ventilatierooster buva topstream 14
8 zonwering Harosol 5060 in RAL 9007
zijgeleiding mbv spankabels

1 aluminium kozijn Reynaers CS38SL
gemoffeld in kleur
2 stalen geveldrager
3 houten binnenspouwblad: Rc ≥ cf EPN
RW-folie, Tyvek o.g.
Verdiepingshoog HSB element,
Minerale wol
Dampremmende folie
Fermacell 15mm o.g.
4 prefab betonnen element vv baksteen:
Kooy 146.P formaat 210 x 98 x 49 mm
5 aluminium zijschot v.v. poedercoating

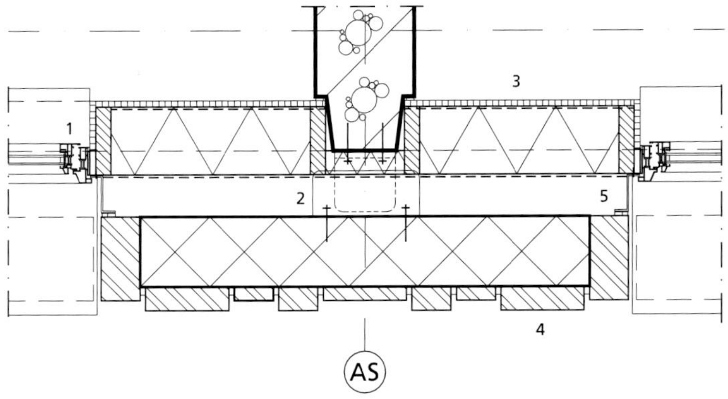

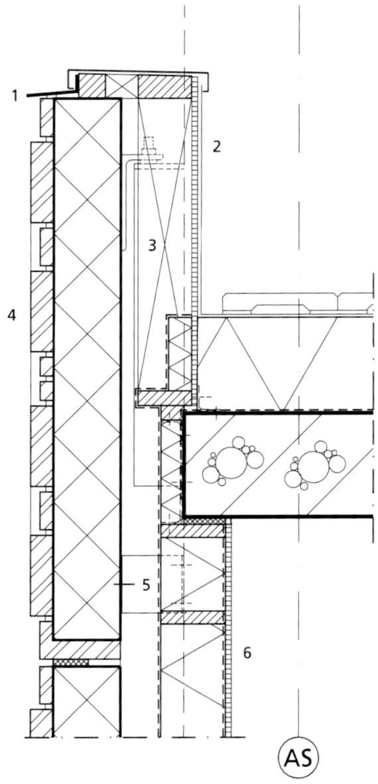

1 staalprofiel, 5mm.
afwerking volgens materiaalstaat
2 cempanel 10 mm
3 stalen baluster tbv verankering
gevelelement cf opgave constructeur
4 prefab betonnen element vv baksteen:
Kooy 146.P formaat 210 x 98 x 49 mm
5 stalen console tbv afhangen prefab
gevelelement cf opgave aannemer
6 houten binnenspouwblad: Rc ≥ cf EPN
RW-folie, Tyvek o.g.
Verdiepingshoog HSB element,
Minerale wol
Dampremmende folie
Fermacell 15mm o.g.

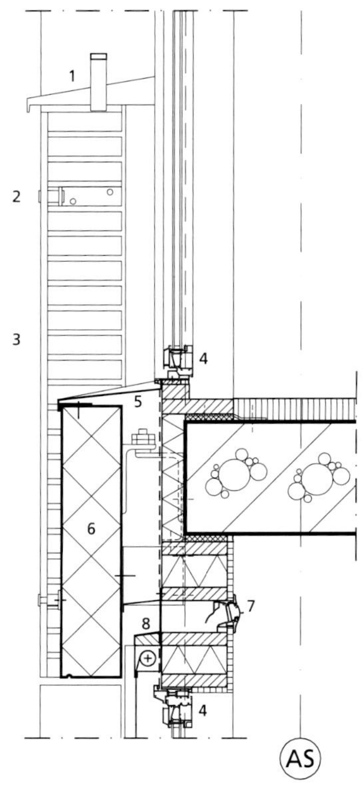

1 stalen doorvalbeveiliging
dimensionering cf opgave constructeur
strip 15x 40 mm gepoedercoat in kleur ntb.
2 bevestiging glas op afstandhouder
stalen hoek t.b.v. montage glaspaneel
3 glaspaneel screenprint vlg. gevelfragment
4 aluminium raamkozijn Reynaers CS38SL
gemoffeld in kleur
5 beloopbare alum. waterslag
vv. poedercoating, kleur ntb
vilt onderzijde
6 doorgaand prefab beton
in kleur geschilderd
cf opgave constructeur
7 ventilatierooster buva topstream 14
8 zonwering Harosol 5060 in RAL 9007
zijgeleiding mbv spankabels

Architect: Francis-Jones Morehen Thorp (fjmt)
Location: 101 Doncaster Ave Kensington NSW 2033
Area: 18750m²
Photographers: Andrew Chung, John Gollings and Mark Donaldson

Capella Apartments, Kensington

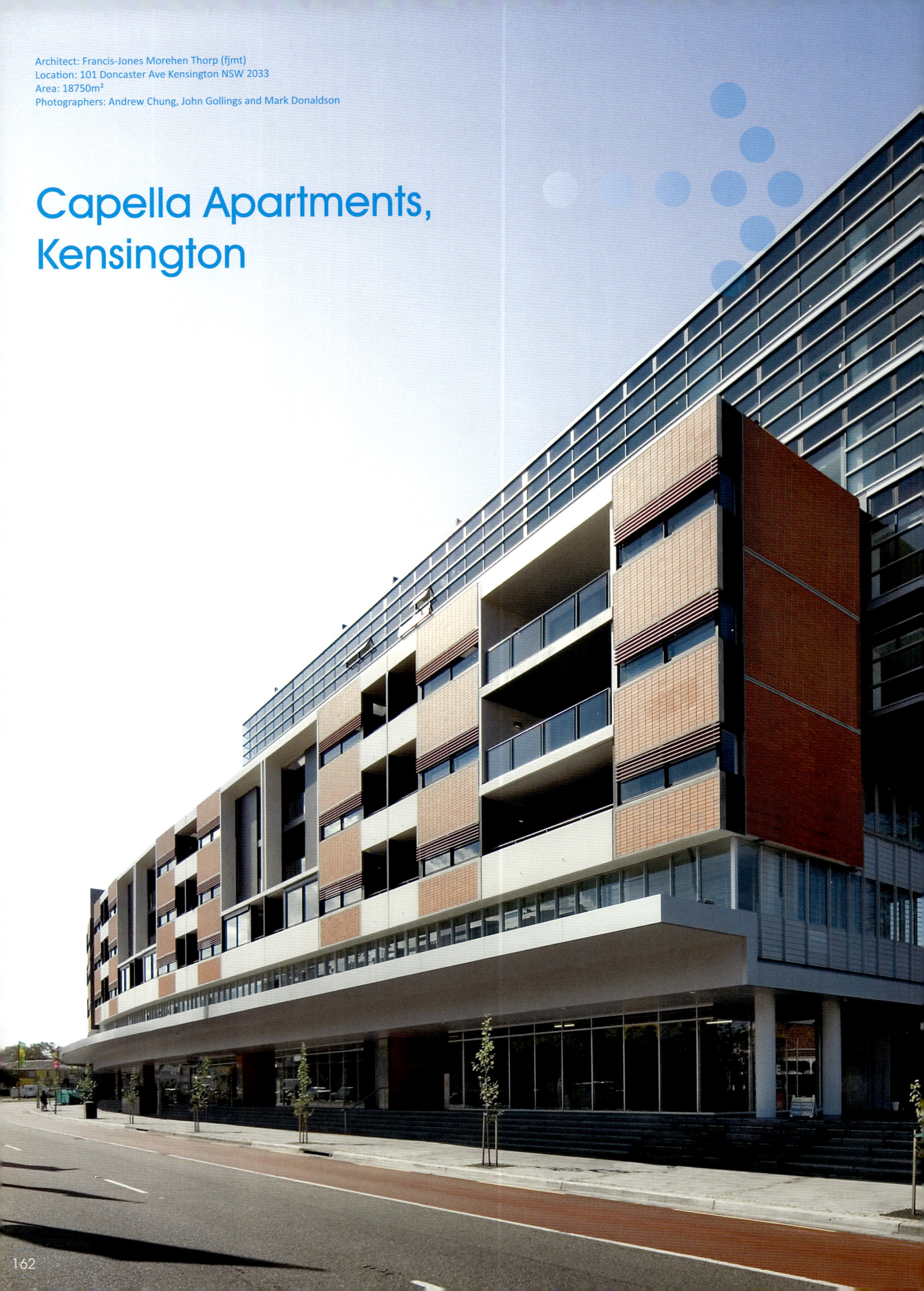

This mix use project for 164 apartments has given the designers the opportunity to develop open space, urban design, and architectural form ideas within an important and sensitive district of Sydney.

The project is a sequence of intersecting layers and volumes: A landscaped podium forms a folding ground plane of open spaces and gardens while absorbing car park loading and back of house retail accommodation. 'Resting' on this platform are two articulated forms of tiled precast and frame. Between podium and street form is the space of the retail and apartment foyers protected by a single awning plane. Free from the scale and modulation of the street are two fully glazed distorted rectangular forms of precise geometry.

At the heart of this project is the creation of a new open space sequence that begins with the broadening of the footpaths through stepped terraces and setback shops and extends into a new landscaped public square that connects two existing streets. This new public square is orientated north to receive sunlight through the sharp angled forms of the new apartment volumes and is bordered by a stepped series of landscaped terraces.

Within the intersecting forms are a wide range of apartment types with varying section and plan forms. The street front form incorporates deeply recessed and protected terraces creating open space 'rooms' while giving daylight to interior spaces. The glazed suspended volumes include a continuous stepped section to create linear thermal chimneys within every room. The garden elevations of these glass forms incorporate a series of continuous vertical blades that are twisted to the north to gently adjust the orientation of the apartments towards the sun and view.

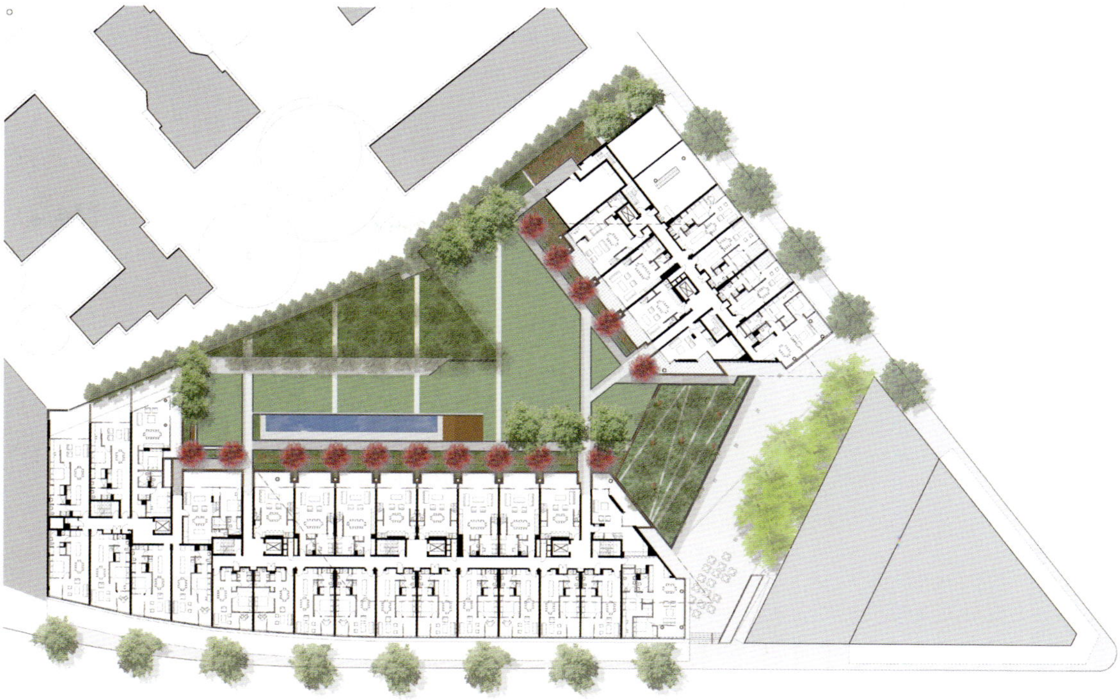

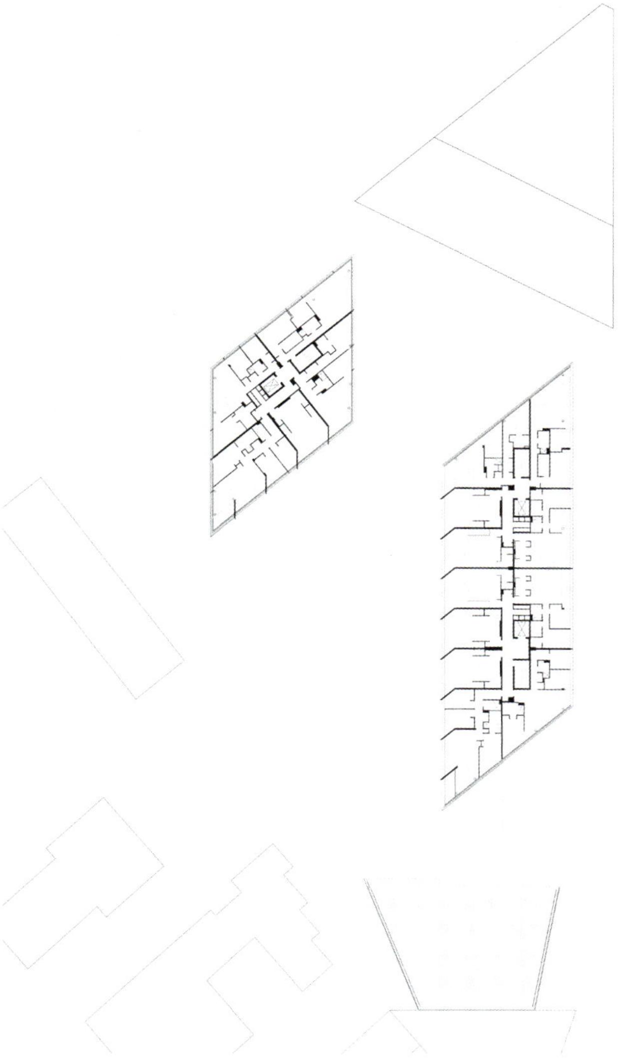

Architect: Rok Oman, Špela Videčnik with Martina Lipicer, Nejc Batistič,
Neža Oman, Florian Frey, Marisa Baptista
Firm: OFIS architects
Location: Izola, Slovenia
Area: 5452m²
Photographer: Tomaž Gregorič

Honeycomb Apartments

The blocks are set out on a hill with a view of Izola Bay on one side and of the surrounding hills on the other. The given urban plot was 2blocks of 60x28 meters.

The brief required 30 apartments of different sizes and structures, varying from studio flats to 3-bedroom apartments. The apartments are small, with minimum-sized rooms according to Slovenian standards. There are no structural elements inside the apartments, thus providing flexibility and the possibility of reorganising things.

The blocks are subject to a Mediterranean climate outdoor space and shade are important elements. The project proposed a veranda for each apartment, thus providing an outdoor space that is intimate, partly connected with the interior, shady and naturally ventilated.

A textile shade protects the balcony and apartment from prying eyes, yet due to its semi-transparency allows the owner to enjoy the views of the bay. Perforated side-panels allow summer breeze to ventilate the space. The strong colours create different atmospheres within the apartments. Boxes at the side of each balcony provide room for air-conditioning units.

The small rooms become visually bigger because the textile shade creates a perspective effect that connects part of the exterior with the interior.

sustainability issues:

Balcony modules are designed as efficient system providing shading and ventilation for the apartments. Textile elements fixed on the front of the balconies block direct sunlight and accumulate "air buffer" zone. In the summer hot accumulated area behind the shadings is naturally ventilated through (10 cm holes) perforated side partitions of the balconies. In the winter the warm air stays in the area and provides additional heating to the apartments.

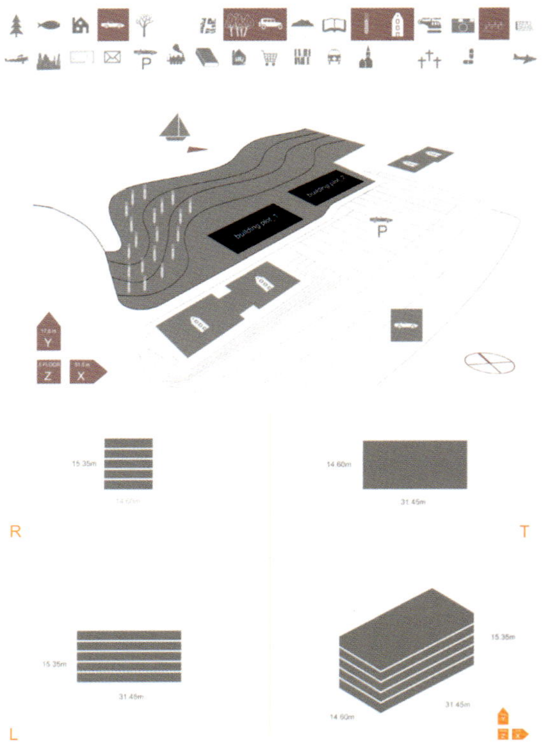

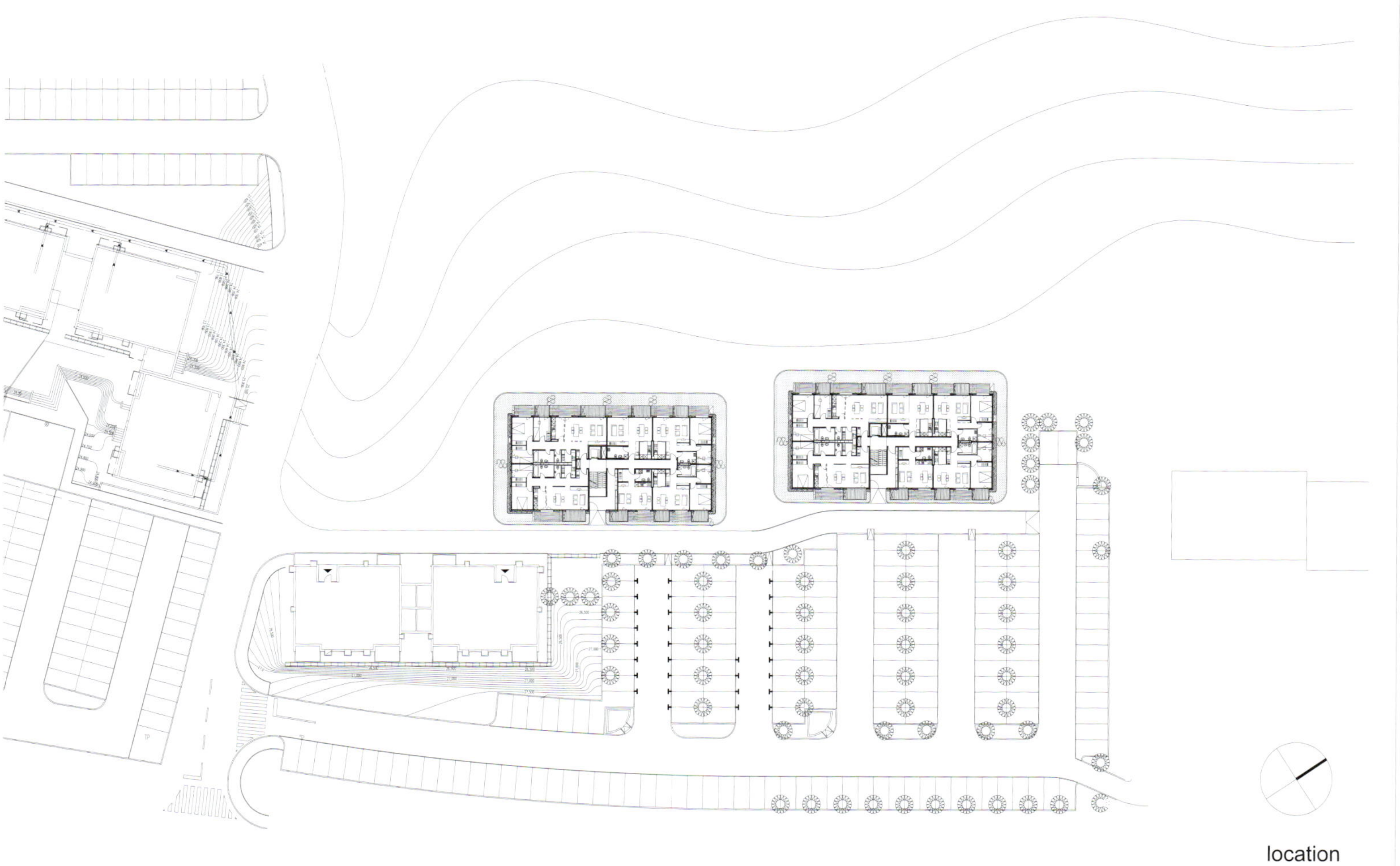

location

northwest facade

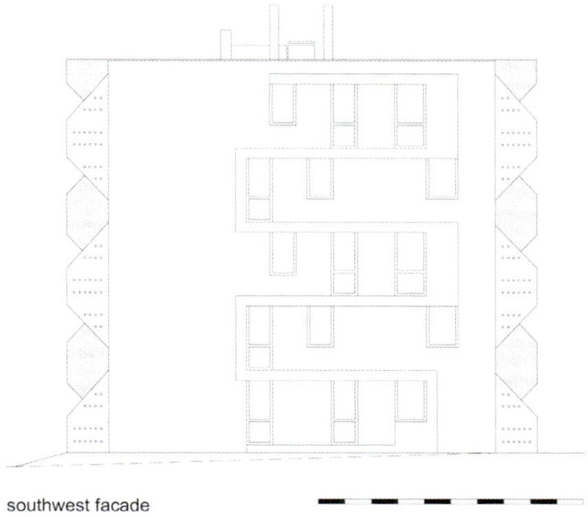

southwest facade

southwest facade

northeast facade

facades

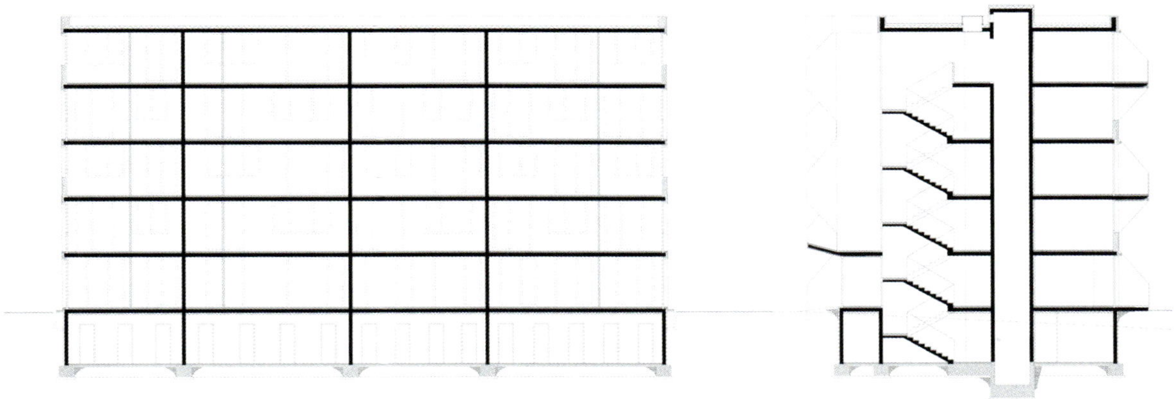

sections

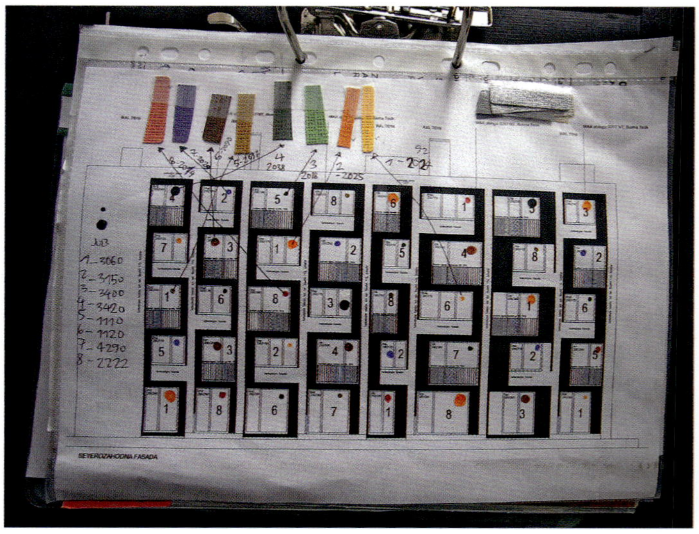

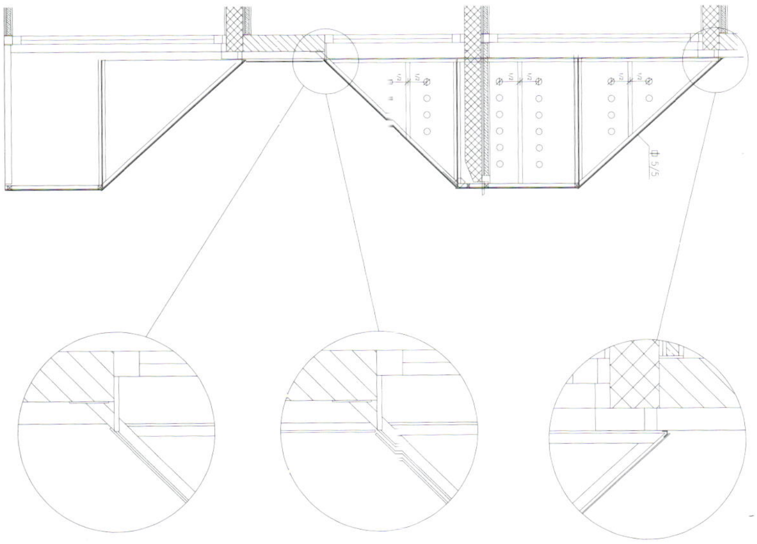

balcony

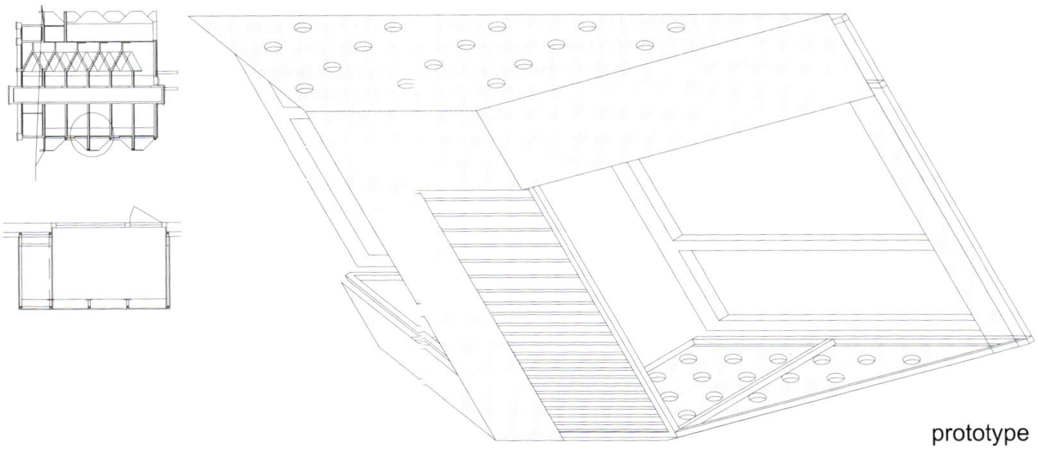

prototype

Architect: Rok Oman, Špela Videčnik
Firm: OFIS architects
Location: Ljubljana, Slovenia
Area: 5000m²
Photographer: Tomaž Gregorič

Tetris Apartments

The location is on the edge of the planned 650 apartments which was finished year ago. Also this apartment block is social type and was sold to Slovenian Housing Fund. The given urban plot of the building was 4 floors high, 58 meters long and 15 in width.

Since the orientation of the building is towards the busy highway the apartment opening together with balconies are shifted as 30 degrees window-wings towards the quieter and south orientated side. In the future also 2 more blocks are planned on both longitudinal sides; therefore there are no direct windows towards east and west. Each apartment has view towards its own balcony, sometimes there is also a glazed loggia. On this way intimacy is created and there will be no direct views from ones apartment directly to the others in the opposite block.

Apartments are of different sizes from 30m² studio flats up to 3 room apartment of 70m². Bigger apartments are developed on the front facades and have nicer views and corner orientation. They are made of economic but quality materials such as wooden oak floors, granite tiled bathrooms and have large windows with external metal blinds. The concept of structure is made in a way, that floor plans are flexible, since only structural walls are those, that separate apartment shell from the rest of the building. All other inner walls are non-structural.

Long after the elevations were planned many people associated them to Tetris game. And so the building got its name. The façade was developed simply – just tracing the floor plan organization.

The inner – structural façade wall has plaster; the external wall that embraces the loggia is glazed or wrapped into precast panel. These panels are wooden of three colors that are tracing the vertical zigzag pattern. Balcony fences are either perforated precast panel or transparent metal fence.

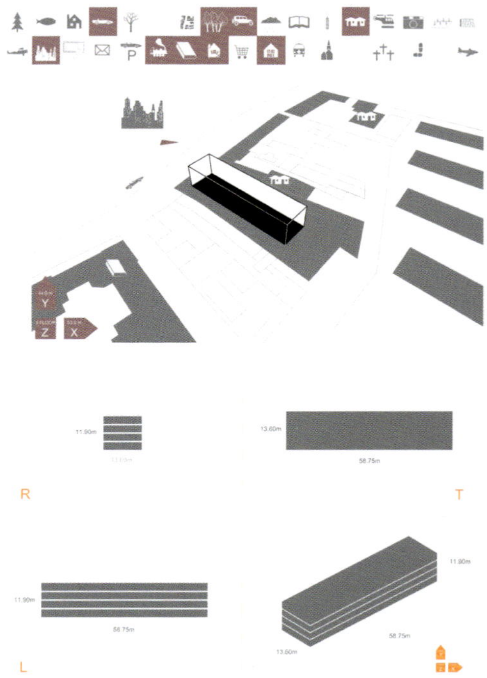

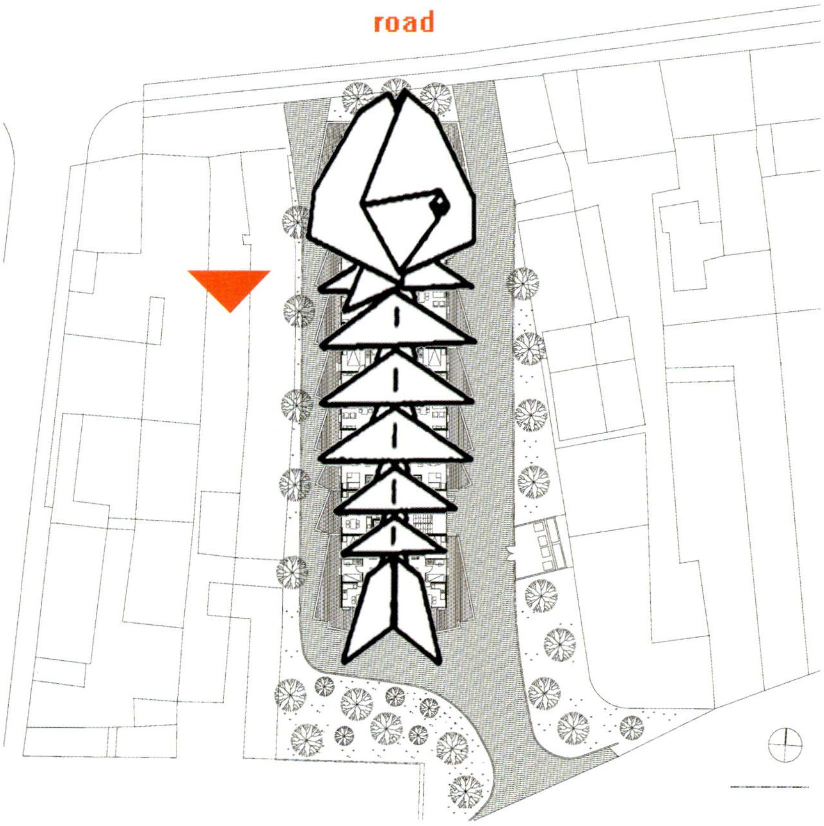
road

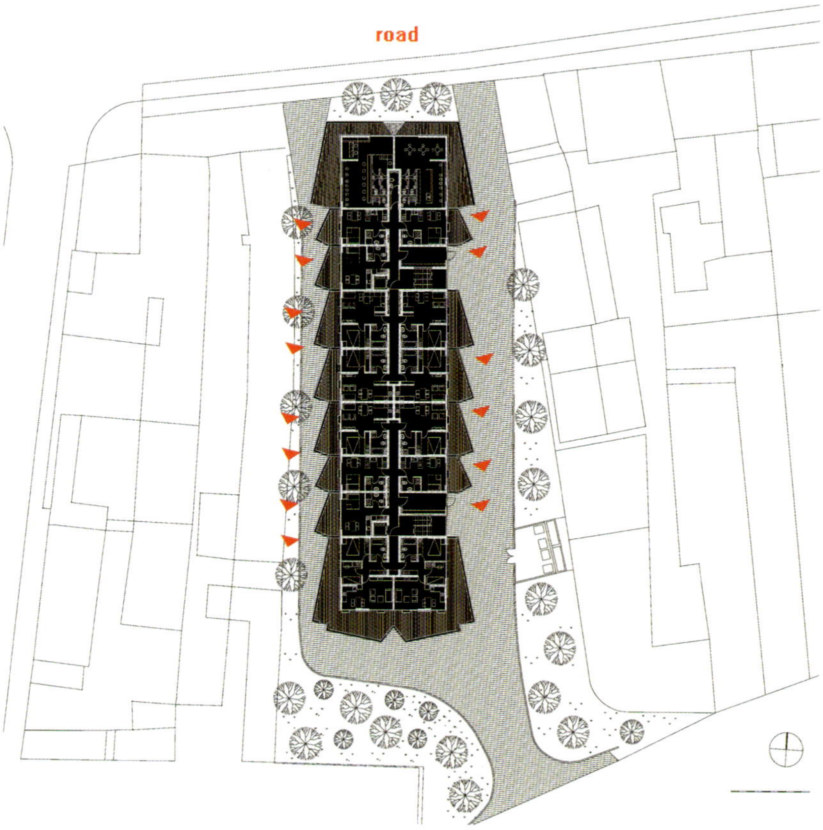
road

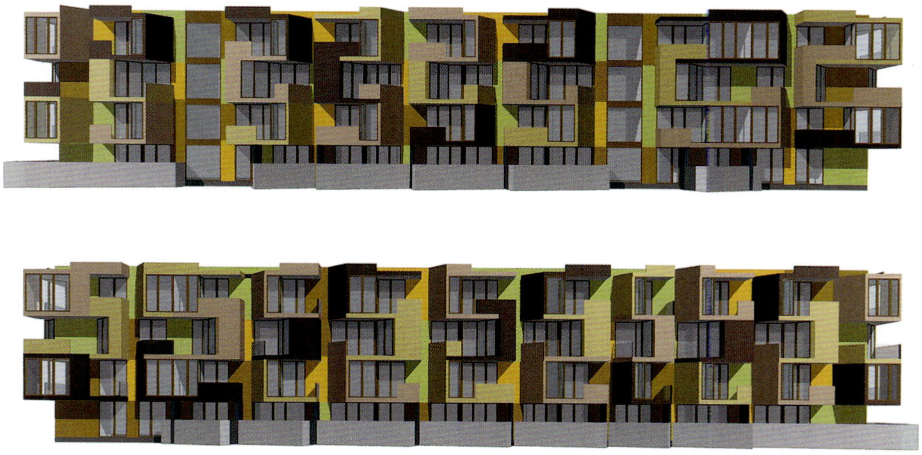

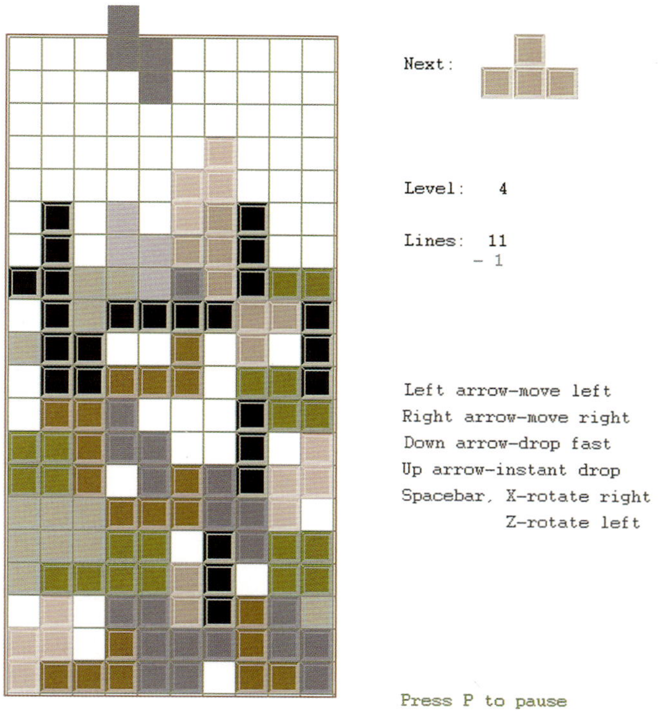

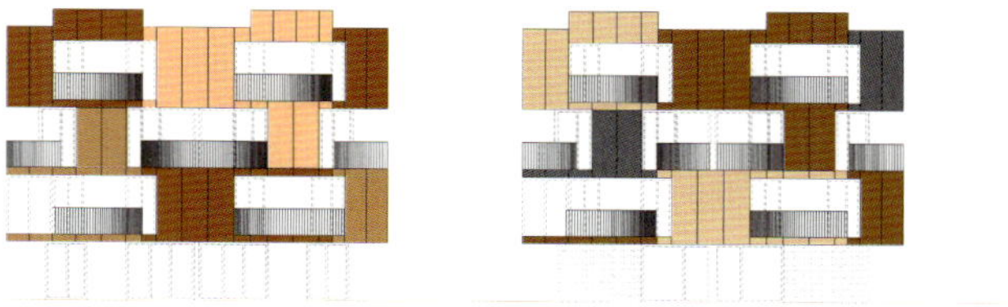

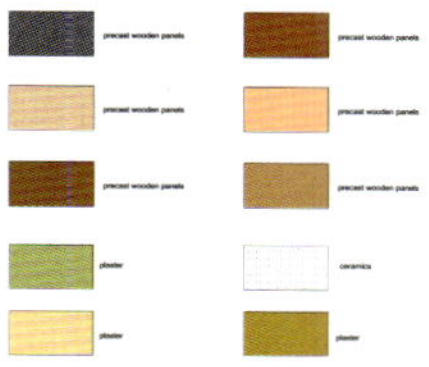

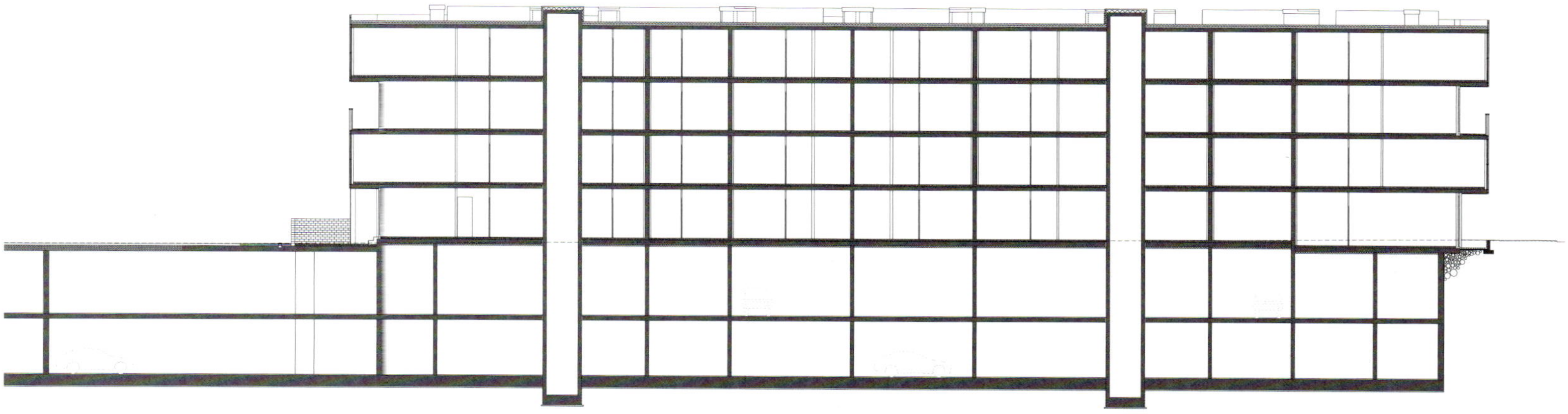

longitudinal section

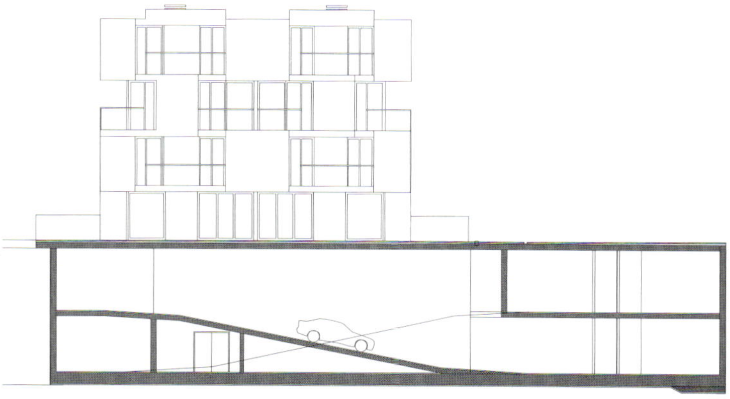

cross section_through garage

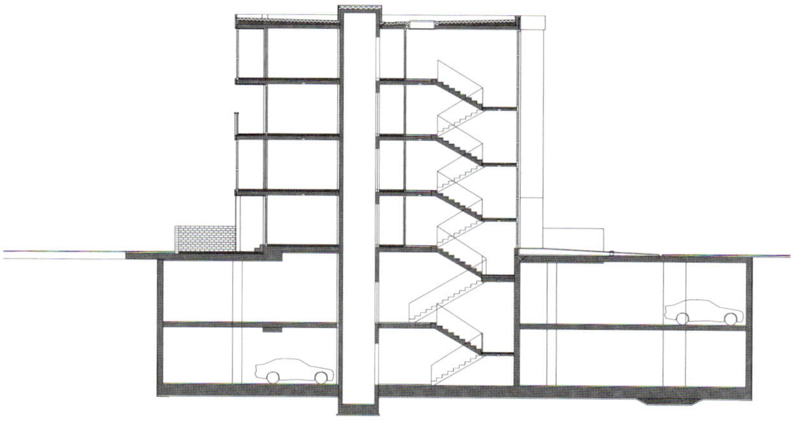

cross section_through staircase

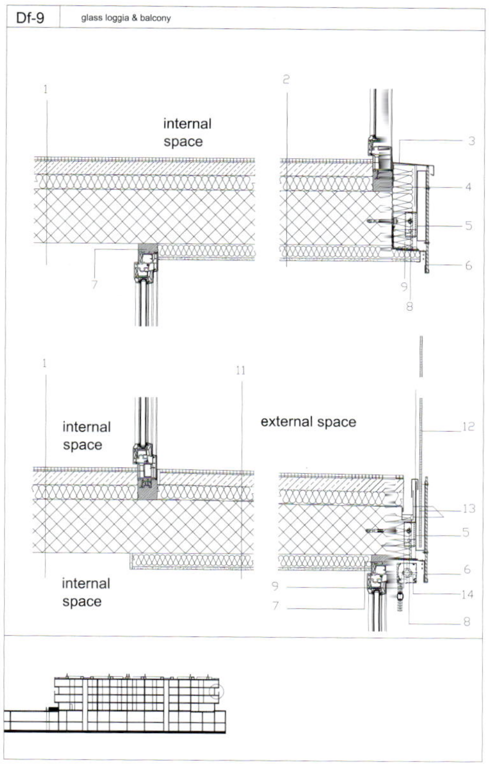

1 living space

-parquet	1,0 cm	
-cement	0,5 cm	
-concrete layer	5,0 cm	
-polyurethane sheeting	0,02 cm	
-sound insulation	5,5 cm	12,0 cm
-reinforced concrete slab	20,0 cm	

2 Loggia

-parquet	1,0 cm	
-cement	0,5 cm	
-concrete layer	5,0 cm	
-polyurethane sheeting	0,02 cm	
-sound insulation	5,5 cm	12,0 cm
-reinforced concrete slab	20,0 cm	
-thermal insulation	6,00 cm	
-plaster	1,25cm	

3 metal sheet

4 supporting profile

5 steel profile fixed in reinforced concrete slab

6 preformed wooden panels

7 filled with propen

8 steel support structure

9 thermal insulation

11 Balcony

-granite tiles	1,0 cm	
-glue	0,5 cm	
-concrete layer	5,0 cm	
-polyurethane sheeting		
-thermal insulation	5,5 cm	12,0cm
-reinforced concrete slab	20,0 cm	
-thermal insulation	6,0 cm	
-plaster	1,00cm	

12 tempered glass

13 drainage for rainwater

14 space for external shading sistem

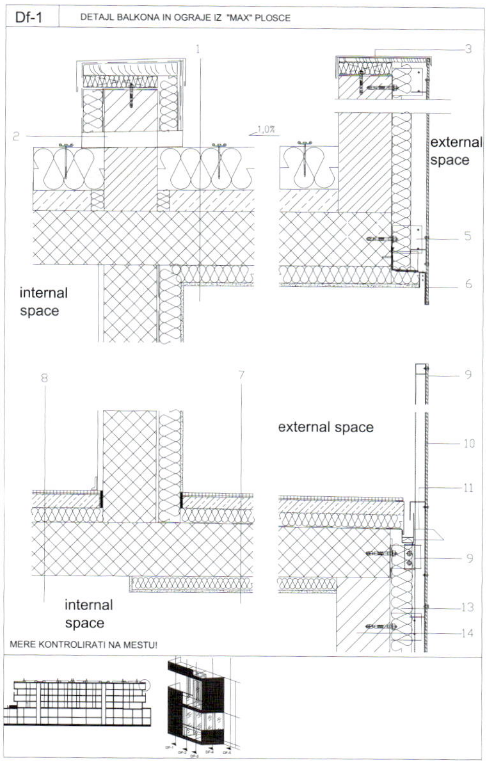

1 flat roof

-hydroinsulation		
-thermal insulation	16,0 cm	
-vapour barrier		
- concrete layer 1,5 %	25,0 cm	
-raiforced concrete slab	20,0 cm	
- thermal insulation	6,50 cm	
-plaster		

2 drainage for rainwater

3 metal sheet

5 steel profile fixed in reinforced concrete slab

6 preformed wooden panels

7 Balcony

-granite tiles	1,0 cm	
-glue	0,5 cm	
-concrete layer	5,0 cm	
-polyurethane sheeting		
-thermal insulation	5 cm	12,0cm
-reinforced concrete slab	20,0 cm	
-thermal insulation	6,0 cm	
-plaster	1,00cm	

8 living space

-parquet	1,0 cm	
-cement	0,5 cm	
-concrete layer	5,0 cm	
-polyurethane sheeting	0,02 cm	
-sound insulation	5,5 cm	12,0 cm
-reinforced concrete slab	20,0 cm	

9 steel profile

10 preformed wooden panels

11 drainage for rainwater

13 steel support structure

14 thermal insulation

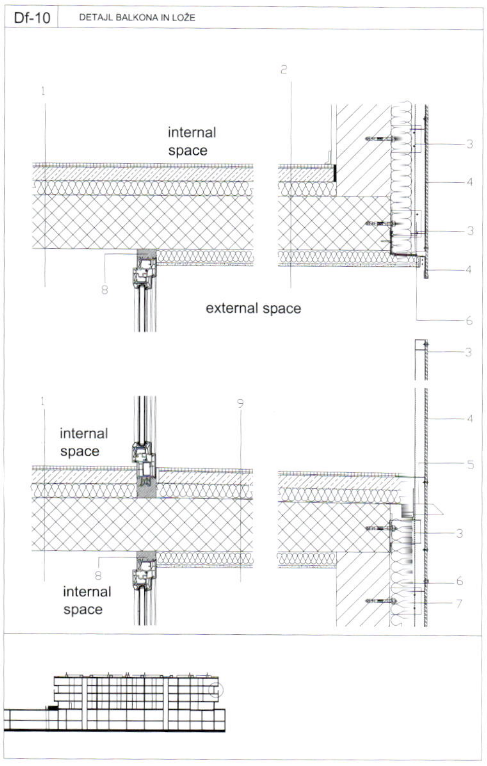

1 living space

-parquet	1,0 cm	
-cement	0,5 cm	
-concrete layer	5,0 cm	
-polyurethane sheeting	0,02 cm	
-sound insulation	5,5 cm	12,0 cm
-reinforced concrete slab	20,0 cm	

2 Loggia

-parquet	1,0 cm	
-cement	0,5 cm	
-concrete layer	5,0 cm	
-polyurethane sheeting	0,02 cm	
-sound insulation	5,5 cm	12,0 cm
-reinforced concrete slab	20,0 cm	
- thermal insulation	6,00 cm	
-plaster	1,25cm	

3 steel profile

4 preformed wooden panels

5 drainage for rainwater

6 thermal insulation

7 steel support structure

8 filled with propen

9 Balcony

-granite tiles	1,0 cm	
-glue	0,5 cm	
-concrete layer	5,0 cm	
-polyurethane sheeting		
-thermal insulation	5,5 cm	12,0cm
-reinforced concrete slab	20,0 cm	
-thermal insulation	6,0 cm	
- plaster	1,00cm	

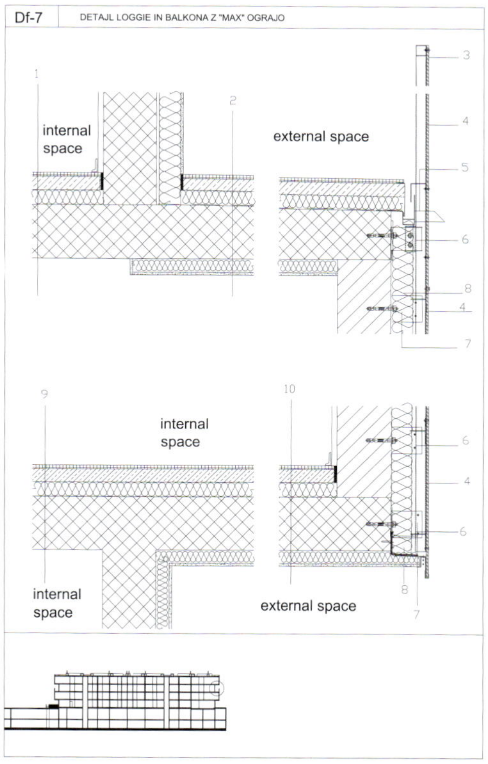

1 living space

-parquet	1,0 cm	
-cement	0,5 cm	
-concrete layer	5,0 cm	
-polyurethane sheeting	0,02 cm	
-sound insulation	5,5 cm	12,0 cm
-reinforced concrete slab	20,0 cm	

2 Balcony

-granite tiles	1,0 cm	
-glue	0,5 cm	
-concrete layer	5,0 cm	
-polyurethane sheeting		
-thermal insulation	5,5 cm	12,0cm
-reinforced concrete slab	20,0 cm	
-thermal insulation	6,0 cm	
-plaster	1,00cm	

3 -steel profile

4 preformed wooden panels

5 drainage for rainwater

6 steel profile

7 steel support structure

8 thermal insulation

9 living space

-parquet	1,0 cm	
-cement	0,5 cm	
-concrete layer	5,0 cm	
-polyurethane sheeting	0,02 cm	
-sound insulation	5,5 cm	12,0 cm
-reinforced concrete slab	20,0 cm	

10 Loggia

-parquet	1,0 cm	
-cement	0,5 cm	
-concrete layer	5,0 cm	
-polyurethane sheeting	0,02 cm	
-sound insulation	5,5 cm	12,0 cm
-reinforced concrete slab	20,0 cm	
- thermal insulation	6,00 cm	
-plaster	1,25cm	

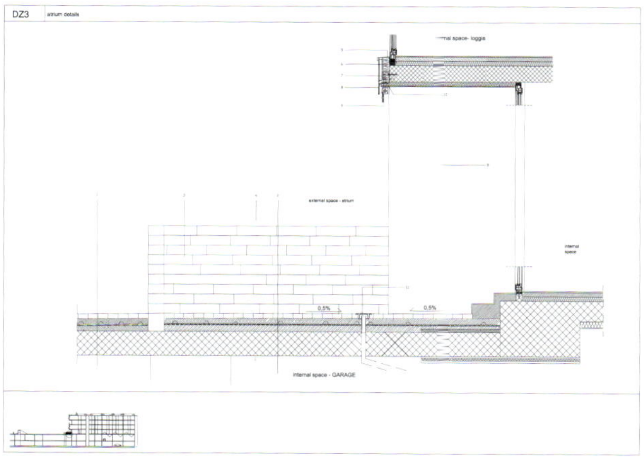

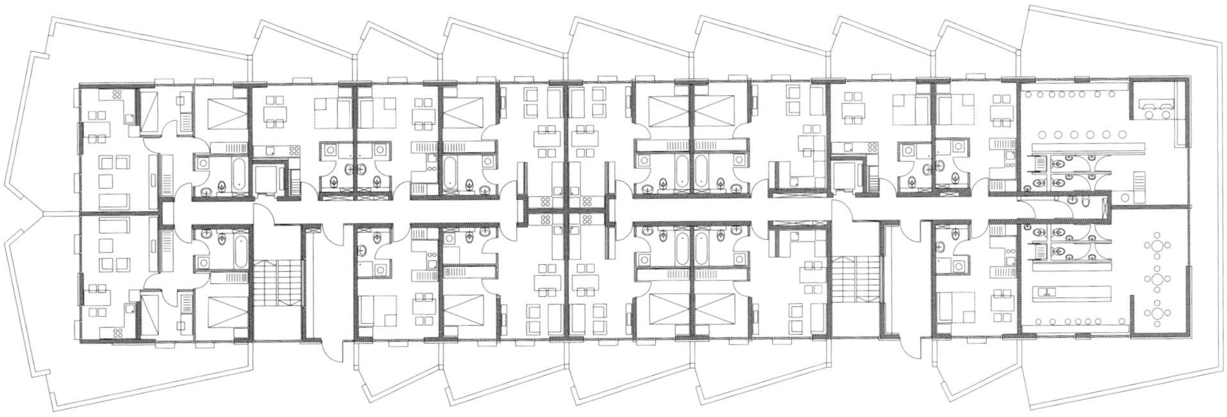

0

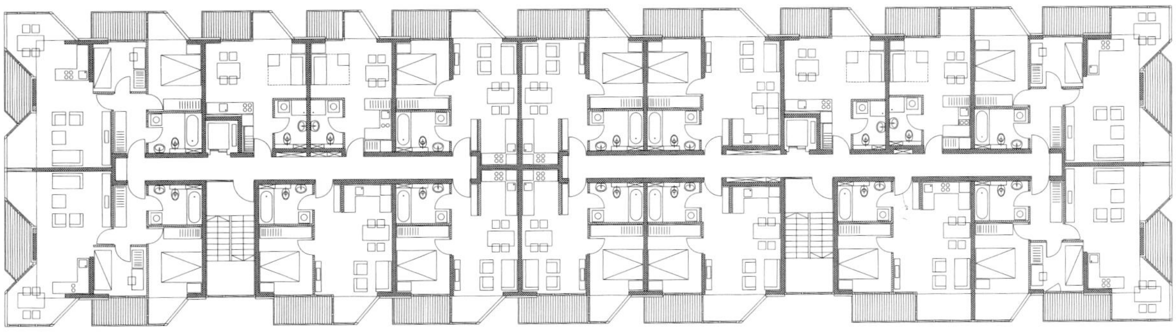

1,3 FLOOR

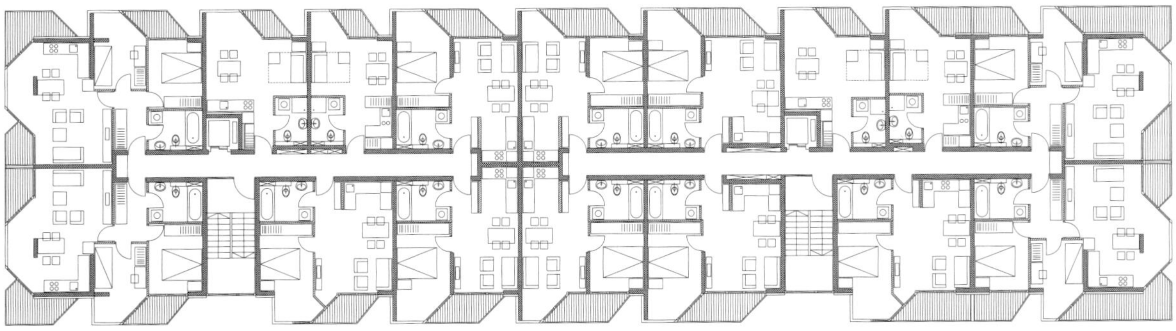

2 FLOOR

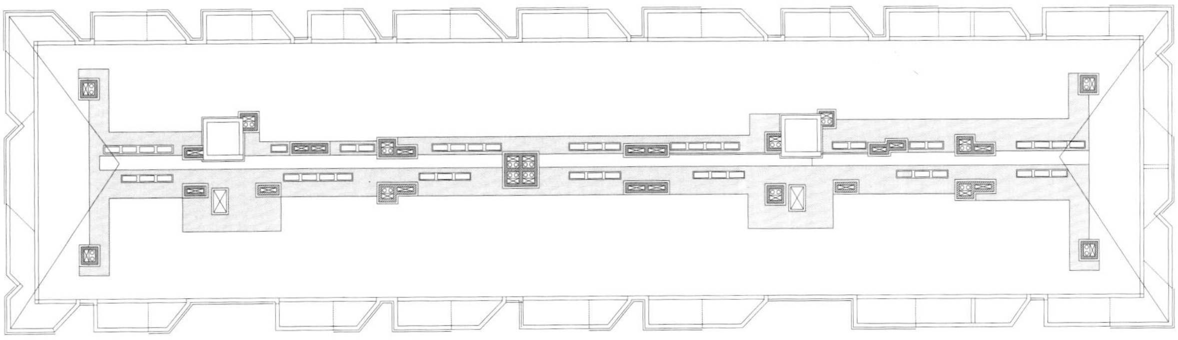

ROOF

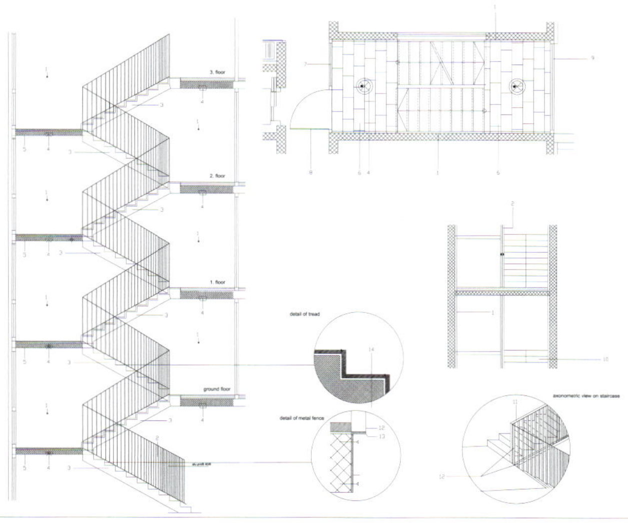

1 washable oil coating
2 fence: metal rails
3 painted metal sheet
4 ceiling lamp
5 ceiling: washable oil coating
6 stone tiles
7 single glass panel aluminium frame
8 door: single glass aluminium frame
9 glass panel
10 staircase
11 top metal rail
12 metal rail
13 weld
14 glue 0,5cm reinforced concrete slab 12cm

Architect: Rok Oman, Špela Videčnik
Firm: OFIS architects
Location: Ljubljana, Slovenia
Area: 47400m² (housing); 42000m² (underground parking); 400m² (commercial areas)
Photographer: Tomaž Gregorič

650 Apartments, Ljubljana

The project won an invitation competition for developing four apartment blocks from 125 to 140 meters in length. The client chose this proposal for economic reasons: the designers managed to provide more saleable surface on the predefined urban set-up and a lower floor space index than all the other proposals.

The scheduled time for preparing plans and for resolving the building work was very tight: 650 apartments in less than eighteen months, which meant more than one apartment a day. The budget was also extremely limited: the aim was to build low-cost apartments and the designers had to bring the final production cost in at around 500 €/m².

To make planning and construction simple and to allow for the use of such prefabricated elements as bathrooms, windows and facade panels the buildings were designed in module form.

Each building is divided into four identical modules, each with its own vertical communications core. There are 42 apartments in each module, varying from small 30 m² studios to 1.5-bedroom 60 m² apartments on four identical floors, and larger duplex apartments from 85 to 105 m² on the top two floors. The module is repeated four times with slight variations at the far ends of the building.

The facade is designed in two layers, the inner facade and outdoor space being formed by items such as glazed loggias, balconies, terraces and verandas. The second skin is constructed with pre-formed wooden panels, glass and metal rails. The structure of the apartments is such that each apartment gets at least one balcony and loggia that connects outdoor and indoor space. Like the modules, the facade layer is also repeated four times, but given the different geometry of the elements the repetition passes virtually unnoticed.

There are two parking levels beneath the site. The landscape provides a contrast to the geometrical facade through the use of gently curving rails and other features. It breaks up the sightlines through the complex and creates a difference between public and private space.

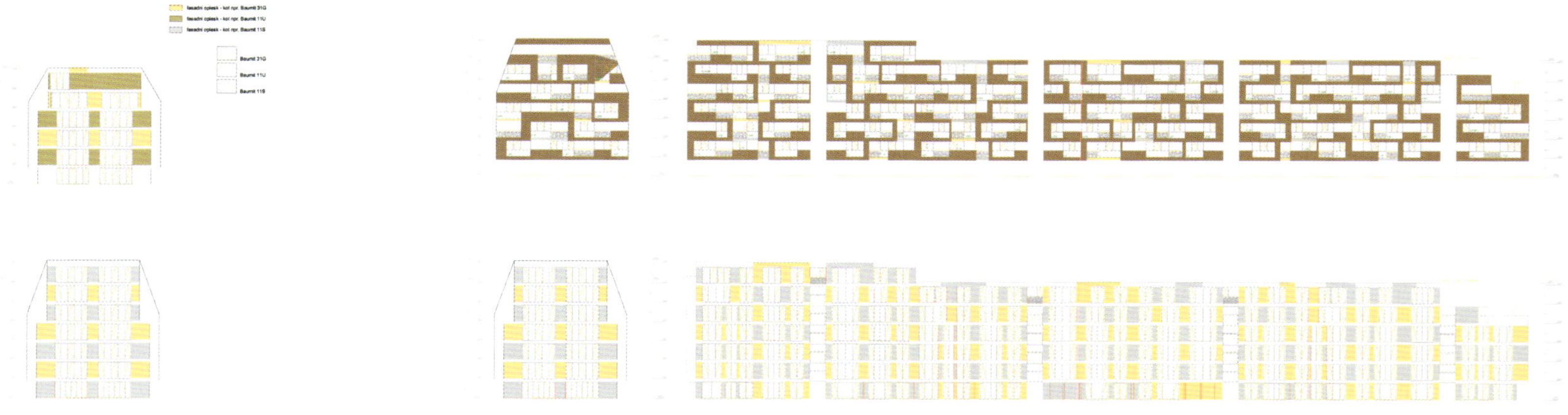
fasadni oplesk - kot npr. Baumit 31G
fasadni oplesk - kot npr. Baumit 11U
fasadni oplesk - kot npr. Baumit 11S
Baumit 31G
Baumit 11U
Baumit 11S

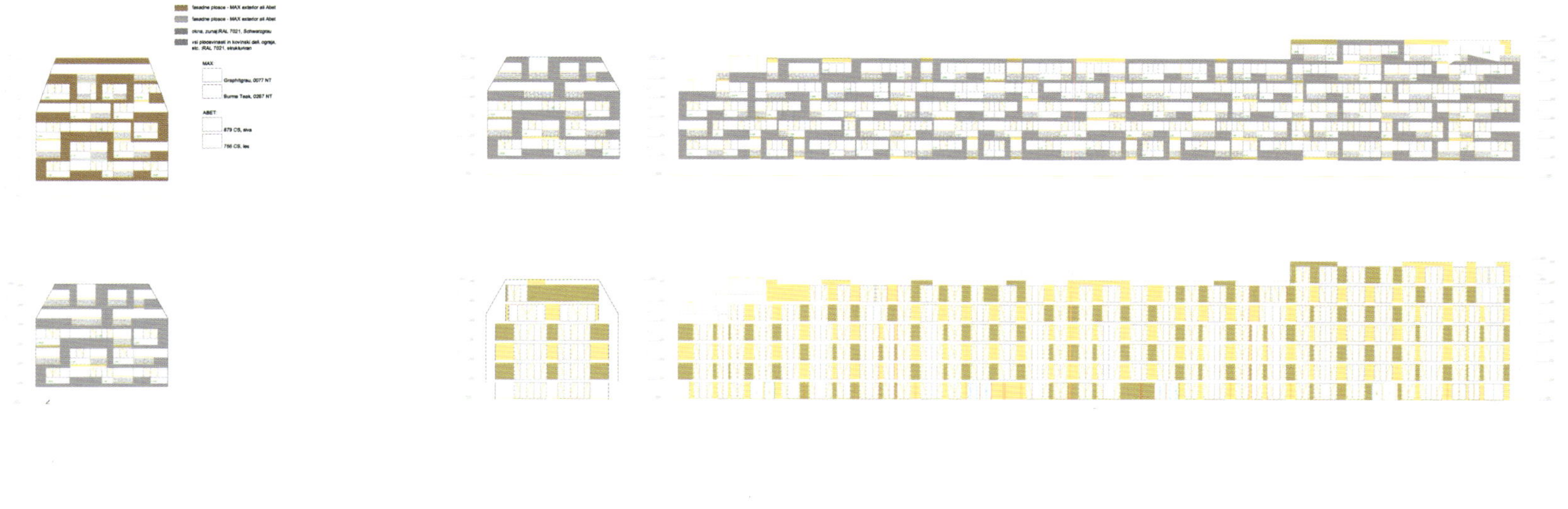
fasadne plosce - MAX exterior ali Abet
fasadne plosce - MAX exterior ali Abet
okna, zunaj:RAL 7021, Schwarzgrau
vsi plocevinasti in kovinski deli, ograje, etc. :RAL 7021, strukturiran
MAX
Graphitgrau, 0077 NT
Burma Teak, 0267 NT
ABET
879 CS, siva
756 CS, les

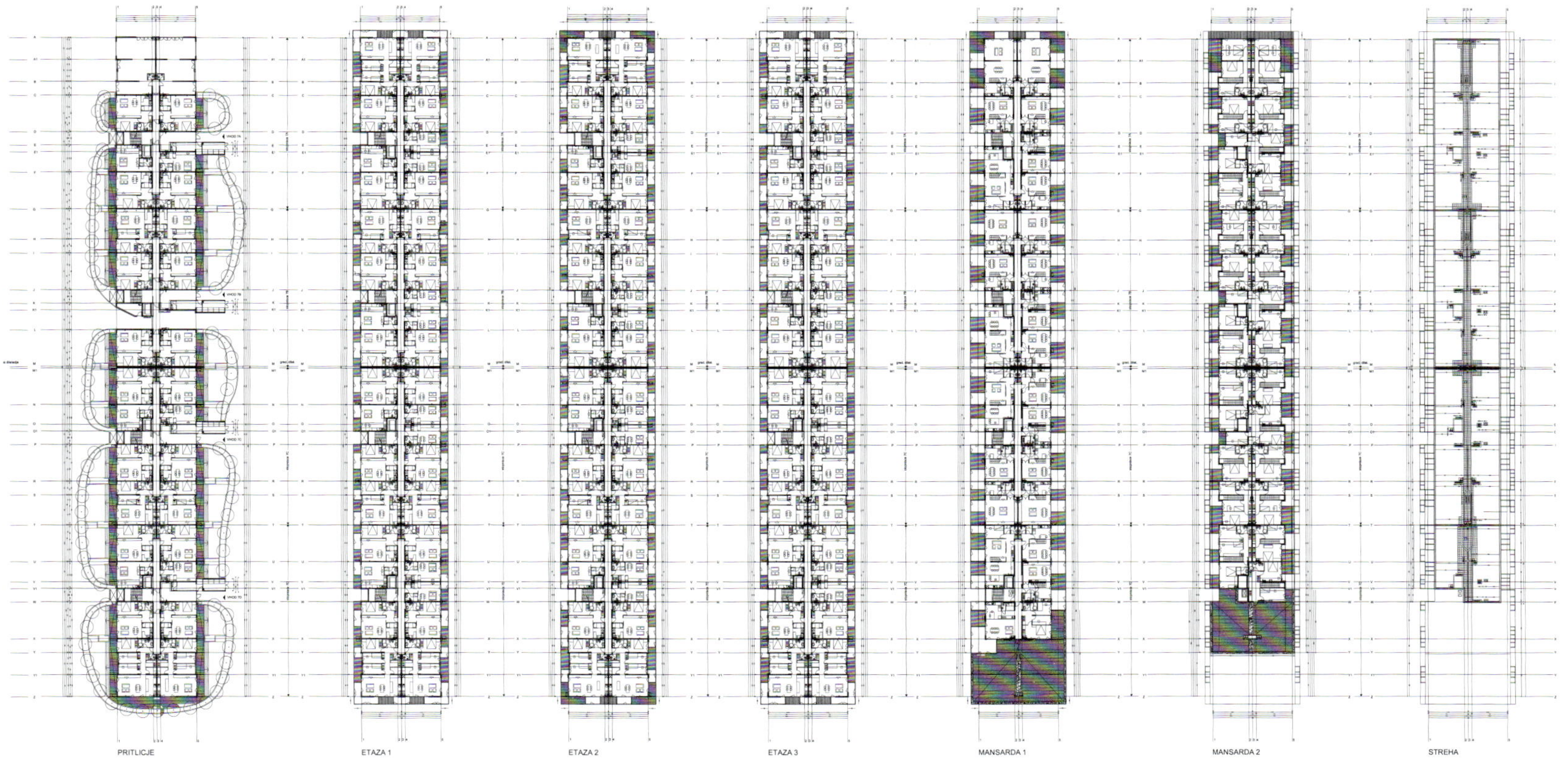
PRITLICJE
ETAZA 1
ETAZA 2
ETAZA 3
MANSARDA 1
MANSARDA 2
STREHA

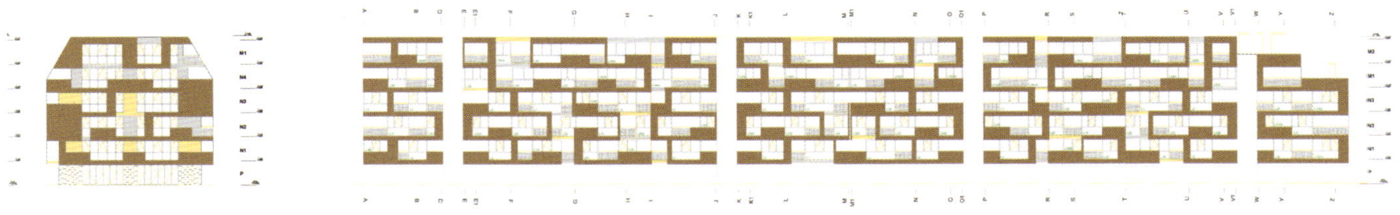

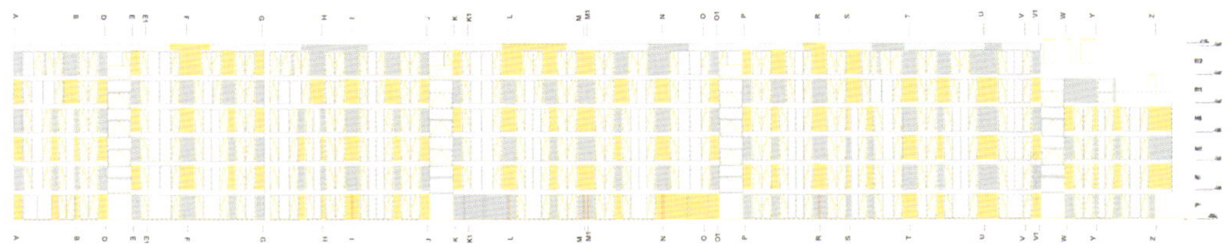

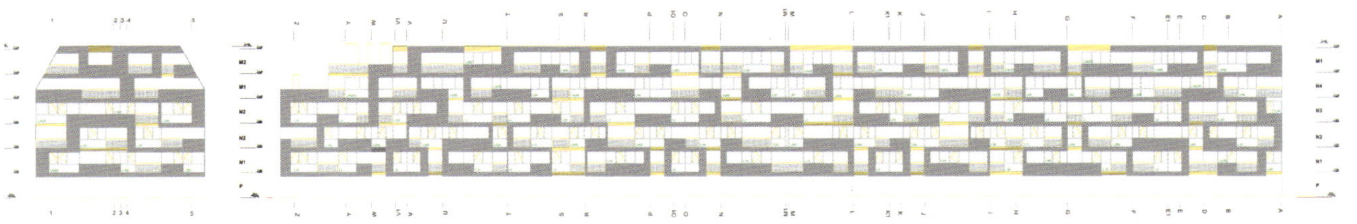

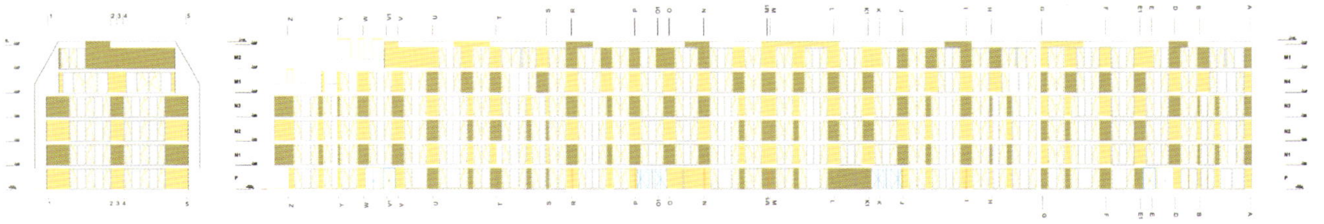

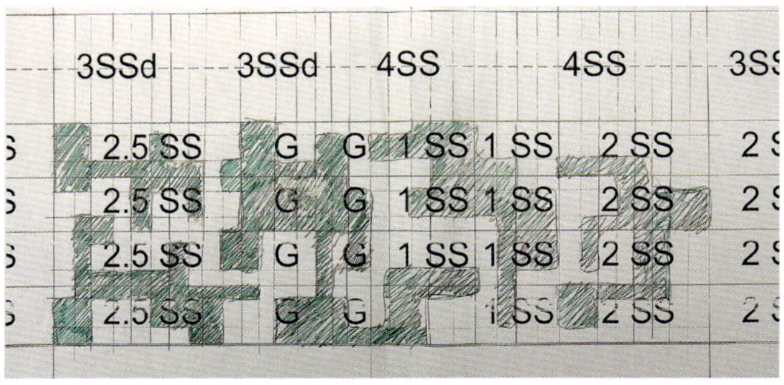
3SSd 3SSd 4SS 4SS
2.5 SS G G 1 SS 1 SS 2 SS
2.5 SS G G 1 SS 1 SS 2 SS
2.5 SS G G 1 SS 1 SS 2 SS

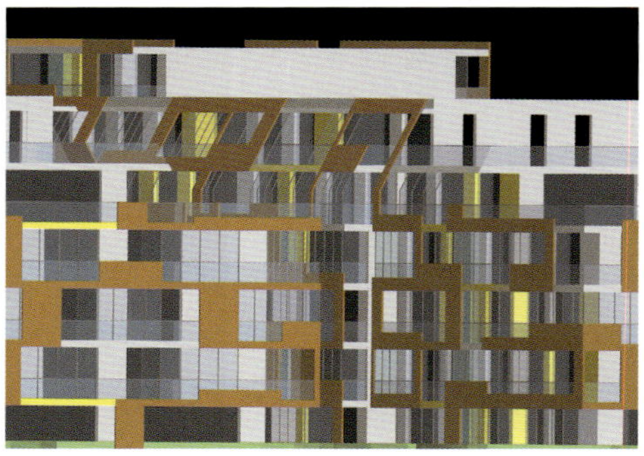

17.85m
R
T
17.85m
30.00 m
L

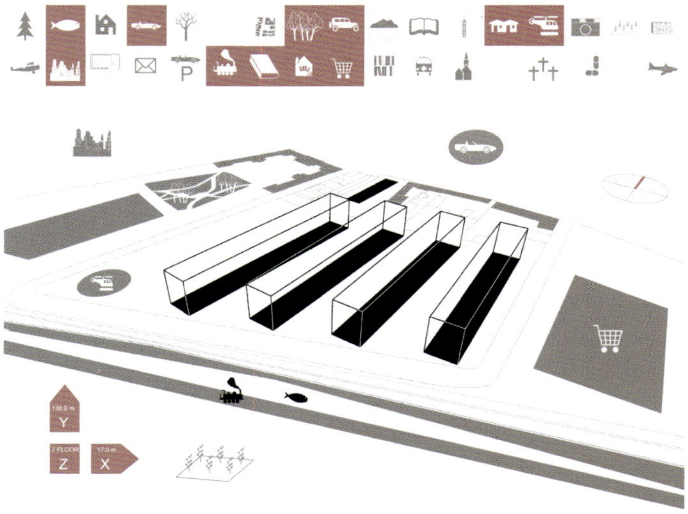
Y
Z
X

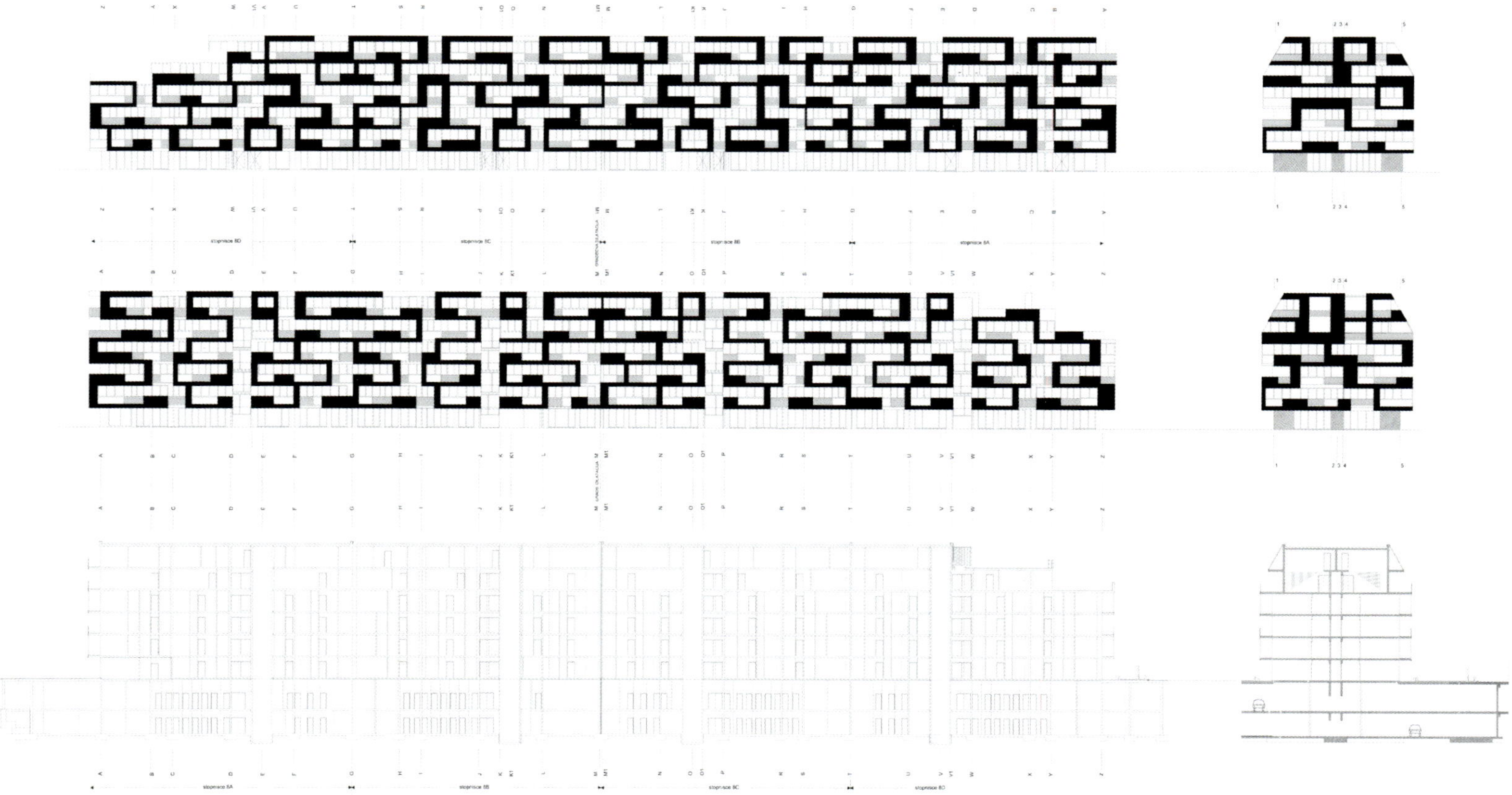

POGLED A

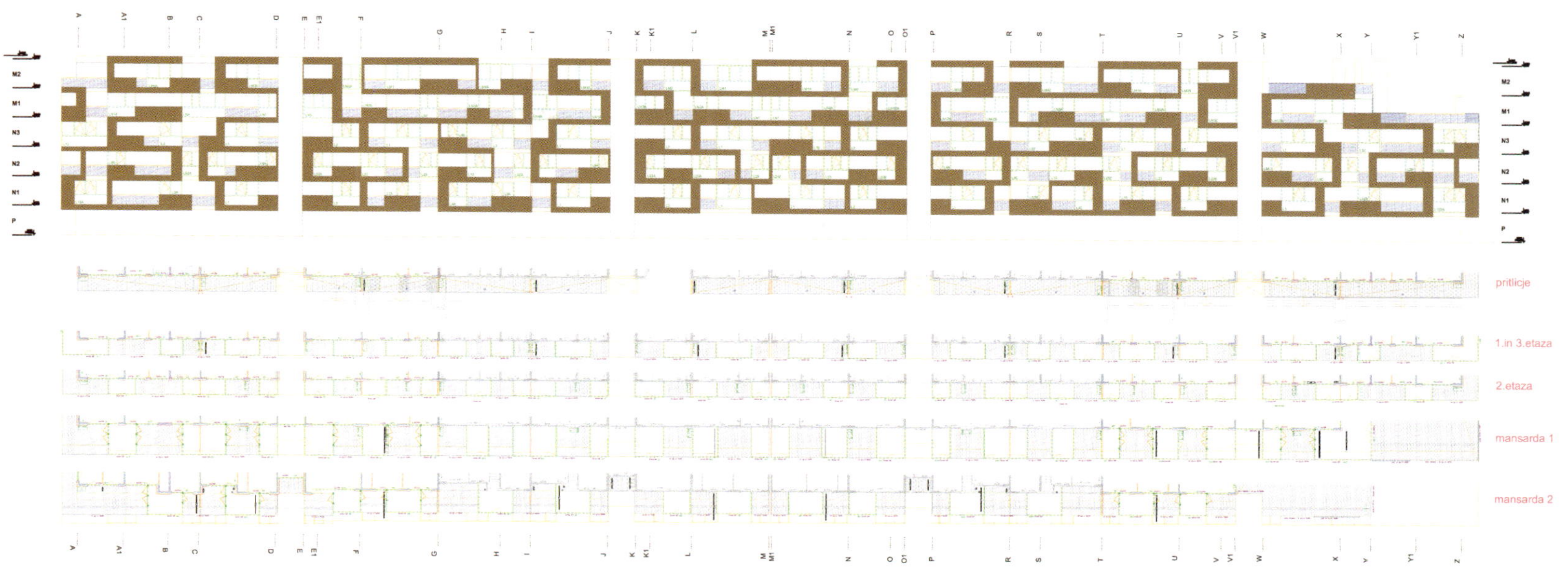

POGLED B

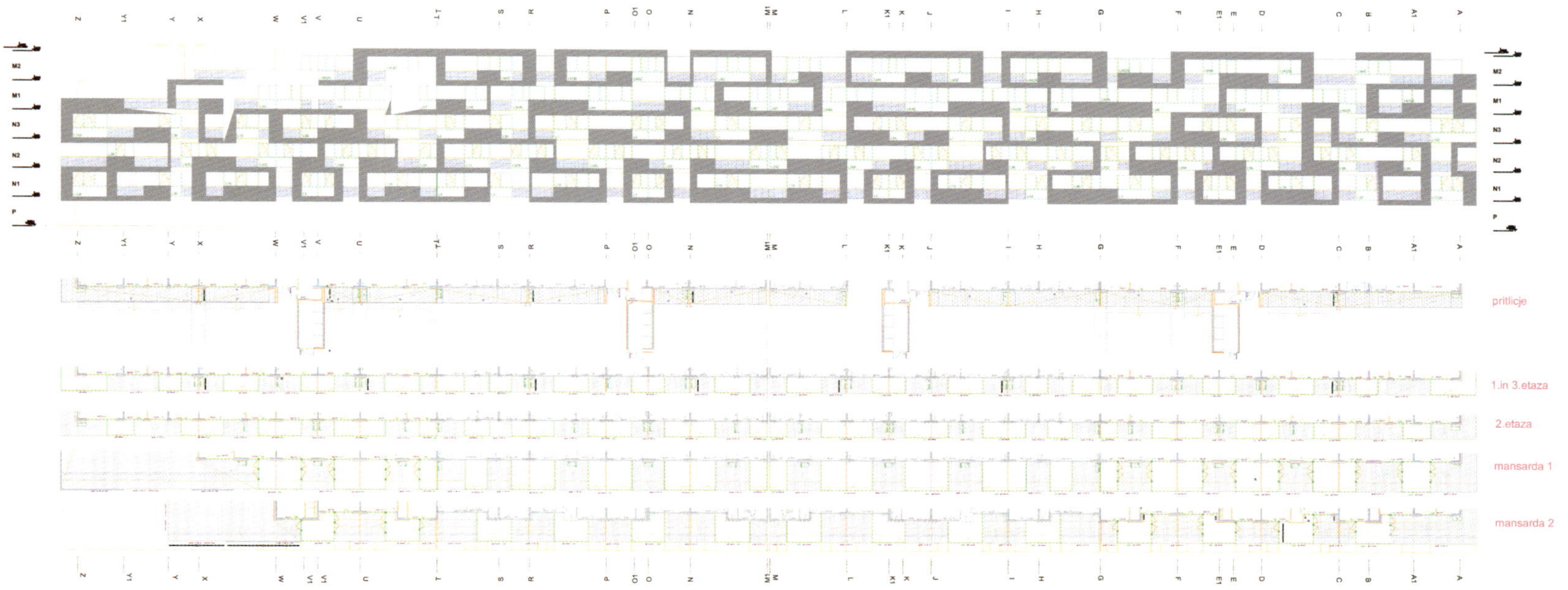

POGLED C

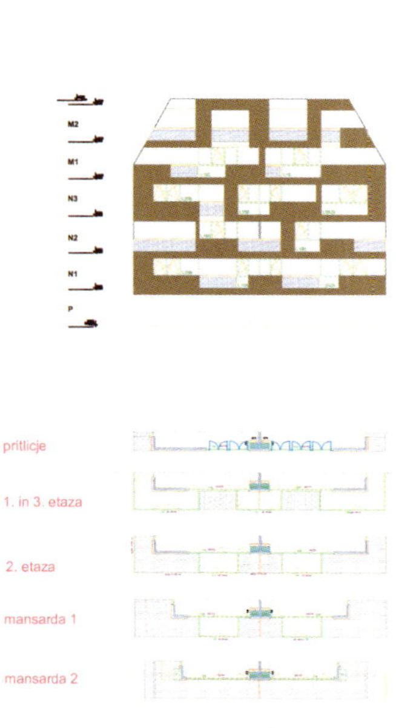

POGLED D

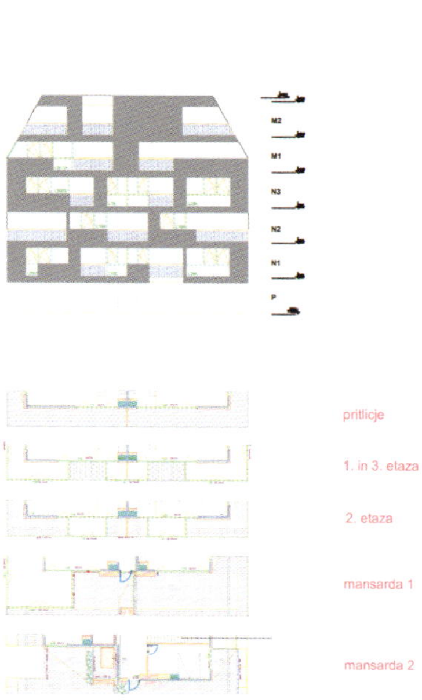

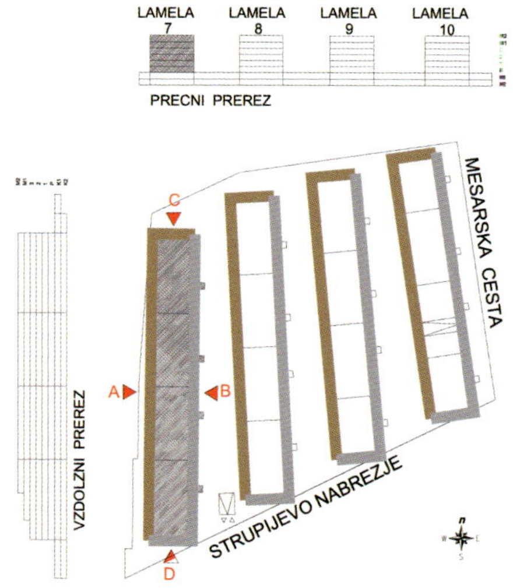

LEGENDA MATERIALOV:

- fasadne plosce - "MAX" - rjava barva
- fasadne plosce - "MAX" - siva barva
- ograja iz kovinske mreze

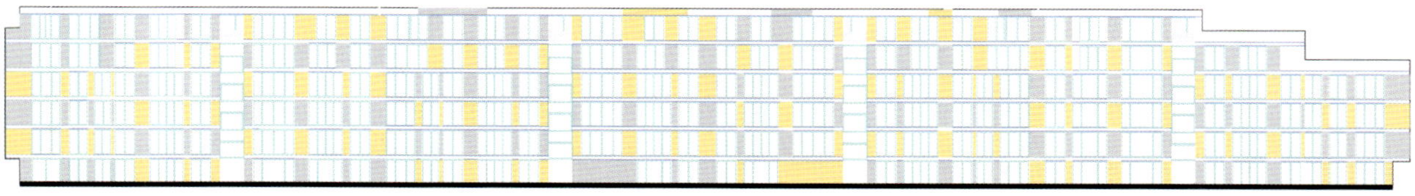

1. layer_plaster facade

+

2. layer_loggias

+

3. layer_laminates+steele fences

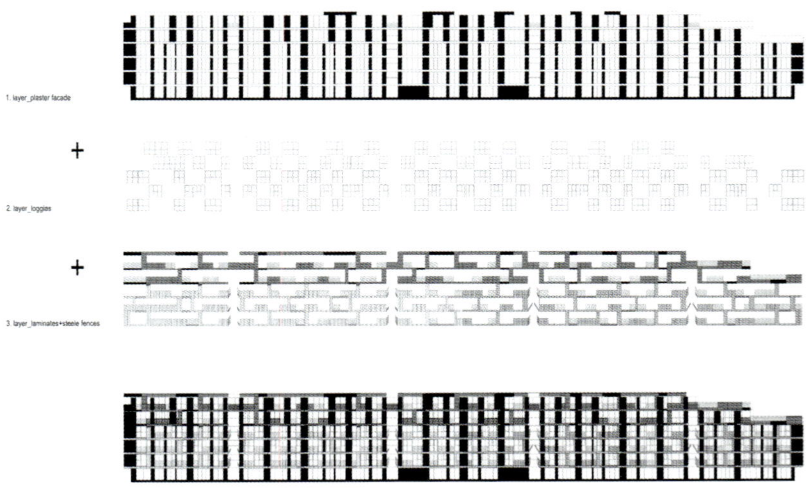
1. layer_plaster facade
+
2. layer_loggias
+
3. layer_laminates+steele fences

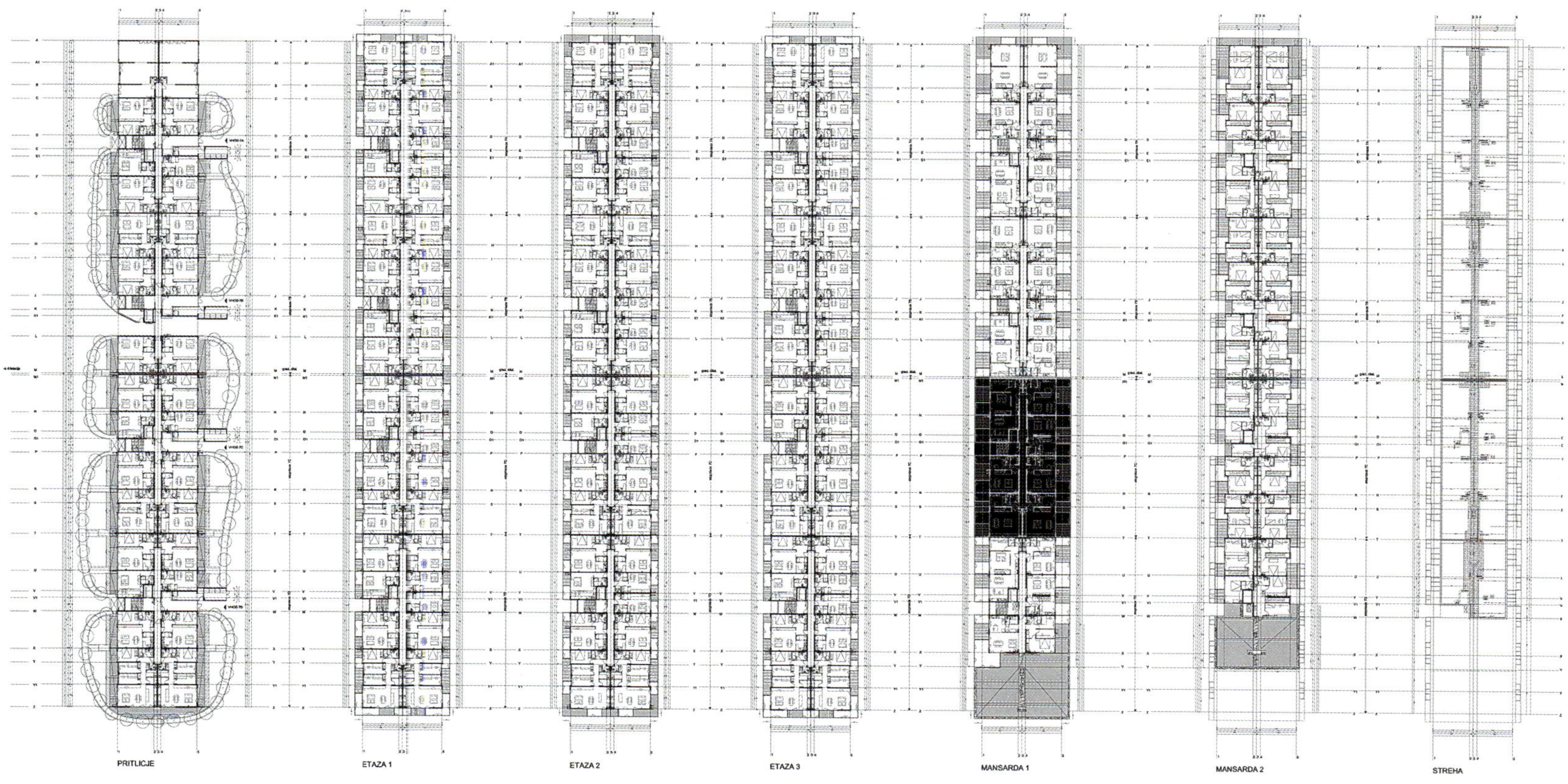
PRITLICJE
ETAZA 1
ETAZA 2
ETAZA 3
MANSARDA 1
MANSARDA 2
STREHA

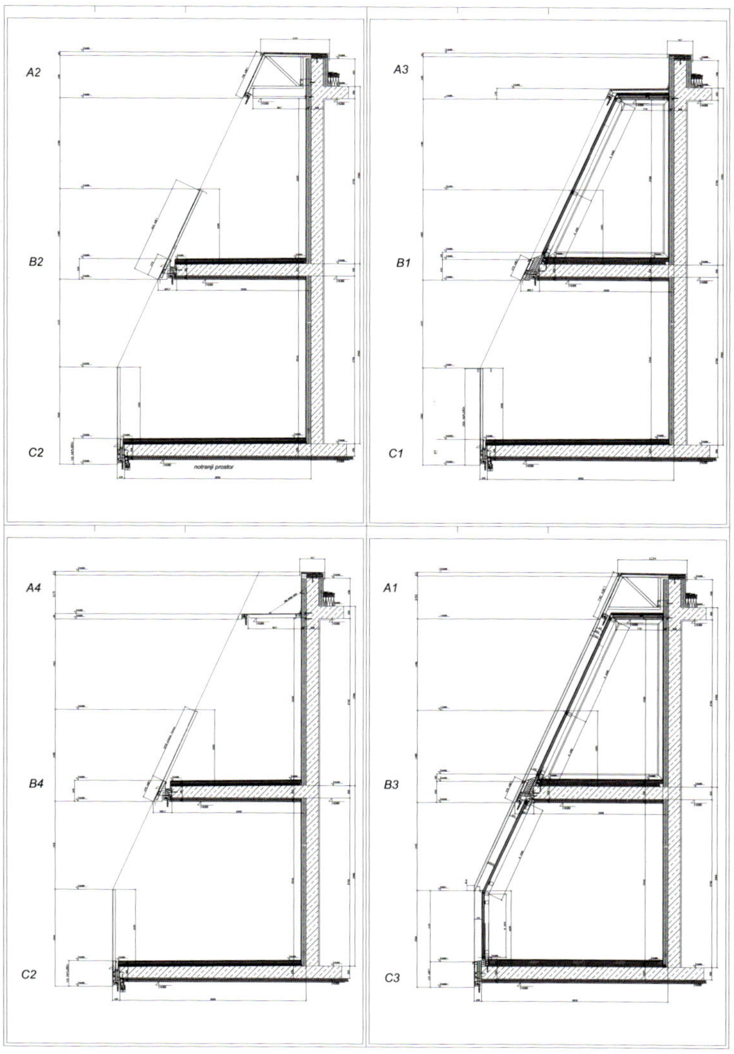
A2
B2
C2
notranji prostor
A3
B1
C1
A4
B4
C2
A1
B3
C3

Architect: Rok Oman, Špela Videčnik
Firm: OFIS architects
Location: nova Gorica, Slovenia
Area: 1200m²
Photographer: Tomaž Gregorič

Lace Apartments

Nova Gorica has very specific climate conditions – it is renowned as the hottest town in Slovenia in summer and also has very strong winds in winter.

The city with its climate, vegetation and way of living has Mediterranean character, where external shady space has high importance. Therefore the client demand was to design rich external spaces of different characters. Studying the different external spaces that existing houses in the area have the designers proposed types such as: balcony and terraces – both opened and covered with roof or pergola, loggias that are closed from the side and fully or partly glazed and fences of different characters – transparent with glass or metal fence, full and of different heights. Many balconies have integrated cupboards for storing external furniture. Using all this traditional elements apartments gain different characters that are more exposed or intimate and that offer opened or partly enclosed views. Using the elements building reinstates three-dimensional lace which embraces its volume.

The fixed urban plot of the building had to be orthogonal block 48 x 16m x 5 floors. Clients brief was also clear with apartments' sizes and typologies that actually are simple and repeated. Over this simple structure the second skin of terraces provides each apartment different character and possibility to buyer to choose the one that responds to his lifestyle.

The facade color patterns are taken from typical color elements of the area (typical Gorica valley soil, wine and brick roof-tops) ...nevertheless the locals started to call it "pyjamas" (reminds on man night wear pattern).

Besides 5 floors two underground garages are build. In the ground floor offices and shops/services are located.

Air condition external elements are partly on the roof and partly behind the metal grid, which forms façade of the mezzanine above ground floor.

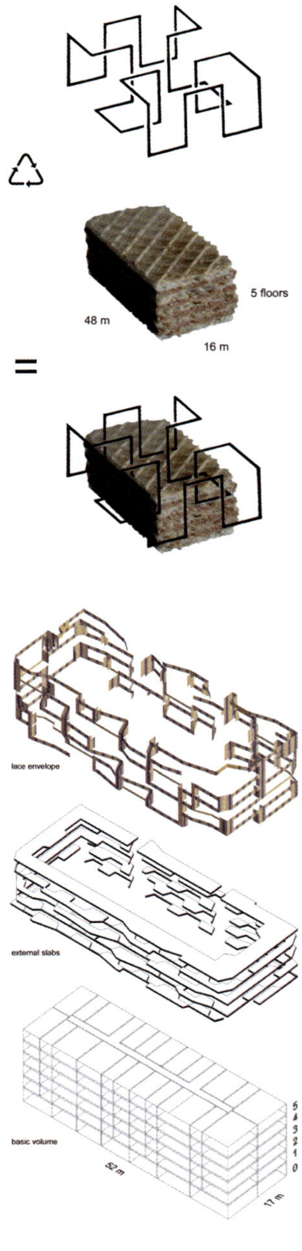

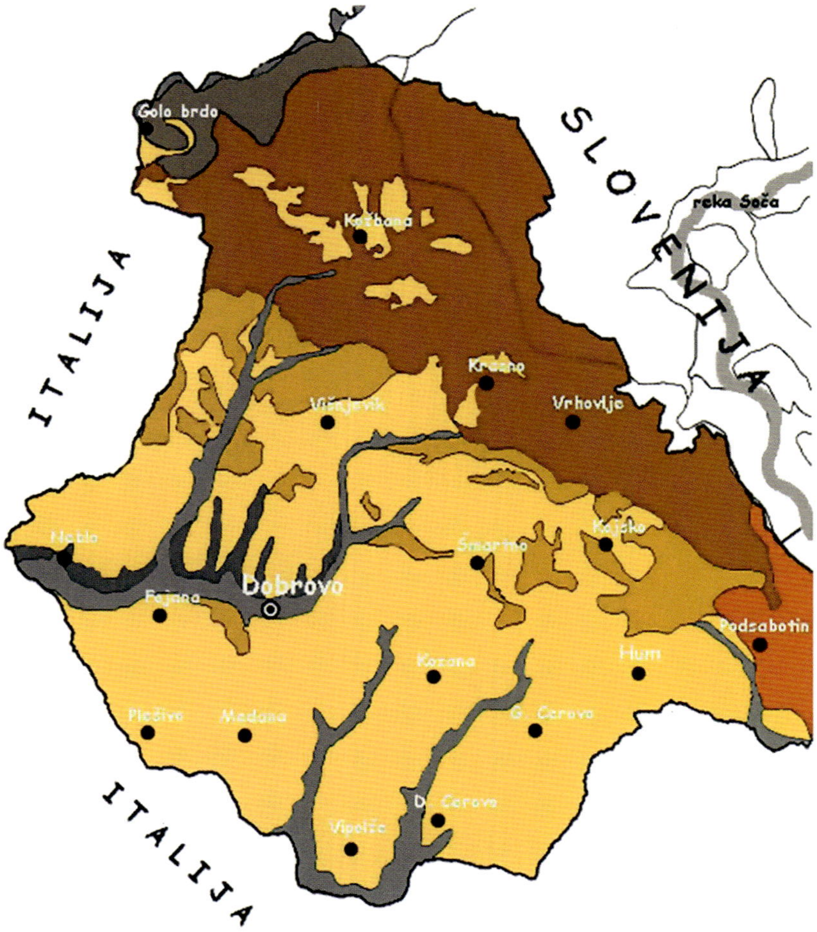

soil type colours = facade colours

70% rjava pokarbonatna apnenčasta tla
20% rendzina / apnenčasta sprsteninasta tla
10% pokarbonatna apnenčasta izprana tla

50% evtrična rjava tla / eoceanski fliš, lapor s primesjo apnenih breč
30% ranker evtrična regolitična tla
20% evtrična rjava tla / fliš, lapor s primesjo apnenca

80% evtrična rjava tla / eoceanski fliš
20% rendzina / fliš, sprsteninasta

60% evtrična rjava tla / eoceanski fliš
40% ectrična rjava / eoceanski fliš, antropogena

70% rendzina / apnenčasta, sprsteninasta tla
30% rjava pokarbonatna tla / apnenčasta tla

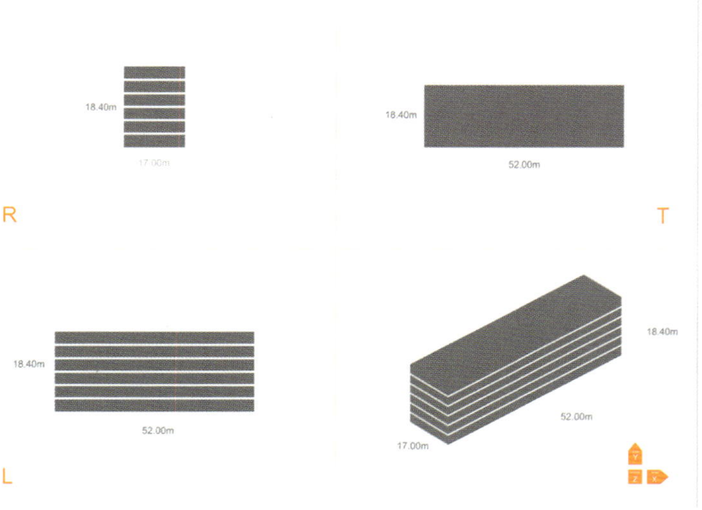

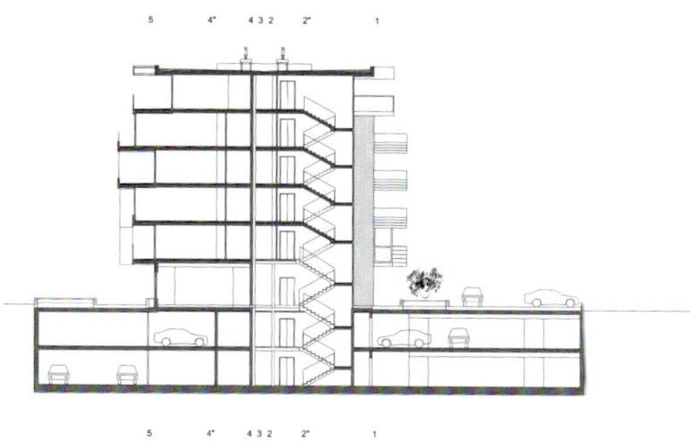

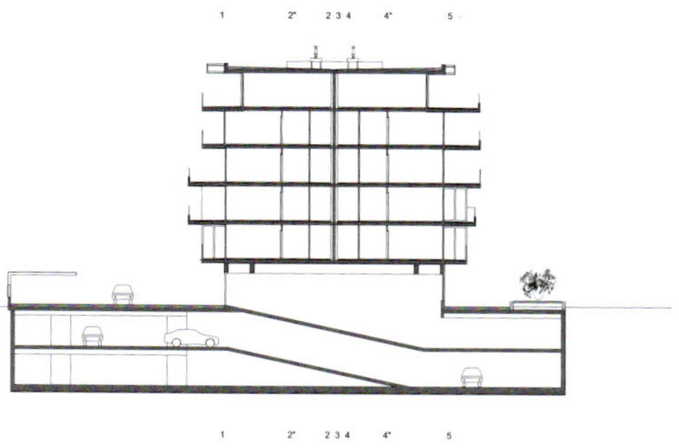

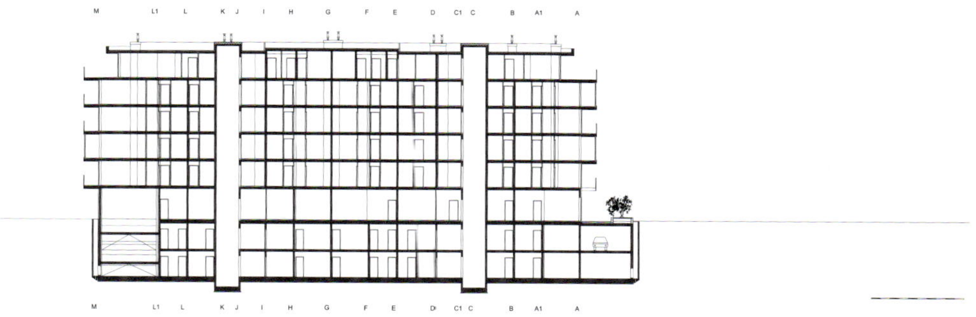

WEST

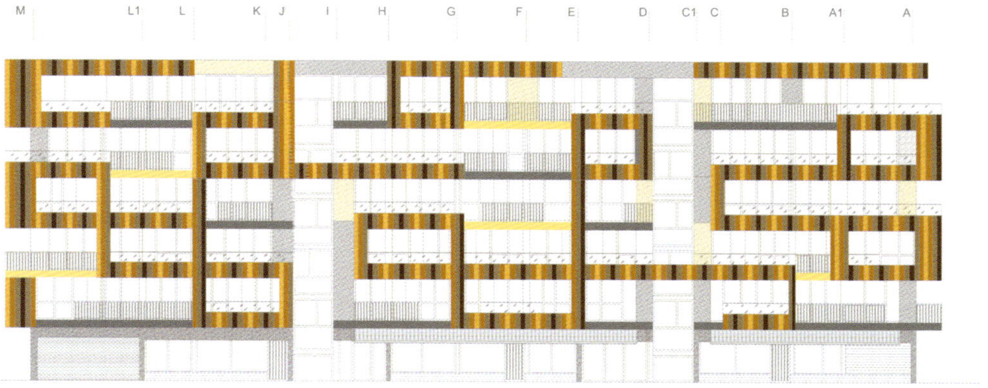

EAST

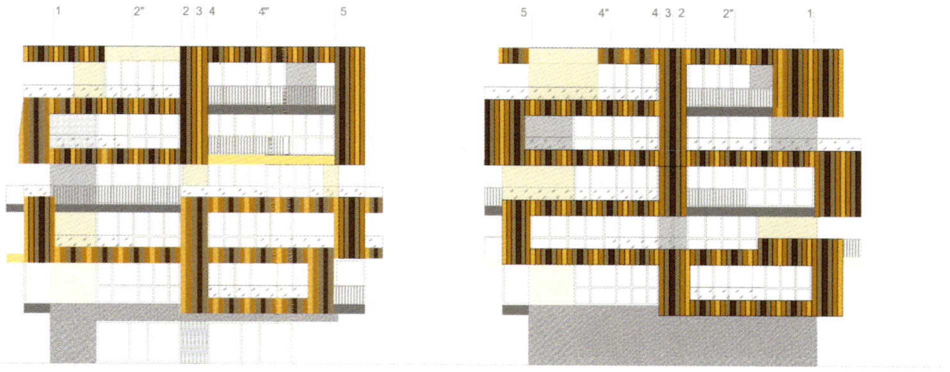

NORTH

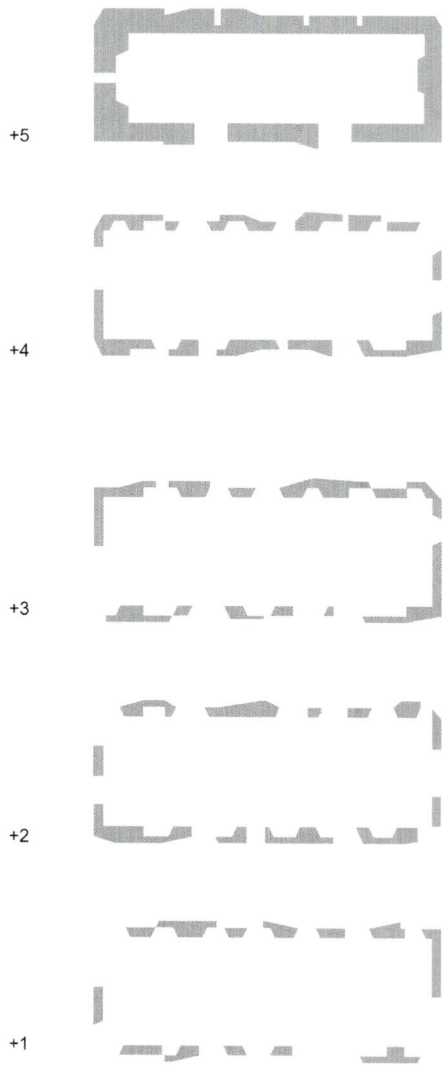

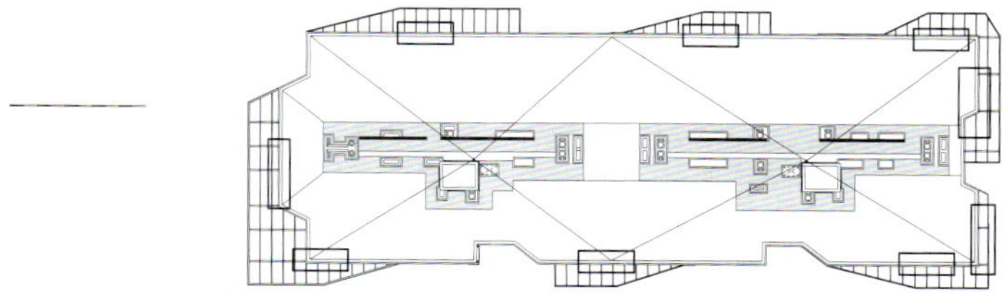

roof level

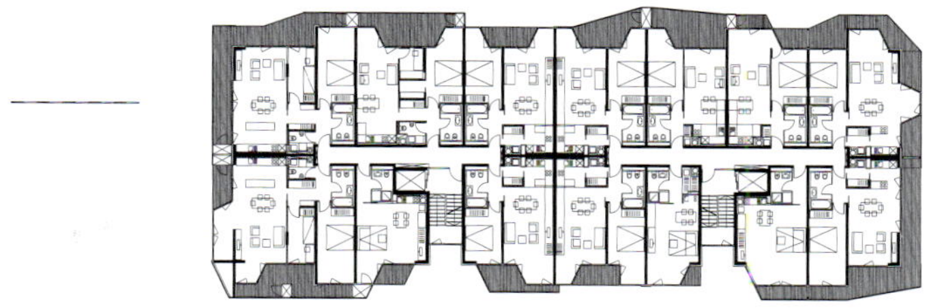

level 3

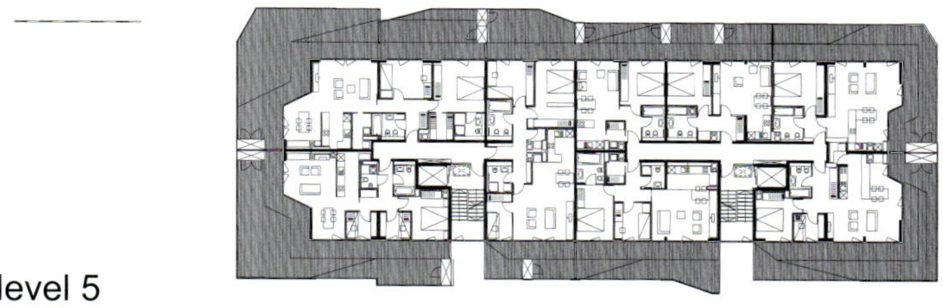

level 5

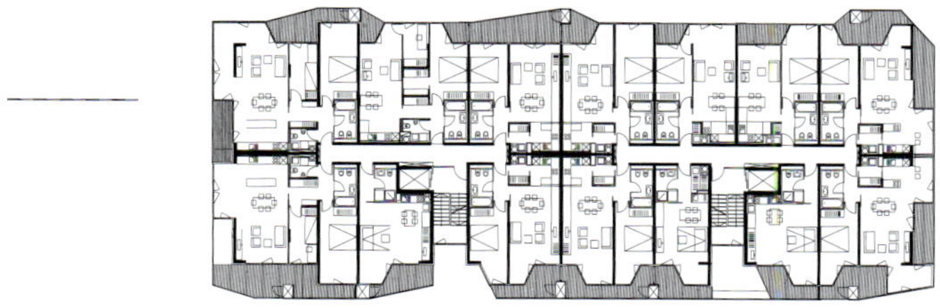

level 2

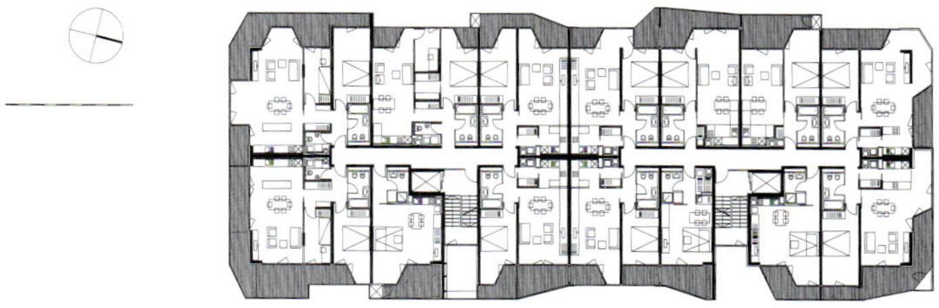

level 4

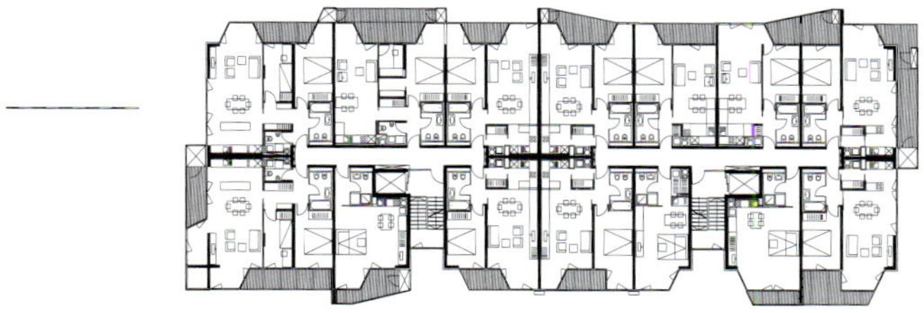

level 1

Architect: HSH architects
Firm: HSH architects
Engineer: Ingenieurbüro Unterberg
Location: Gormannstr. 8-9, 10119 Berlin
Area: 1050m²
Photographer: Tomek Kwiatosz

Apartment Building Gormannstreet Berlin

The building site in Gormannstreet in Berlin-Mitte already showed a long planning-history when HSH architects got involved: Two 20 meters high firewalls confined the premises. Nevertheless preceding conceptions proposed solely horizontal organised floors. All of them failed. Nobody wanted to live in the dark lower apartments.

HSH architects structured the building vertically. Three individual apartment houses span four storeys each, starting from access at street level up to the top floor with an own roof terrace. The inner configuration defines the outer facade. Yet in an urban scale the single houses are perceived as one building, enfolded by a slightly coloured aluminium skin.

Inside the apartments are organised to allow the occupants to view each other within and accross several storeys: Every floor is assigned to specific functions, but simultaneously linked in vertical direction. Spacious stairways connect the different levels. Voids, patios in the two lateral and a loggia in the central apartment mediate as communicative elements between the storeys and play with the relationship between inside and outside. The patios create very private but open interior spaces amidst the city.

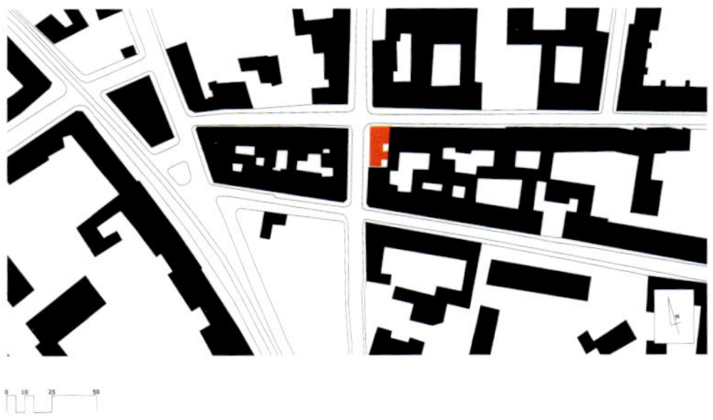

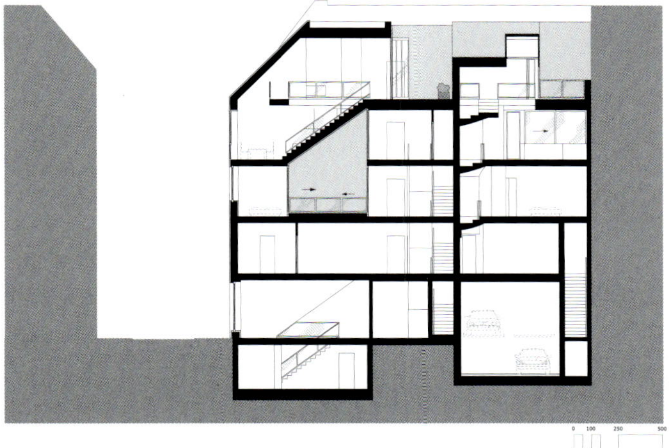
0 100 250 500

0 100 250 500

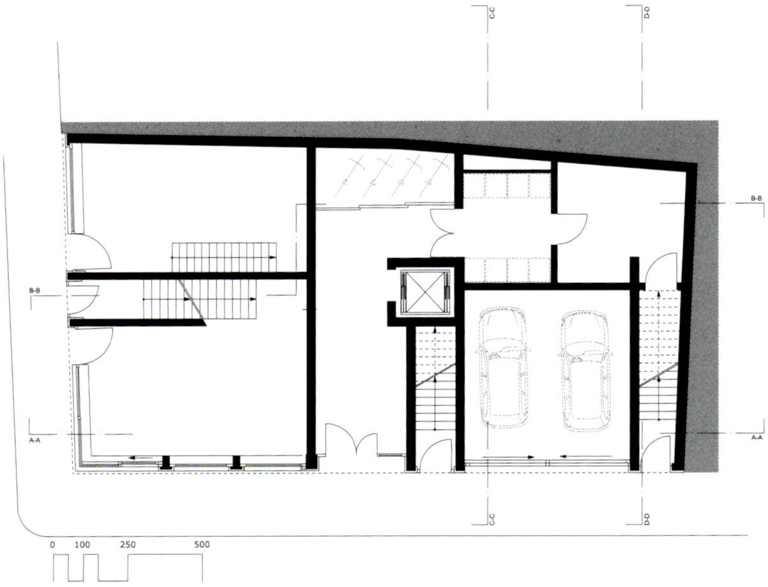
C-C
D-D
B-B
A-A
0 100 250 500

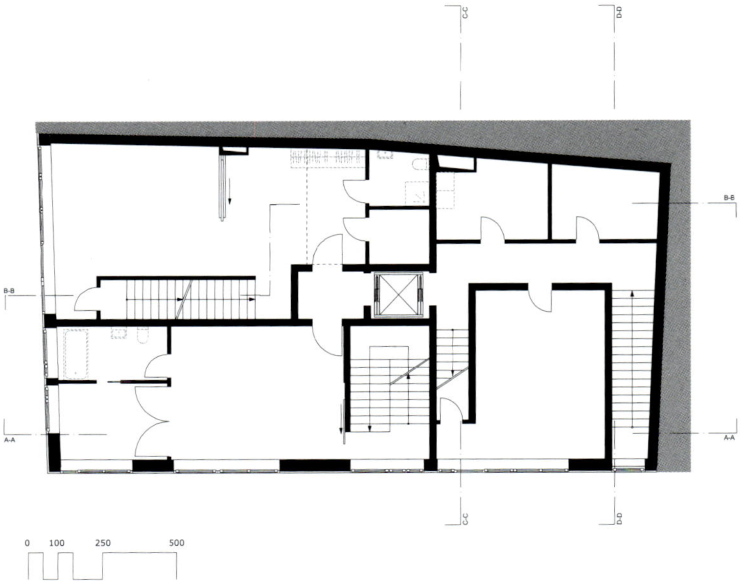
C-C
D-D
B-B
A-A
0 100 250 500

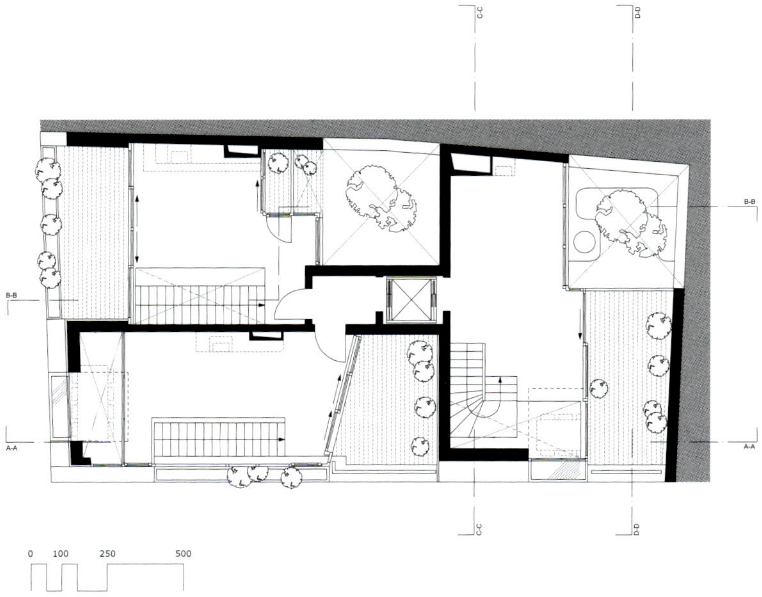
C-C
D-D
B-B
A-A
0 100 250 500

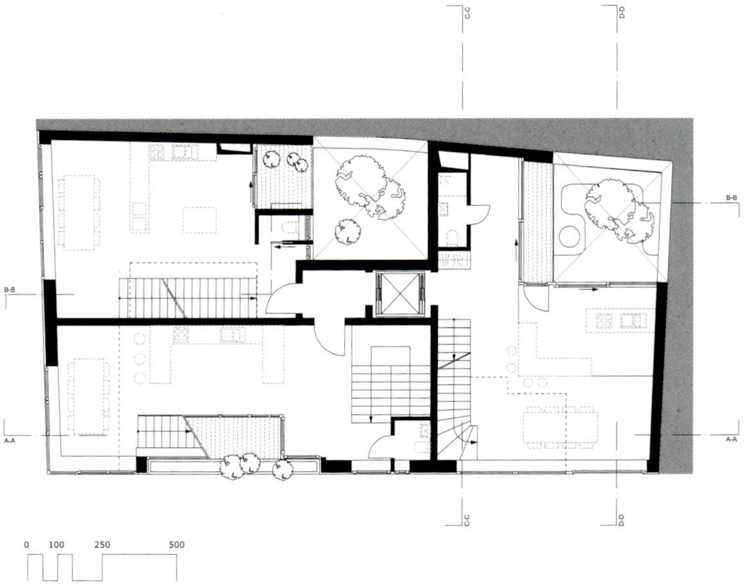
C-C
D-D
B-B
A-A
0 100 250 500

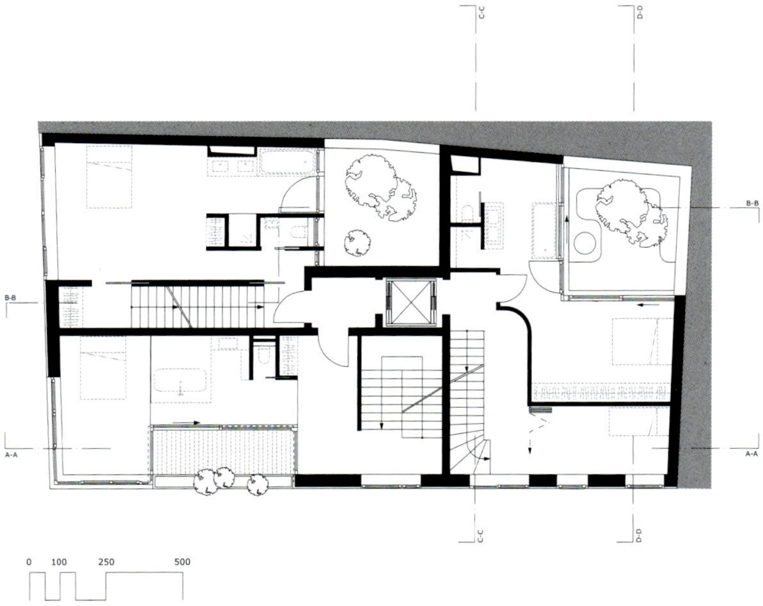
C-C
D-D
B-B
A-A
0 100 250 500

Architect: Igor Franić
Firm: Studio za arhitekturu d.o.o.
Collaborators: etc.
Contractor: Tehnika d.d.
Structure: Skoro d.o.o.
Mechanical installations: Projektbiro Tolic d.o.o.
Electrical installations: Elektro-ekspert d.o.o.
Sewage and water supply: Projektbiro Tolic d.o.o.
Fire protection: Solido d.o.o.
Location: Zagreb
Area: 40275m²

Zagreb Street Center

"Centar Zagrebačka" truly represents a new spatial species: a central-peripheral, inclusive -exclusive, supercontextual -alienated, hermaphrodite. Here a series of apartments on several storeys (by their nature, differentiated and "long") have been raised up and away from the noisy and unpleasant suburban road, over a mass shopping area. Such spatial disposition is very common on a smaller scale in the case of a built-in or detached suburban house. What makes this special is its size (forty thousand square meters of gross surface area over three underground storeys and nine storeys above the ground), its scale (because it represents discontinuity in a mostly homogeneous suburban mass), and its suburban situation of non-place. In this situation a logical answer to the hybridity of the program is not the addition of different horizontal layers or the fixing; of the borderlines of an "urban" space that does not exist, but deduction, from the common and closed volume, in a spatial operation of minimalist sculpture, focused on the visual effects of reflexion, diffraction and refraction of light. The architectural program is in this way logically fulfilled and – equally logically - fully abstracted.

In the dynamic physical landscape, on the way to a changing city, the abstract flow of money materialises in separate situations, at the same time proto- and post-urban. The "architecture" of a corner structure without continuous blocks, especially when it insists on its "urban" character is too often lost in them. Here this is not the case – on the contrary. Abstracted by the omission of details, reduced to a colourless volume with a very thin red skin, made of profiled metal sheets like countless warehouses scattered through the outskirts of perhaps all contemporary cities. Some portions of this skin have fine perforations, so that the program packed behind them is visible only at night. As a witty trace of the designer's quantification of "suburban" scale instead of the composition of "city" facades, floor levels and cardinal points are marked in white on red metal sheets. Cardinal points are a confirmation of such "architecture's" autonomy and its geometrical depersonalization In this way "Centar" is simultaneously non-existent and can be anywhere.

Mercator

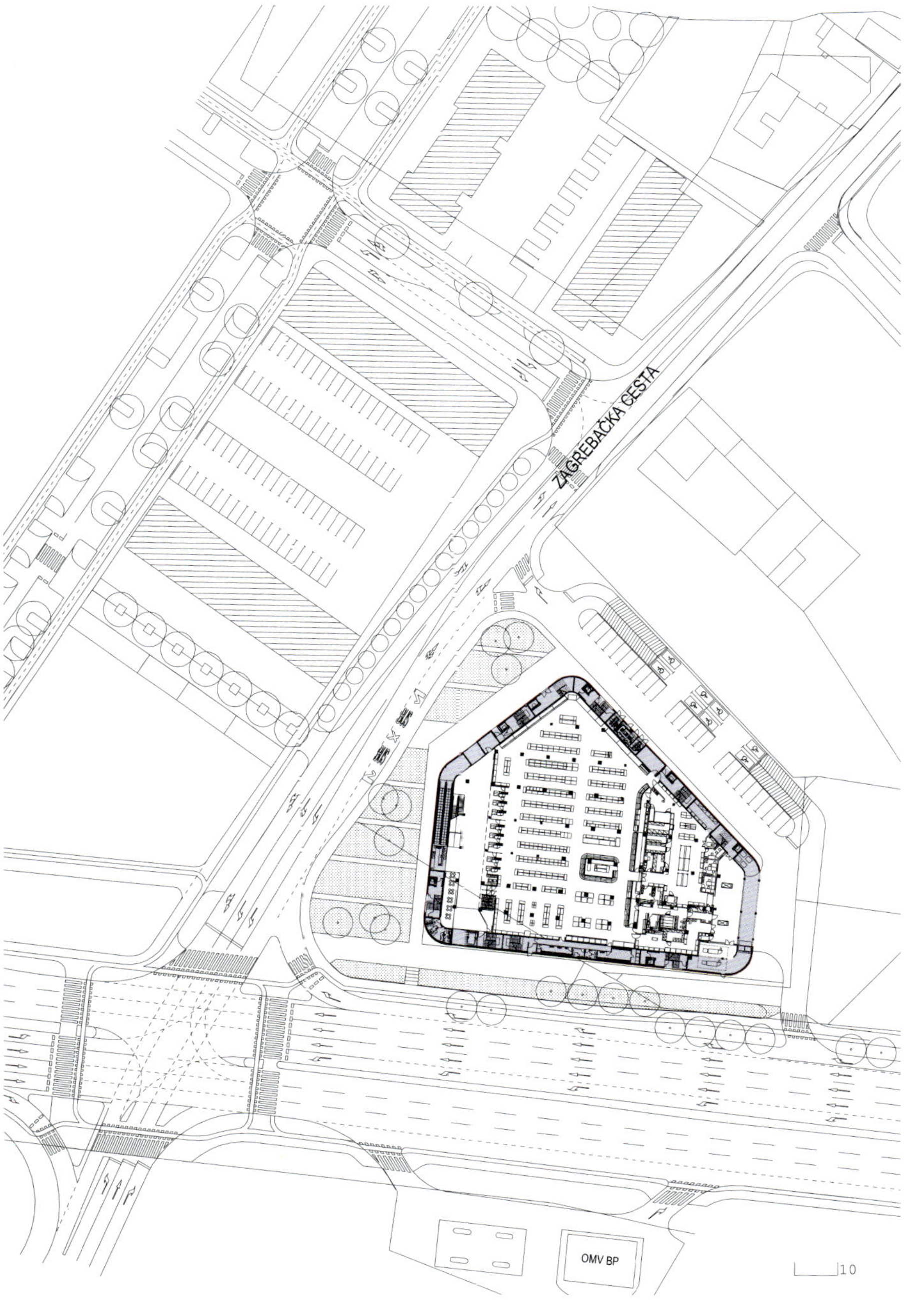
ZAGREBAČKA CESTA
OMV BP
10

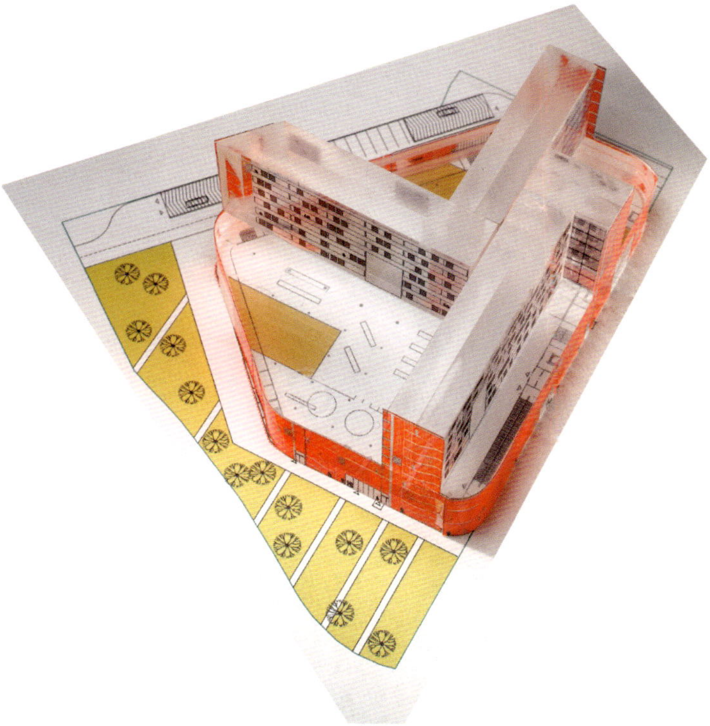

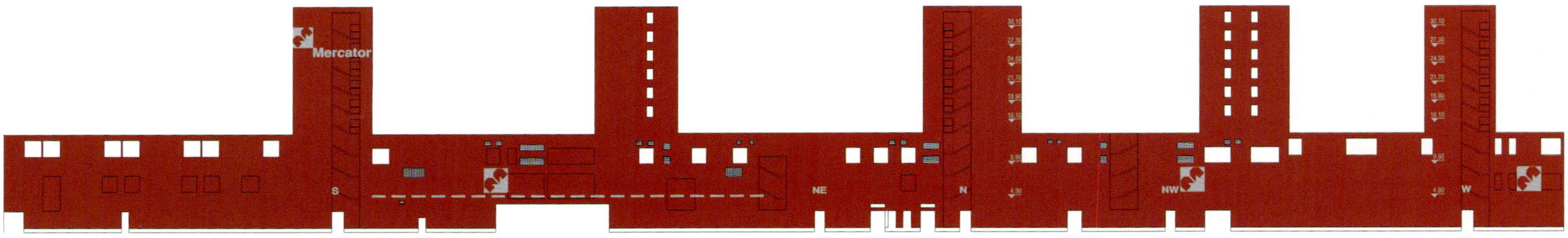
Mercator
S
NE
N
NW
W

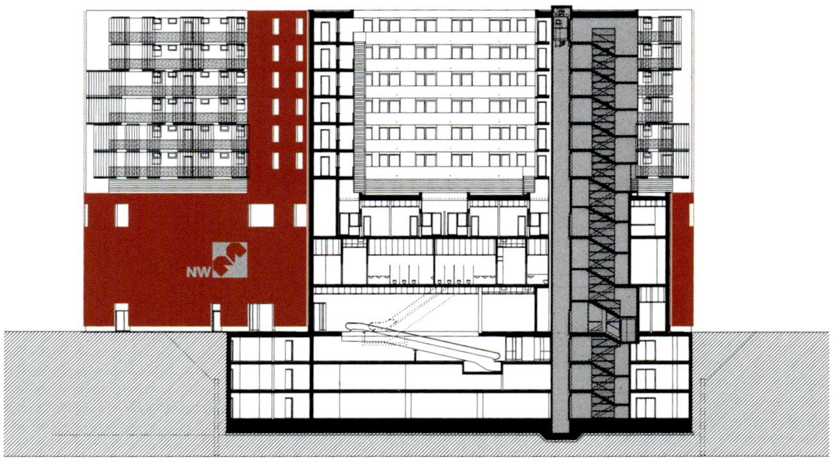
NW

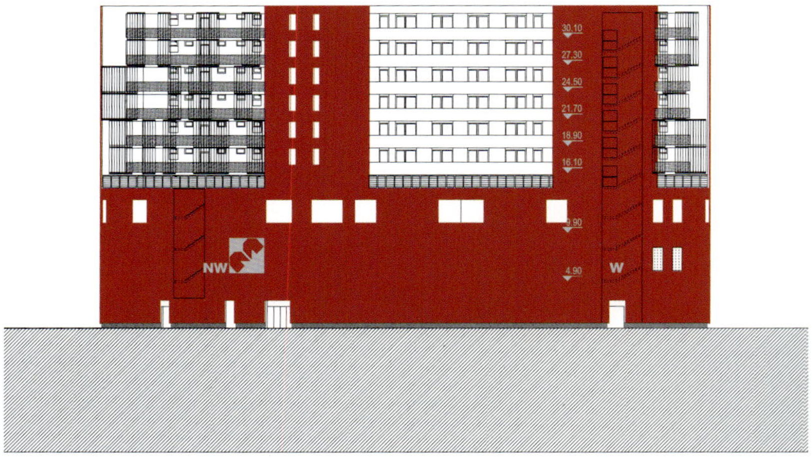
30.10
27.30
24.50
21.70
18.90
16.10
4.90
NW
W

NW
Panasonic
www.intelkom.hr
Mercator Centar

30.10
27.30
24.50
21.70
18.90
16.10
9.90
4.90
NW

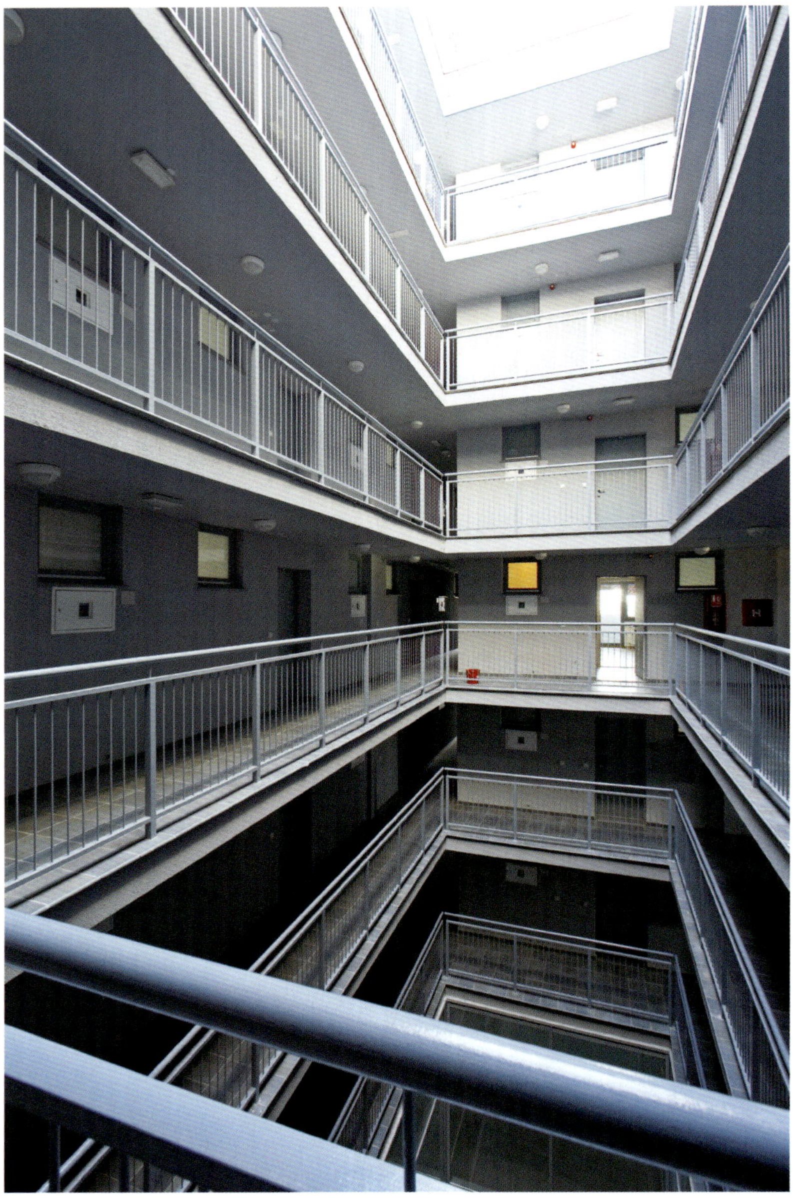

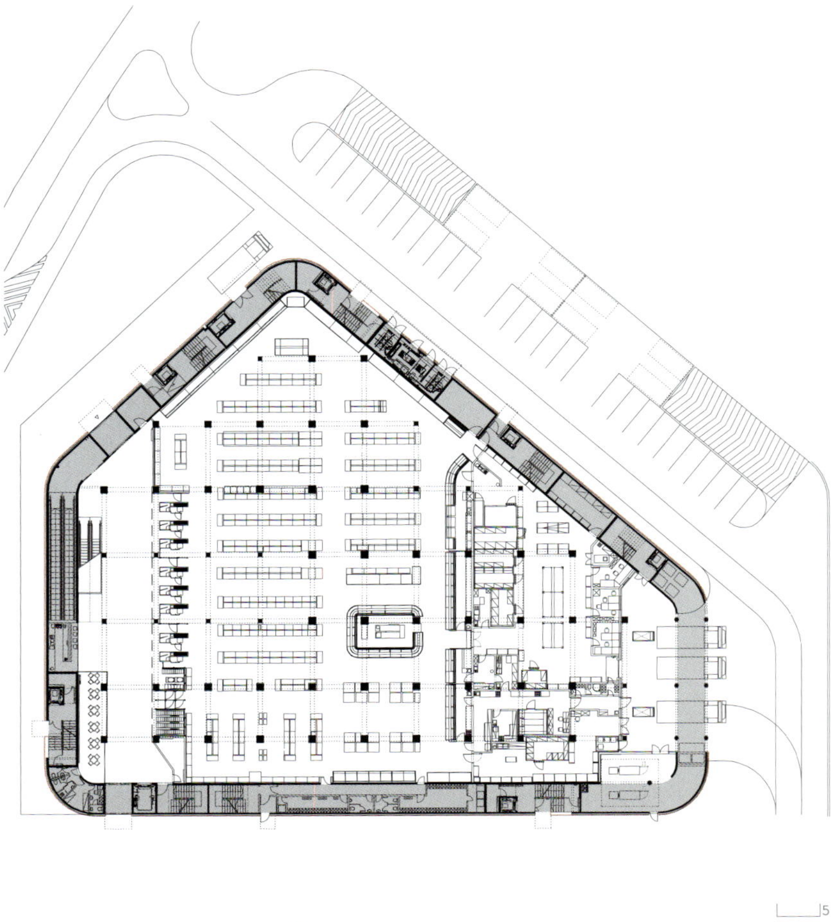

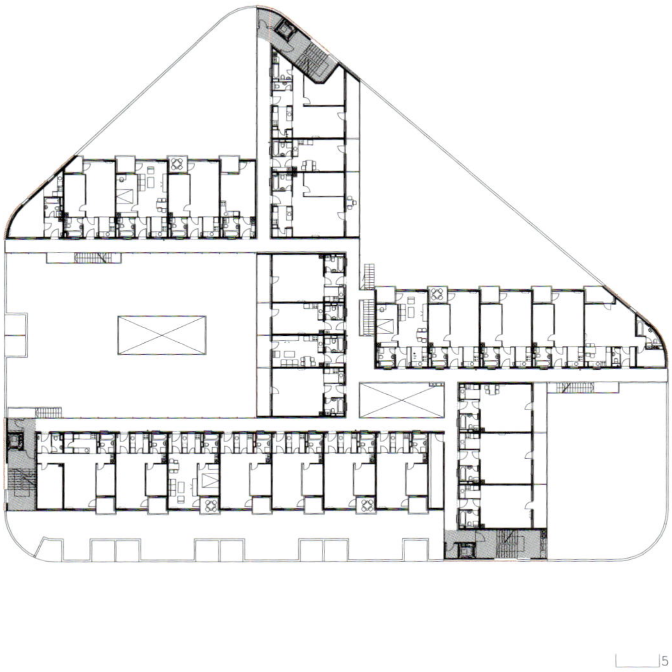

La Griff
STUDIO ZA NJEGU KOSE
MODERNE FRIZURE
ŠIŠANJE
RAZNE TEHNIKE PRAMENOVA
NJEGA KOSE
SAVJETOVANJE
tel: 386 42 96, 1. kat
www.lagriff.hr

RITUALS...

Architect: Paul Diederen, Bert Dirrix
Firm: diederendirrix
Location: Eindhoven, the Netherlands
Area: 1875m²

Central Residential Complex

The location is situated between characteristic elements, consisting of a monumental park, a cemetery, an old villa neighbourhood and apartment complexes from 1980s and the closed façade of the city center. A design contest was held to ensure the quality of the project in this for the city important location. An integral part of the contest objectives was to integrate an altered street layout for the Zwembadweg. The program includes 16 city houses and 18 luxury apartments which share an underground parking garage. The houses and apartments have high quality finishes. The lack of a garden is compensated by the park-like surroundings and an exterior space formed by a large balcony or a patio.
Sixteen townhouses and eighteen apartments are built atop a communal underground parking garage at the edge of Eindhoven's city center, in an environment which is different in every direction.
The apartments are located in a tower built on columns in the city park Dommelplantsoen. Across the repositioned Zwembadweg is a low-rise development of patio dwellings with a strategic view of the park-like surroundings. The bedrooms are on the ground floor, the kitchen is on the first and the lounge area is on the second floor, which also has a terrace.
Central thus provides two options for living in the city while maintaining a certain distance from the turmoil of urban life. This is not only true of the introverted patio houses, whose closed facades guarantee privacy, but also of the tower, which turns its back on the city and opens itself up to the park.

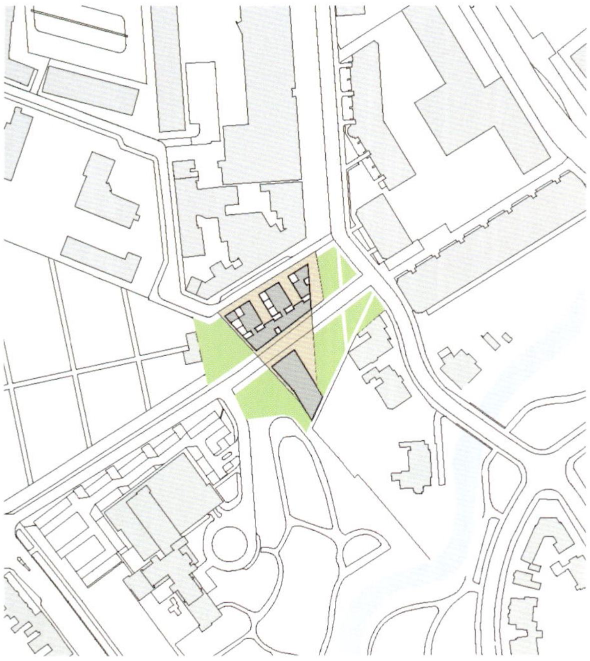

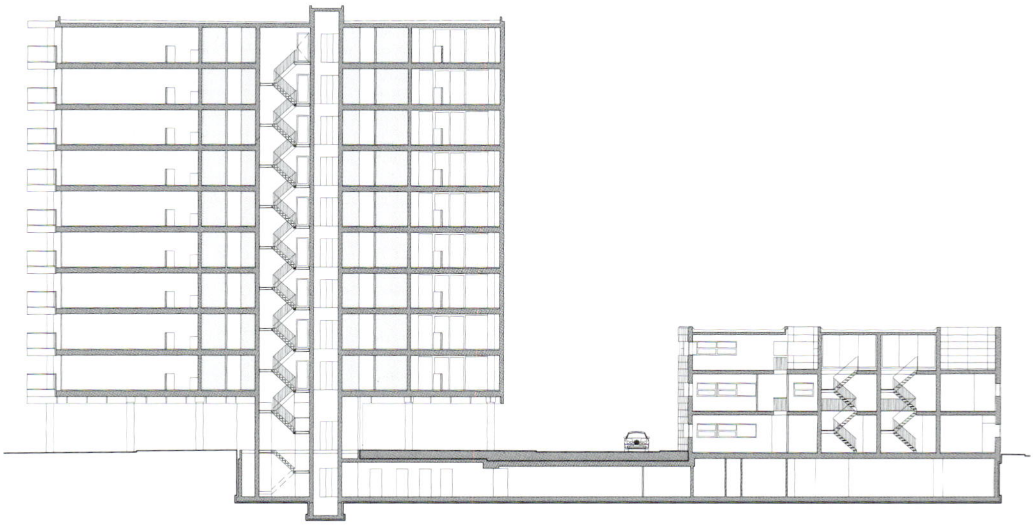

CENTRAL doorsnede 2 1:500

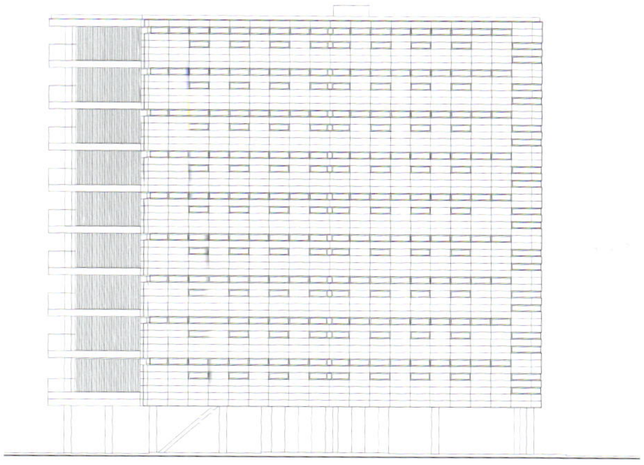

CENTRAL gevel noordoost 1:500

CENTRAL gevel zuidwest 1:500

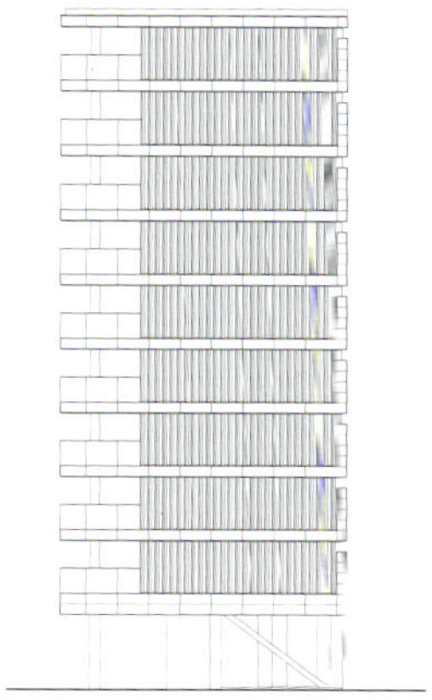

CENTRAL gevel zuidoost (L)

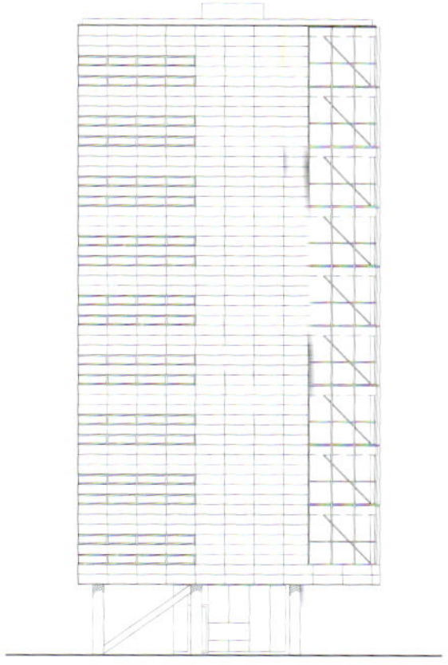

/noordwest (R) 1:500

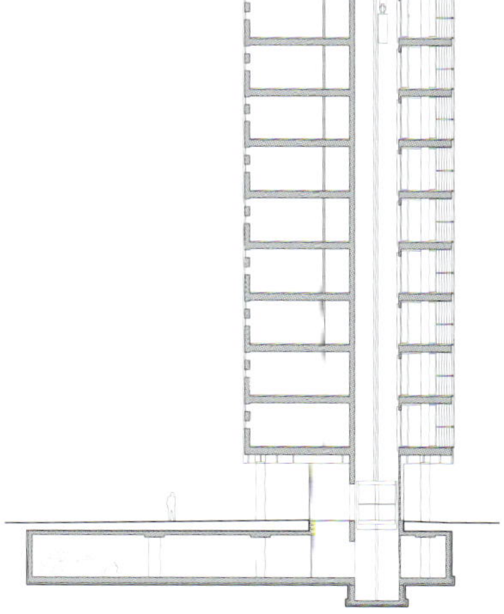

CENTRAL doorsnede 1 1:500

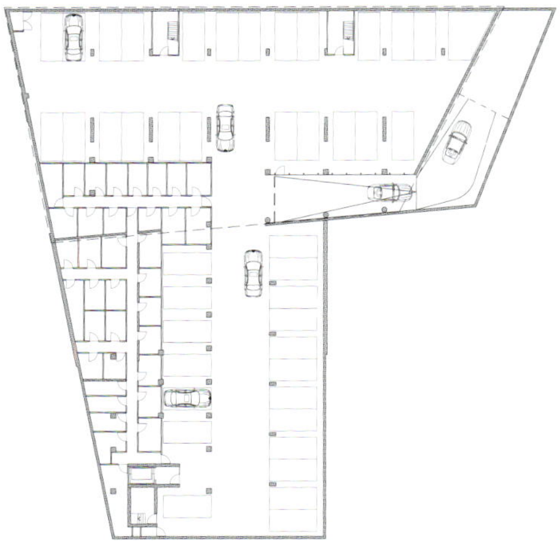

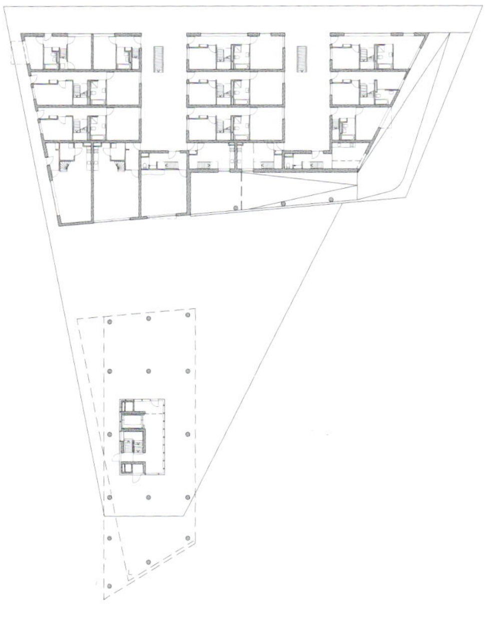

B

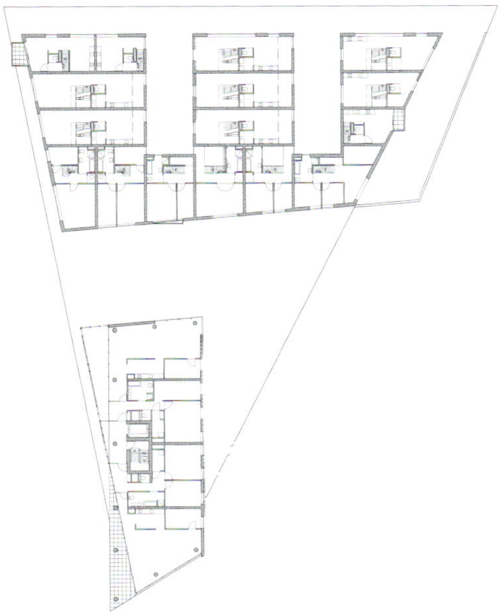

Architect: Theo van de Beek, Marjolein van Eig, Hans van der Heijden, Mathieu Kastelijn, Paul Voorn, Rick Wessels, Manora van Wijck
Firm: biq stadsontwerp bv
Location: Langerakbaan, Utrecht
Photographer: Stefan Müller, Berlijn

Langerak Field 6 + 11

This project of 240 apartments is constructed in the last phase of the city extension area Langerak in Utrecht. Following the urban design of KCAP the main traffic artery, Langerakbaan, has a canal at its south side behind which varying sorts of low-rise neighbourhoods have been built. At it north side, however, more urban forms of housing, such as combined living and working units and large-scale apartment buildings (to the design of architects such as Van Herk & De Kleijn, DKV and ArchitectenCie) have been constructed. Here also variation was a clear ambition. KCAP called this the Polka motif, in which calm parts alternate with spatial accents.
Field 6 and 11 are located just before the end of Langerakbaan. The real end of this avenue is marked by a curved building which is design by Frits van Dongen of ArchitectenCie. The project consists of two orthogonal blocks flanking a continuous strip of urban greenery. Each of those is supplied with an underground car park which forms the plinth for the actual apartments. The Polka motif receives an architectural form by articulating the ends of the block with all-sided towers between which a lower deck access block is oriented at Langerakbaan via its balconies at the south side. The block sits directly at the pedestrian walkways. With the plinth a well-known inner city motif in introduced in the suburb. All round private terraces or French balconies are added on top of the plinth. These porticos are use all round the block. All dwellings have spacious outdoor spaces.
The facade materials are durable and solid. The facade cladding of the apartments is a sand coloured brickwork, used on its edge. This results in minimal reveals framing bronze anodized aluminum window frames. The plinth is articulated in dark bricks forming stately entrance porticos. The buildings do not have front or back sides. The floors and walls of the entrance halls are lined with stone.

GROENEDIJK
LANGERAKBAAN

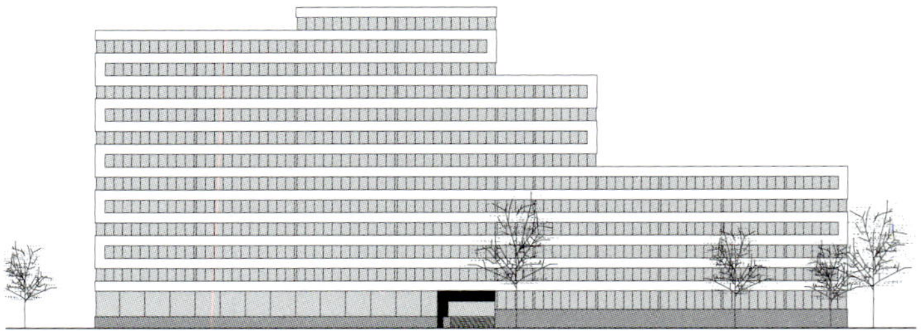

aanzicht langerakbaan

veld 6

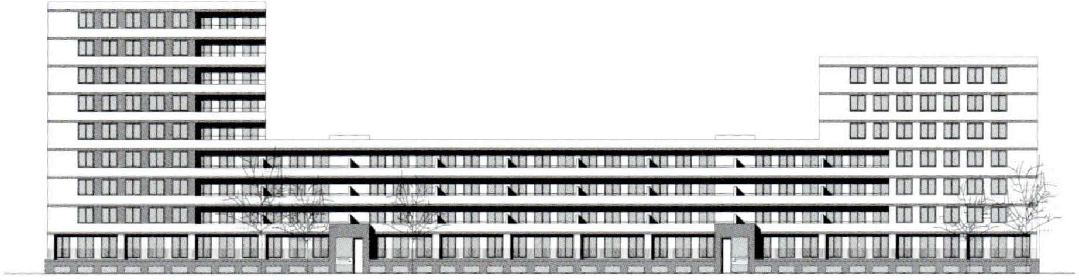

veld 11

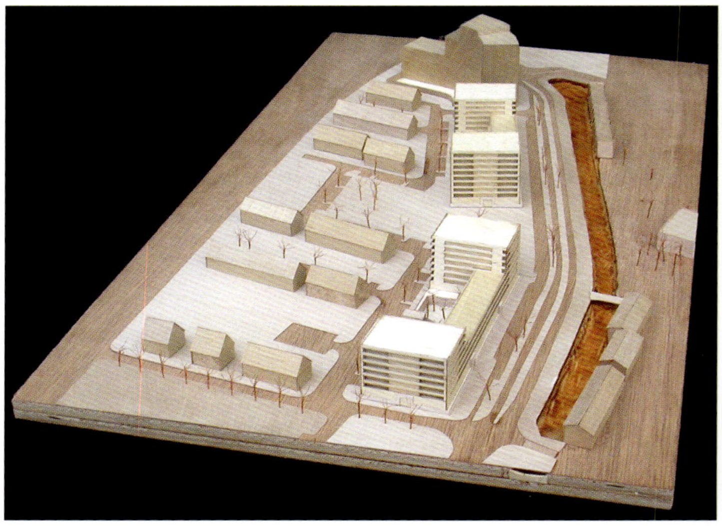

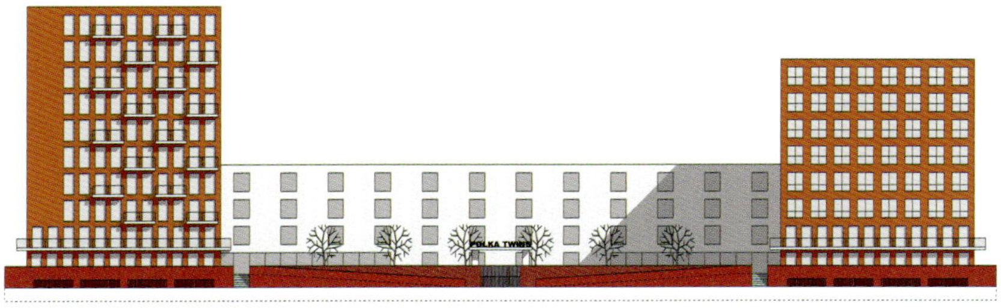

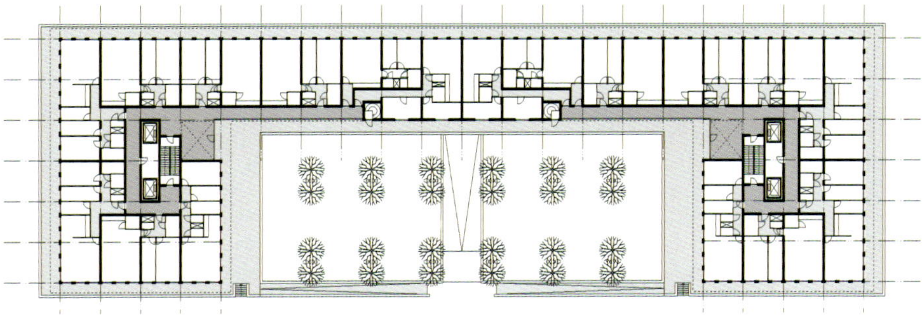

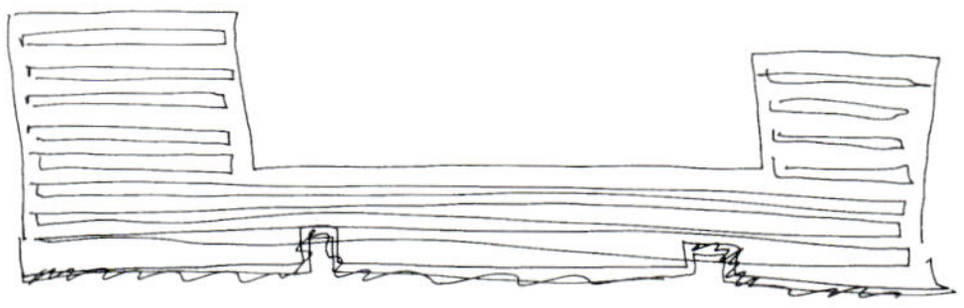

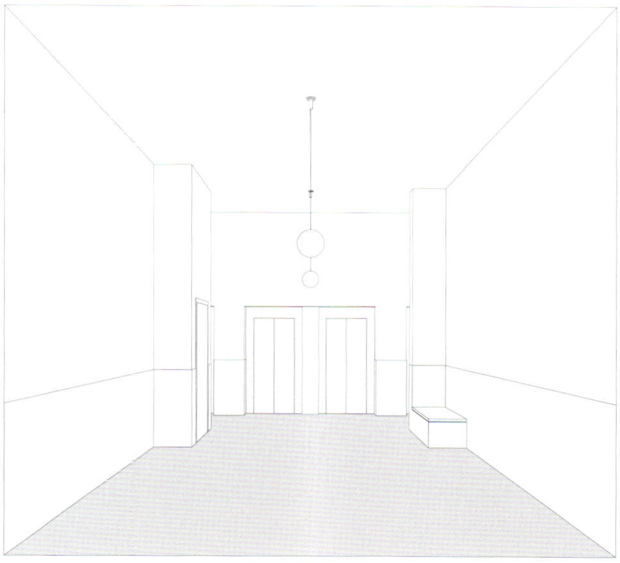

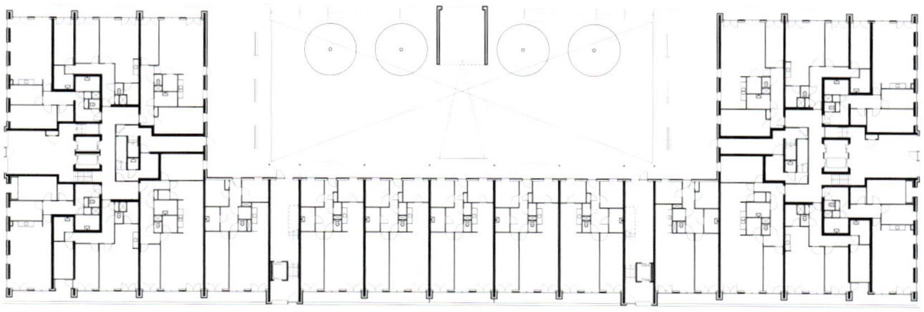

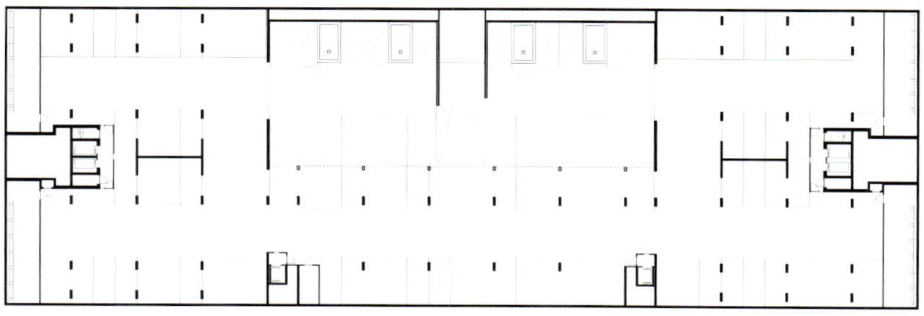

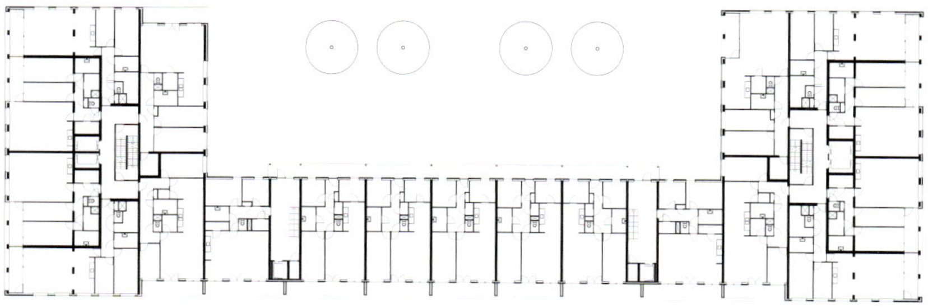

Architect: Gerard Kruunenberg
Firm: Kruunenberg Van der Erve Architecten
Location: Amsterdam, The Netherlands
Area: 6128m²
Photographer: Luuk Kramer
Text: TXT media bv

Furore-blok A

The combined volume of three buildings form the Furore complex which in its entirety is reminiscent of the closed building blocks of the area. Yet the buildings also retain characteristics of autonomous structures and even display the traits of linear urban planning from the 1920s. This double character has led the architects of Blok A to design a massive structure that relates well to the smaller buildings that surround it. The five-meter high base that decorates Blok A at the ground level emanates a solidity that is typical of an autonomous building. Yet in the execution of the design there is refinement expressed in the juxtaposition of clear and rough melted glass. The upper section of this base serves as a balustrade for the French balconies of the first floor residences. Both building units, which stand on the shared base, resemble independent lines but they are connected by walkways and a glass sound screen erected to deflect traffic noise. The brick surface of the two buildings is graced by the rhythm of vertical fenestration and square openings cut into the facade where it continues up to form a high roof ridge. The rhythm is enriched by accents in the form of 'flower frames'; large windows which are shielded from the traffic noise by sliding glass panes set flush to the masonry. The fact that the facade openings form an oblique rather frontal composition enlivens the face of this complex.

Four commercial spaces and three office- residence units are situated on the ground floor level. This is topped by seven floors of owner-occupied residences and rent-controlled units. The two types of residences are interspersed evenly through out the complex. Finally two levels of underground parking have been realized in the basement.

The main stairwell forms a single undivided light-flooded space that stretches from the lowest level of the basement up to the sixth floor. Access to the residences on the first floor is provided via the spacious terraces. The other residences are accessed via the walkways and galleries situated on the north side of the complex.

Thus the inner heart of Amsterdam is enriched with a notable building that beautifully stands in harmony with the surrounding nineteenth century architectures although it is in fact a massive apartment complex.

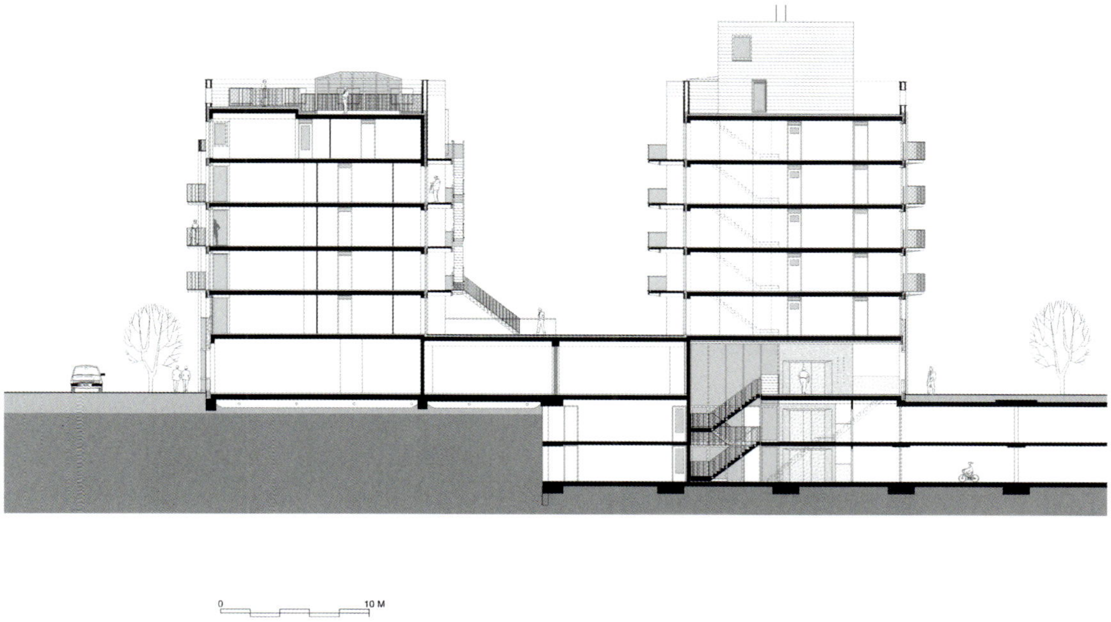
0
10 M

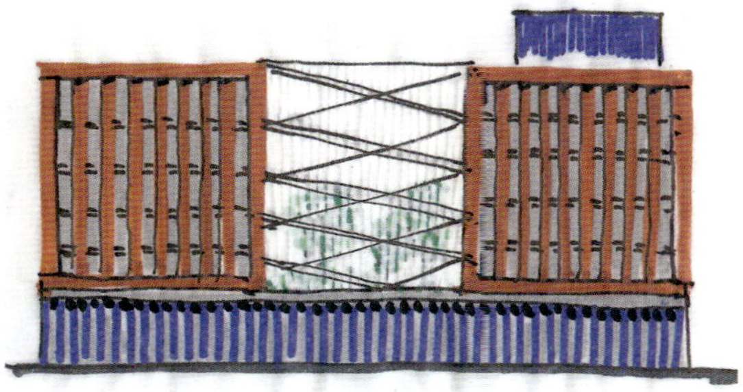

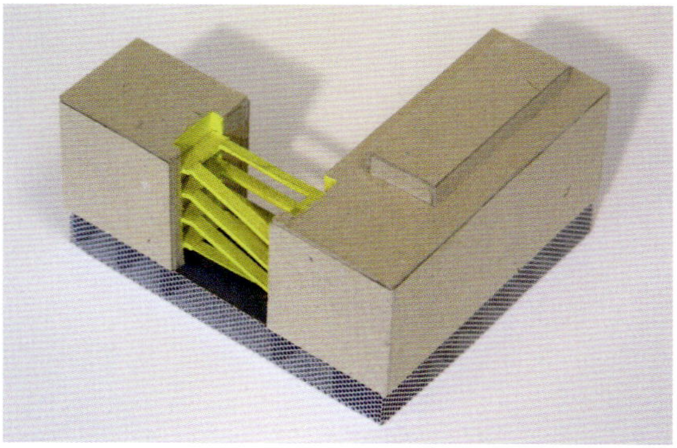

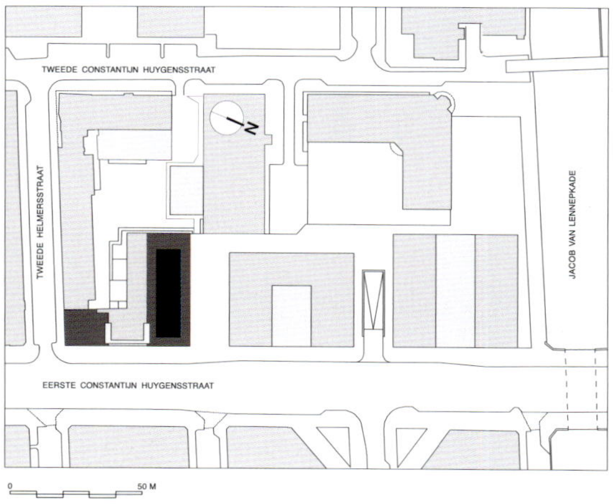
TWEEDE CONSTANTIJN HUYGENSSTRAAT
TWEEDE HELMERSSTRAAT
JACOB VAN LENNEPKADE
EERSTE CONSTANTIJN HUYGENSSTRAAT
0
50 M

48 - 58

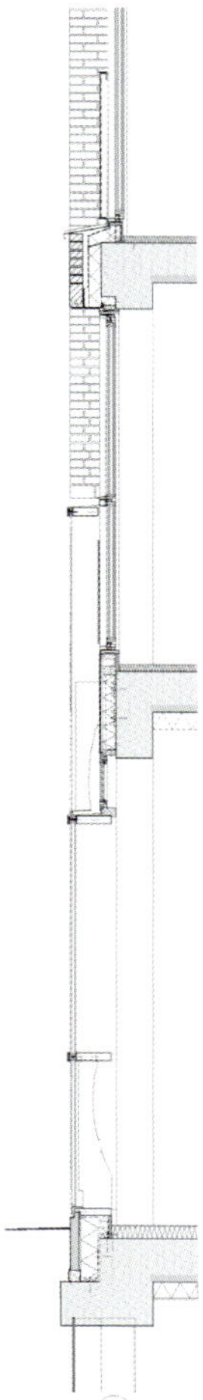

1 Vertikale doorsnede
constructieve wand, fr balkon, etalage

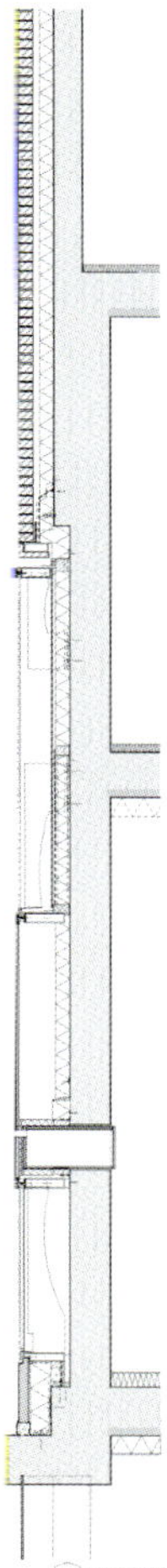

2 Vertikale doorsnede
constructieve wand, brievenbus

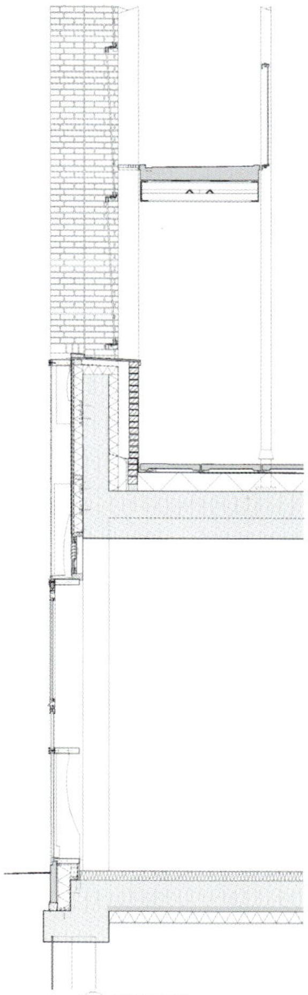

3 Vertikale doorsnede
geluidscherm, raam

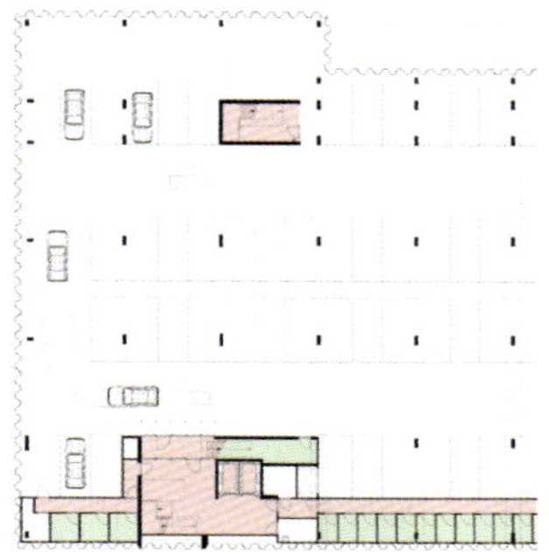

niveau -2

begane grond

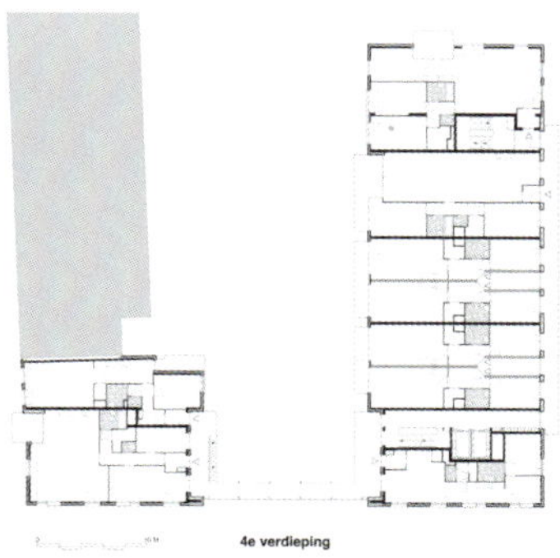

4e verdieping

1e verdieping

6e verdieping

2e verdieping

7e verdieping

3e verdieping

5e verdieping

Architect: Marnix van der Meer, Rene de Korte, Rolf Bruggink, Edwin Deen
Firm: Zecc Architecten
Location: Eemskanaal, Groningen
Area: 25 large dwellings
Photographer: Studio for Architectural Photography

Quay Houses Groningen

On the South-east side of Groningen an industrial area has become indistinct. Here for a large constructing area has become available. A few existing buildings on the waterside stay as they are. On the other area big enclosed building-blocks will be realised. These building-blocks have a large common courtyard and a constructed parking place.

The building-block contains an apartment building designed by Mecanoo, and a row houses along the water, designed by Zecc Architecten. The difference in height between these two elements is distinguished.

The houses have four floors. The houses are being build in a row, but all of them have an individual touch. This individual touch is realised by the sculptural design of the entrance stairs, the vertical articulation of the houses and the different sorting of brick used for the façades.

The lower two levels are realised as a split-level. This is to create enough privacy on the road side and on the same time realise a nice high kitchen on the, private, terrace side.

The houses are strong serially in terms of building system, but offer space for many requirements: additional roof terrace on the south side or view on the water, possibility to choose a larger bay window in the front or a larger entrance area at the ground level.

As corner solution is chosen for weaving two properties together, so both can enjoy the qualities of the Eems canal.

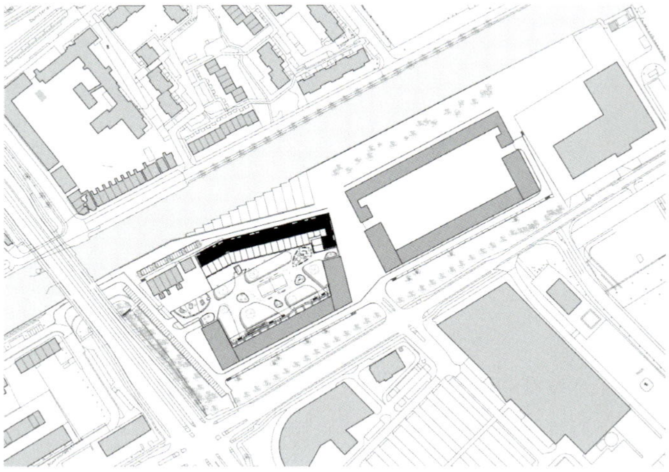

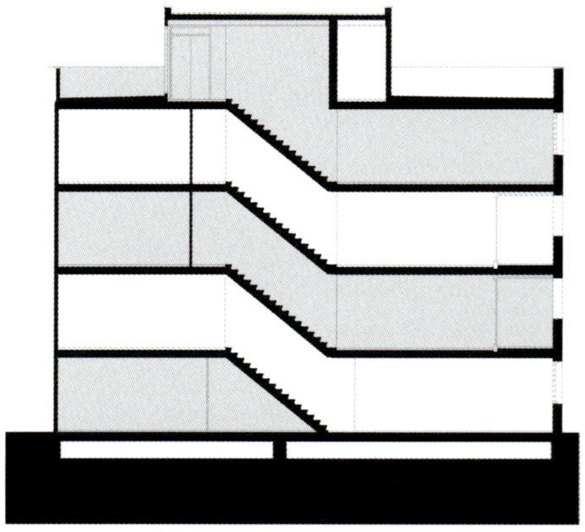

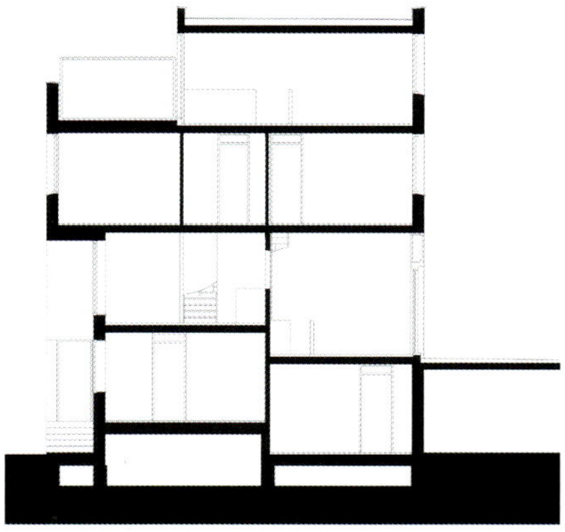

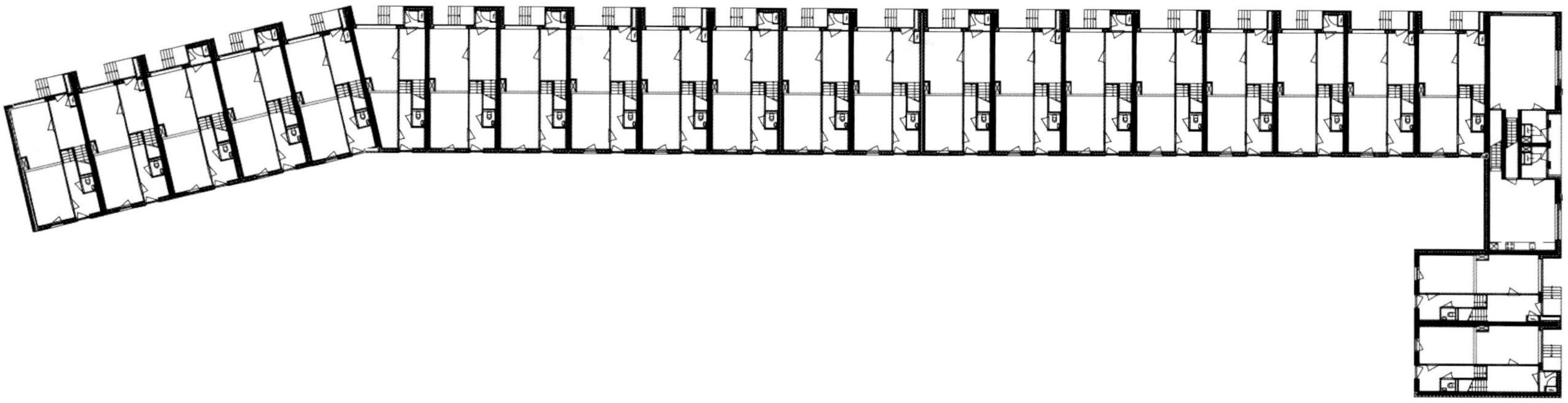

Overzicht begane grond / Ground floor

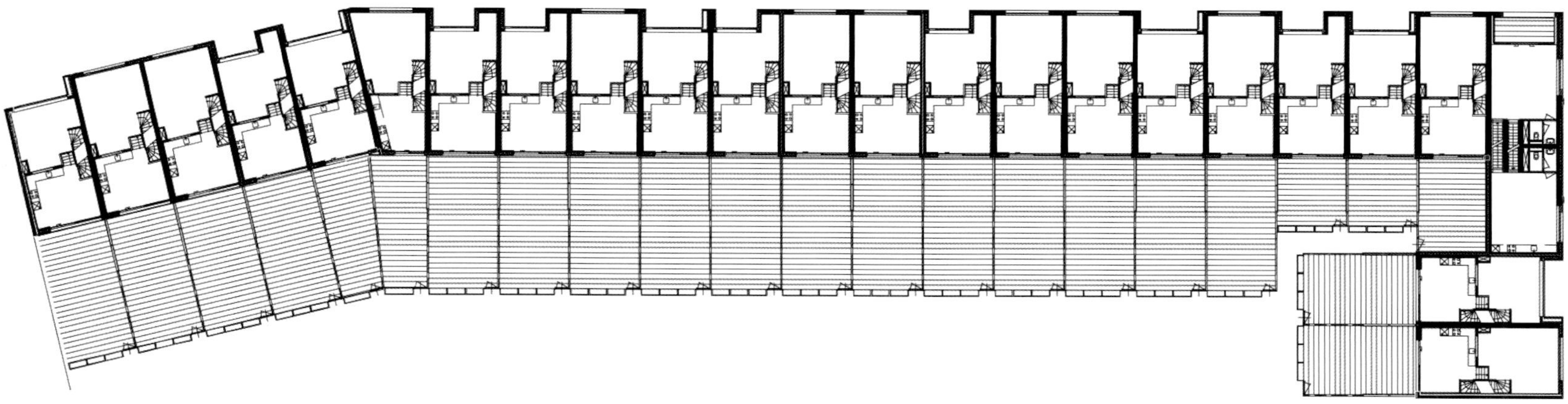

Overzicht 1e verdieping / 2nd floor

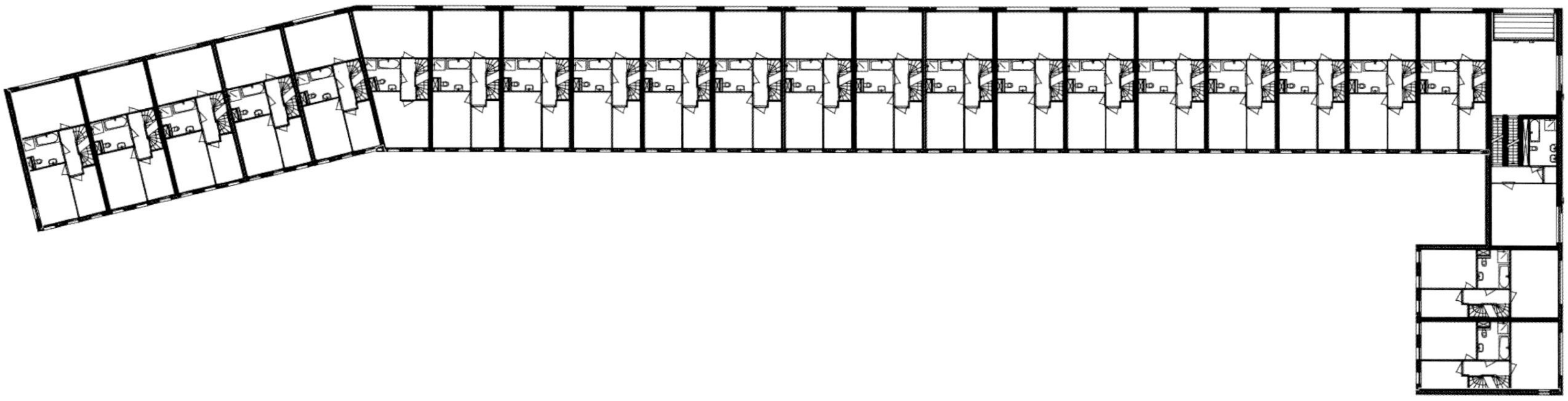

Overzicht 2e verdieping / third floor

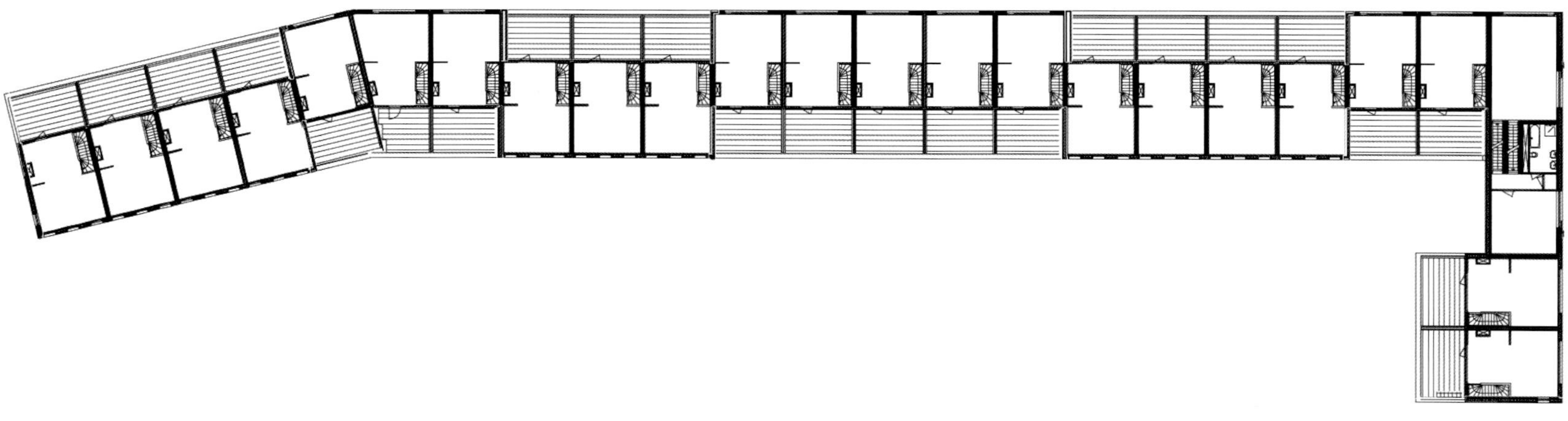

Overzicht 3e verdieping / fourth floor

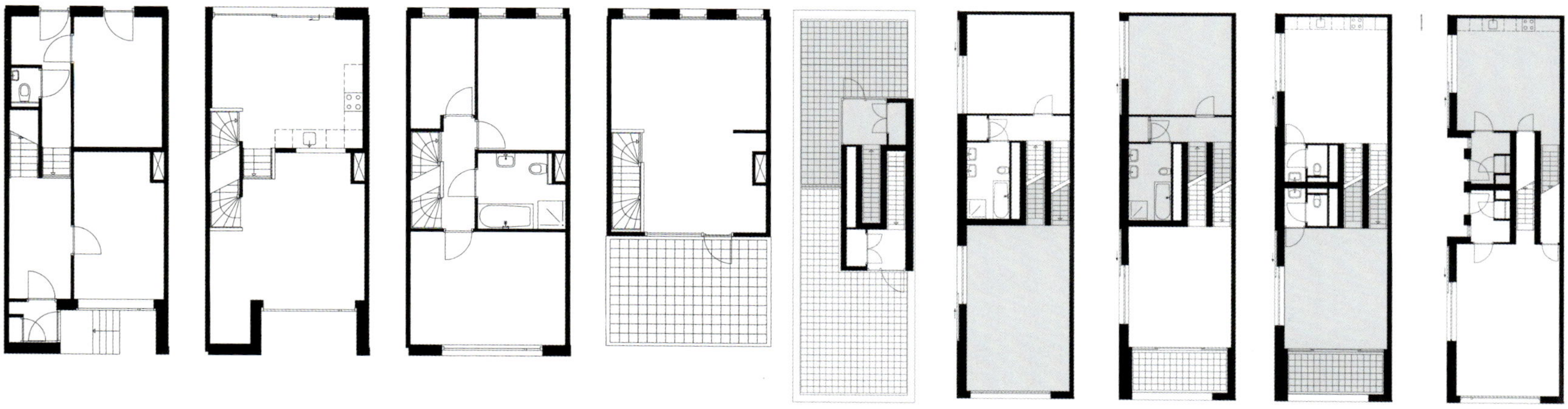

Architect: Silvester Fuller, Tony Owen Partners
Construction: Ceerose Pty Ltd
Location: Boston

Boston University

The eight-level, environmentally-efficient building has been designed to house overseas students visiting from Boston University for a semester in Sydney. It can contain 164 beds as well as a care-taker suite. It also has three lecture halls, a library, an Internet lounge, a rooftop terrace with a timber deck and an adjoining fully-equipped communal kitchen, plus cafe. The unique module houses 4 single bedrooms for every shared lounge and bathrooms.

The design uses large canyon-like slots in the façade which allow sunlight and ventilation to penetrate deep into the building and into each room.

The windows in these slots have a rhomboid shape to maximize efficiency, and deliver a bold architectural façade which is illuminated at night through an ever-changing light show. The windows are oriented to trap sunlight whilst ensuring privacy between rooms.

The end walls of the slots are made from glass louvers that are seven storey's high, and the building also contains a seven-storey glass louvered atrium. Air is drawn through the canyons and passes through the building like gills. It is drawn up through the central void and out the top to ventilate corridor areas, thus allowing the building to breathe naturally.

East-facing operable louvers on each level further help to lower ambient temperatures by drawing in fresh breezes. The design allows more light and ventilation into each bedroom, provides good views, and is a sensible model for residential density in the CBD. At the same time the design is a sensitive infill for a constrained urban site. Regent Street contains a number of heritage buildings. The facades utilize masonry to create a contemporary contextual response that is still contemporary. Color is used on the ground floor structure to celebrate the student life in the street. The corner of the building is set back in the northern corner to widen the lane way. A café and outdoor seating is located here to enliven the lane way and create an ad-hoc public space. At night the fissures are lit up. This will be programmed with changing colours to create a permanent light installation.

Regent St

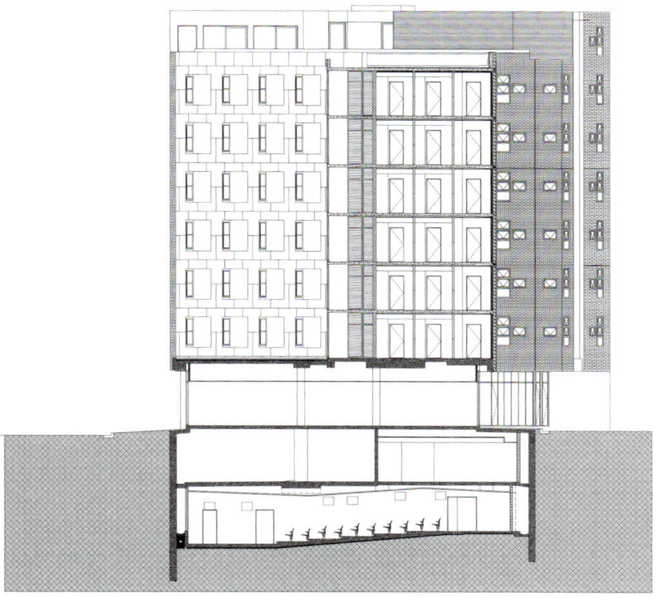

SECTION AA

EAST ELEVATION

Architect: Bjarke Ingels, Finn Nørkjær, Jakob Lange, Jan Borgstrøm
Firm: BIG
Construction manager: Henrick Poulsen
Team: Annette Jensen, Dariusz Bojarski, Dennis Rasmussen, Eva Hviid-Nielsen, Joao Vieira Costa, Jørn Jensen, Karsten V. Vestergaard, Karsten Hammer Hansen, Leon Rost, Louise Steffensen, Malte Rosenquist, Mia Frederiksen, Ole Elkjær-Larsen, Ole Nannberg, Roberto Rosales Salazar, Rong Bin, Sophus Søbye, Søren Lambertsen, Wataru Tanaka
Collaborators: JDS, Moe & Brødsgaard, Freddy Madsen, SLA
Location: COPENHAGEN, DK
Area: 33000m²

The Mountain

How do you combine the splendours of the suburban backyard with the social intensity of urban density?

The Mountain is the 2nd generation of the VM Houses—same client, same size and same street. The program, however, is 2/3 parking and 1/3 living. What if the parking area became the base upon which to place terraced housing, like a concrete hillside covered by a thin layer of housing, cascading from the 11th floor to the street edge? Rather than doing two separate buildings next to each other—a parking and a housing block, the designers decided to merge the two functions into a symbiotic relationship. The parking area needs to be connected to the street, and the homes require sunlight, fresh air and views, thus all apartments have roof gardens facing the sun, amazing views and parking on the 10th floor. The Mountain appears as a suburban neighborhood of garden homes flows over a 10-storey building—suburban living with urban density.

The roof gardens consist of a terrace and a garden with plants changing character according to the changing seasons. The building has a huge watering system which maintains the roof gardens. The only thing that separates the apartment and the garden is a glass façade with sliding doors to provide light and fresh air.

The residents of the 80 apartments will be the first in Orestaden to have the possibility of parking directly outside their homes. The gigantic parking area contains 480 parking spots and a sloping elevator that moves along the mountain's inner walls. In some places the ceiling height is up to 16 meters which gives the impression of a cathedral-like space.

The north and west facades are covered by perforated aluminium plates, which let in air and light to the parking area. The holes in the facade form a huge reproduction of Mount Everest. At day the holes in the aluminium plates will appear black on the bright aluminium, and the gigantic picture will resemble that of a rough rasterized photo. At night time the facade will be lit from the inside and appear as a photo negative in different colours as each floor in the parking area has different colours.

The Mountain is located in Orestad city and offers the better of two worlds: closeness to the hectic city life in the center of Copenhagen, and the tranquillity characteristic of suburban life.

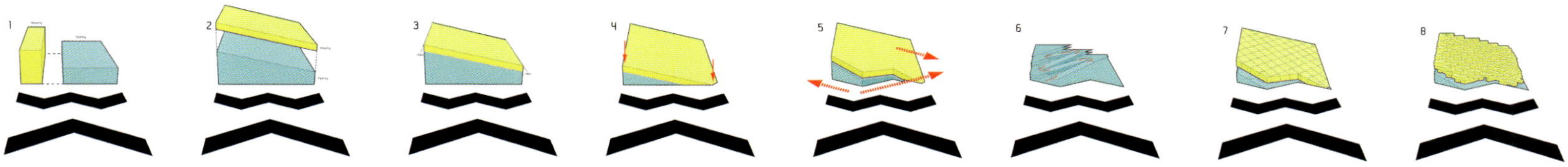

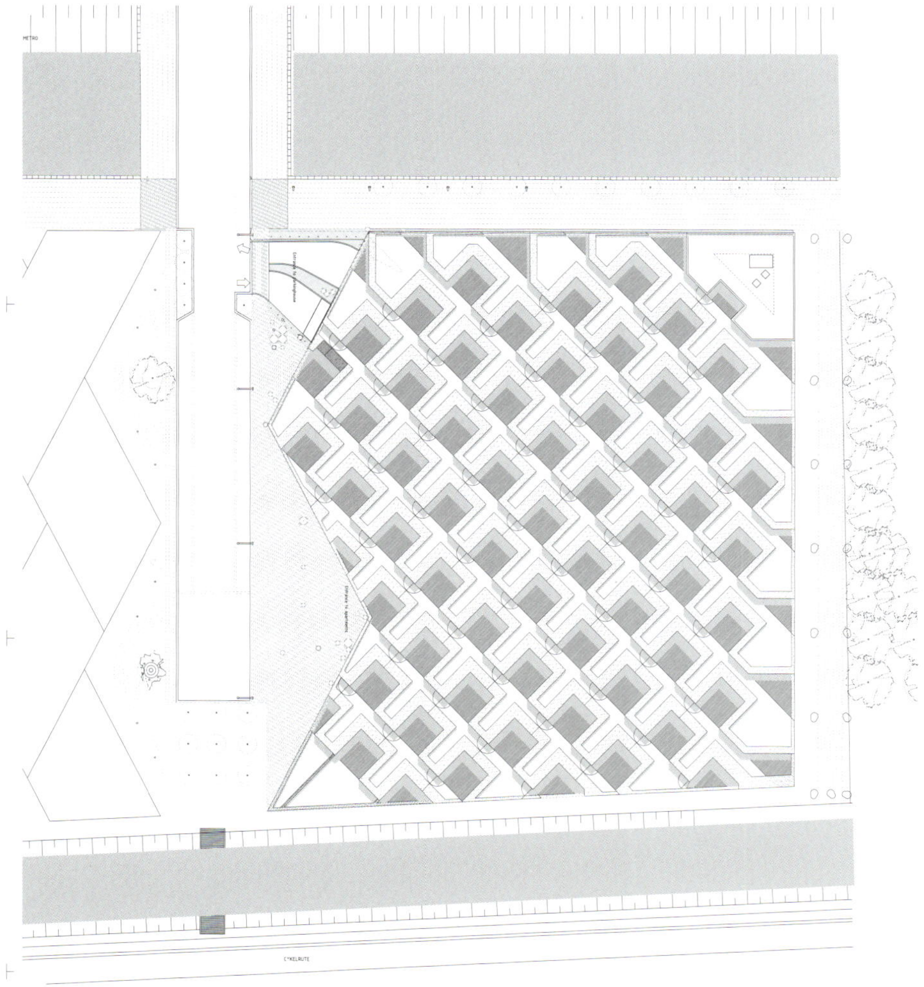

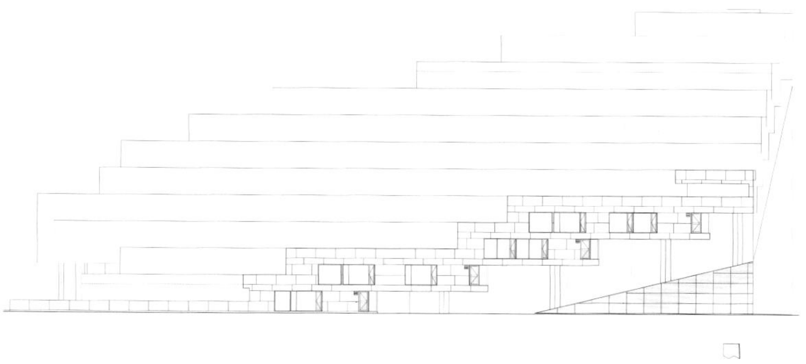

East

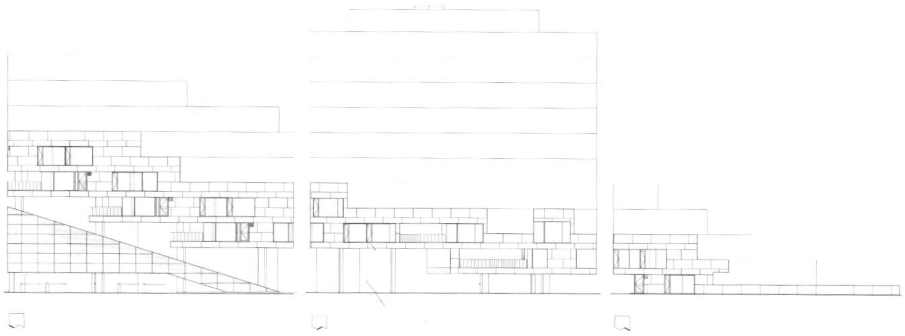

South

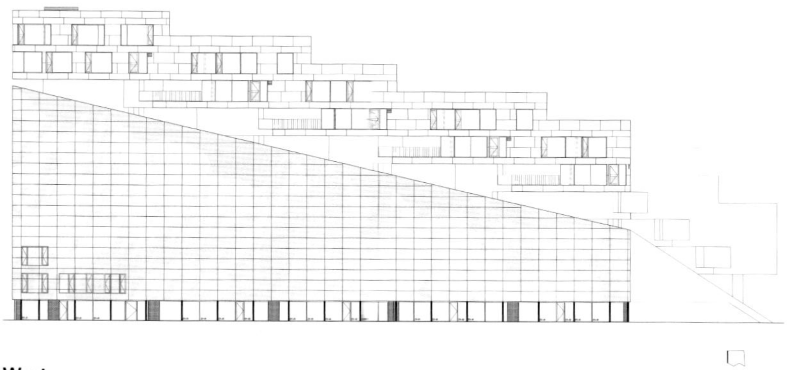

West

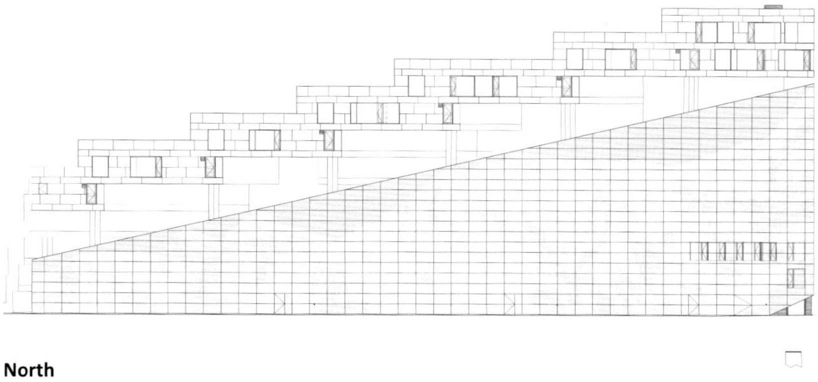

North

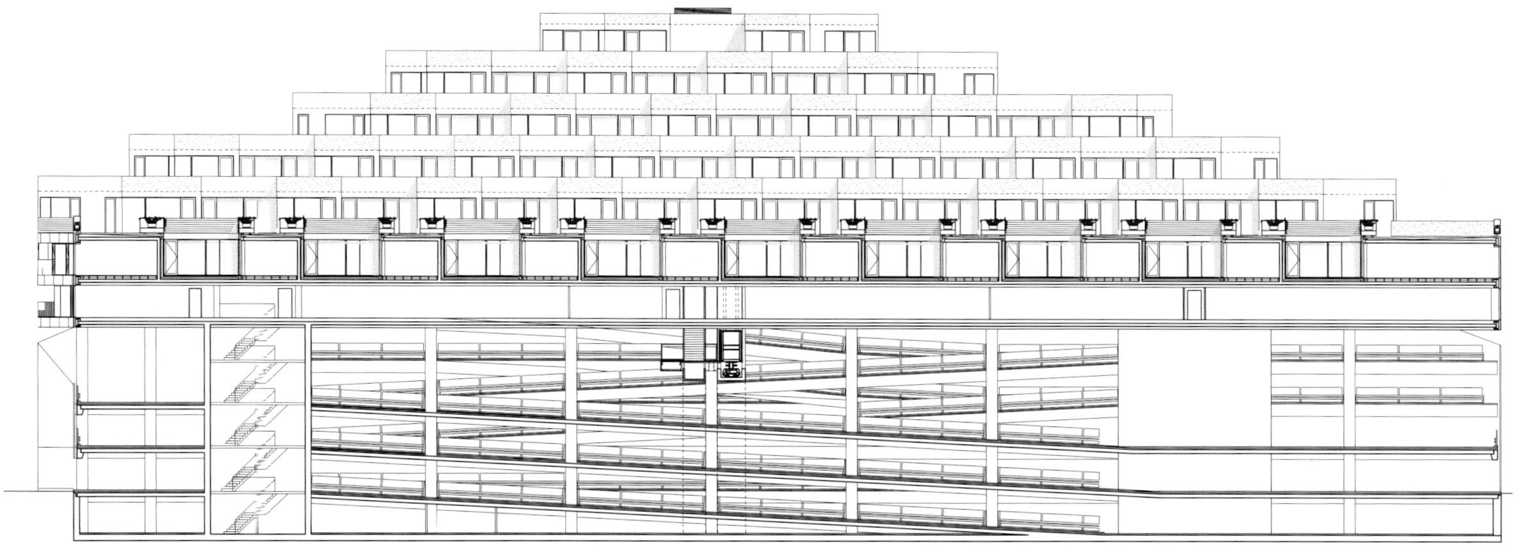

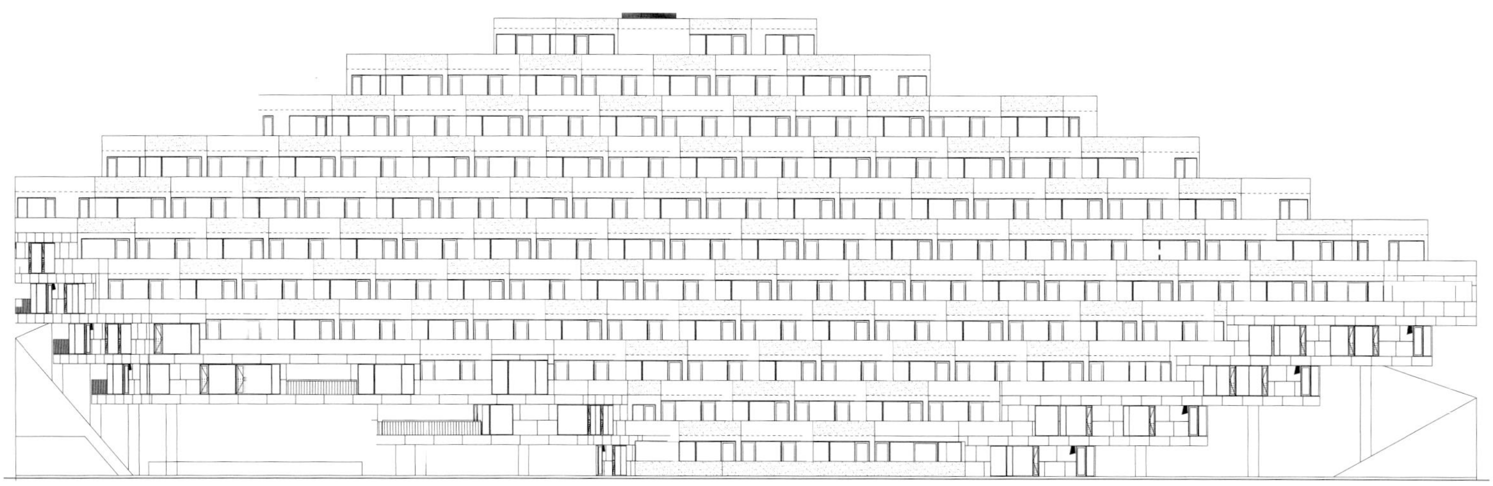

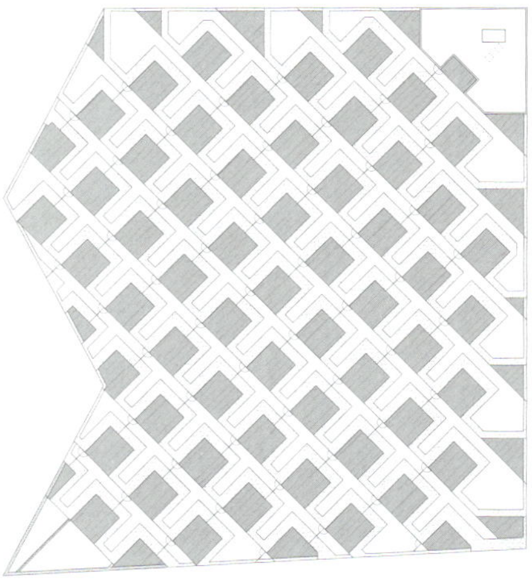

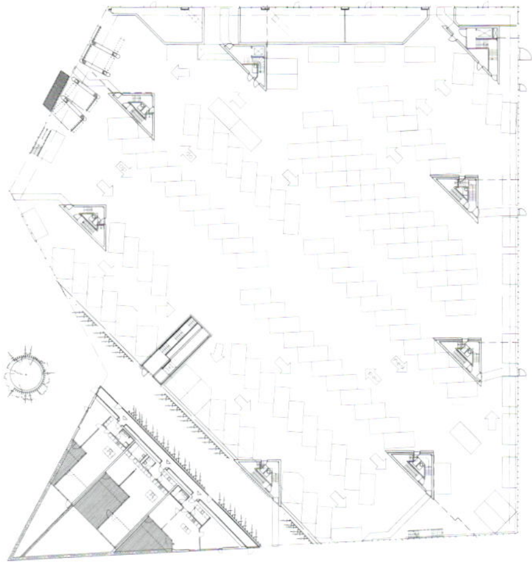

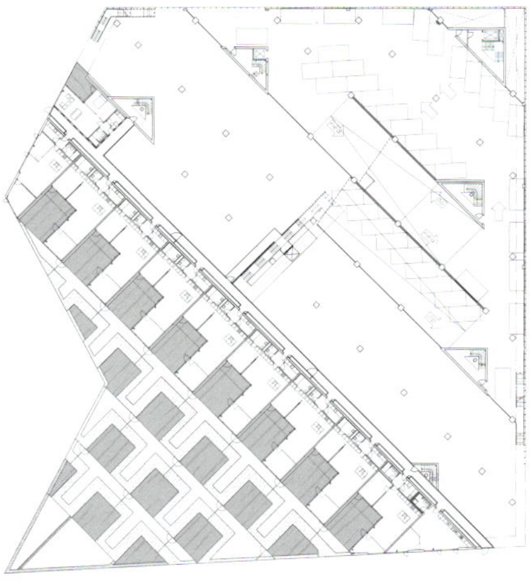

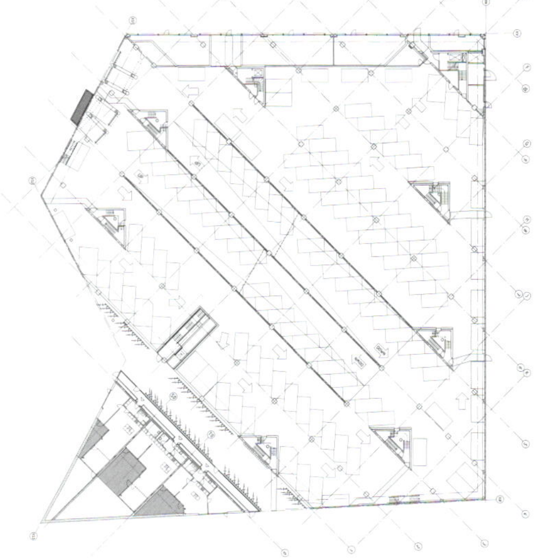

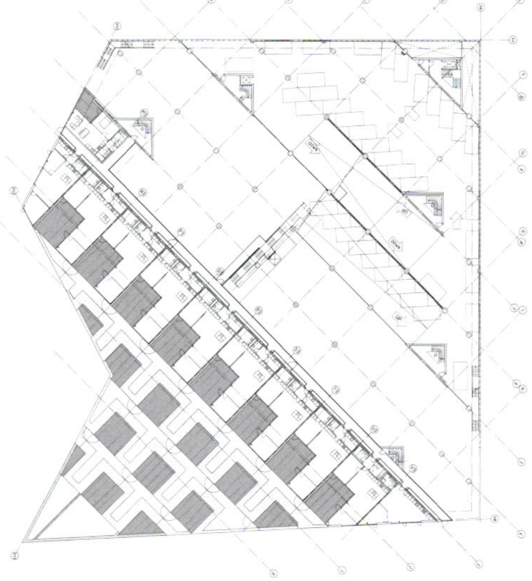

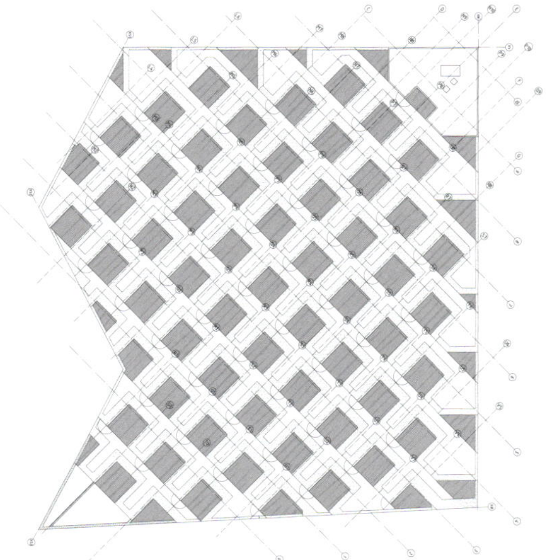

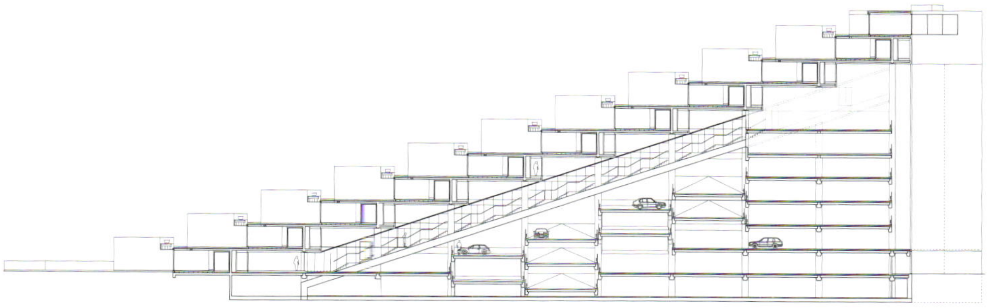

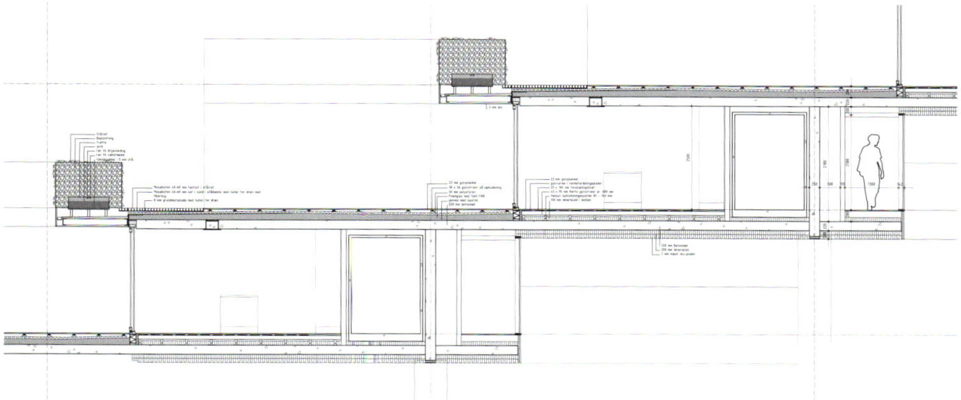

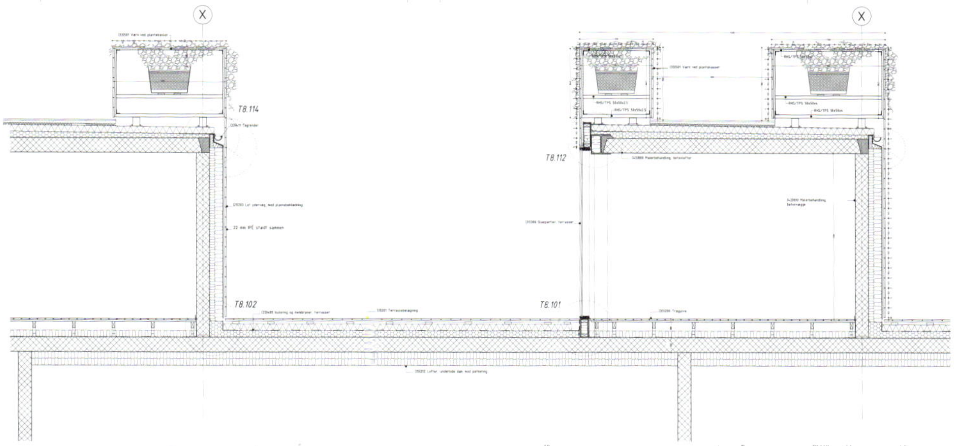

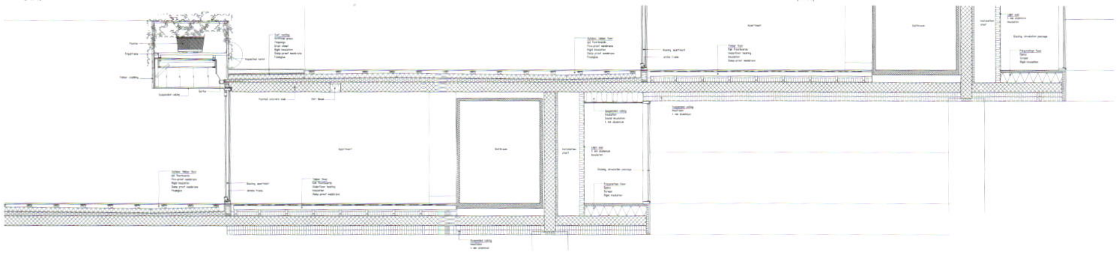

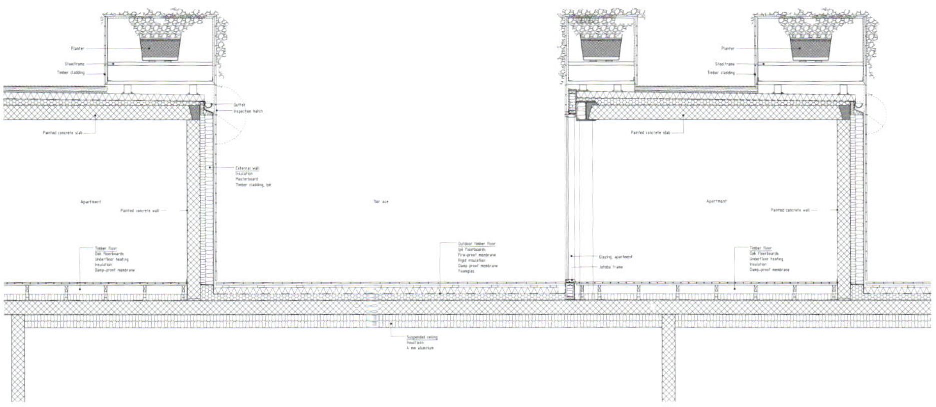

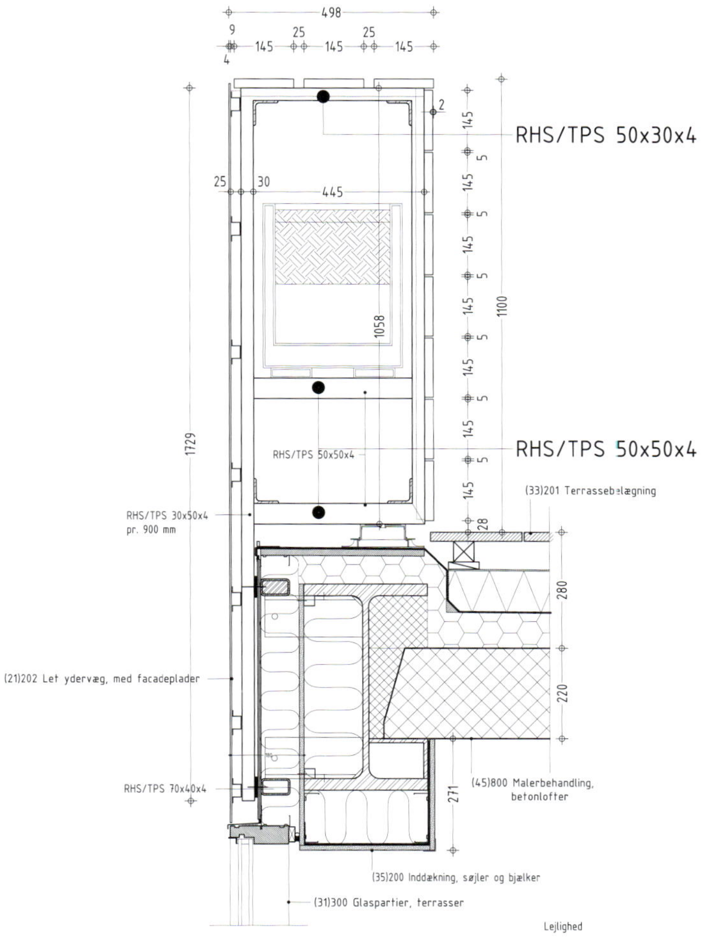

498
145
145
145
RHS/TPS 50x30x4
445
1058
1100
1729
RHS/TPS 50x50x4
RHS/TPS 50x50x4
(33)201 Terrassebelægning
RHS/TPS 30x50x4
pr. 900 mm
280
220
(21)202 Let ydervæg, med facadeplader
RHS/TPS 70x40x4
271
(45)800 Malerbehandling,
betonlofter
(35)200 Inddækning, søjler og bjælker
(31)300 Glaspartier, terrasser
Lejlighed

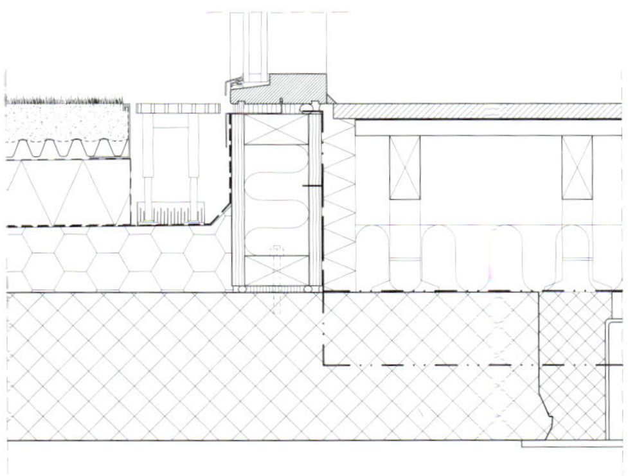

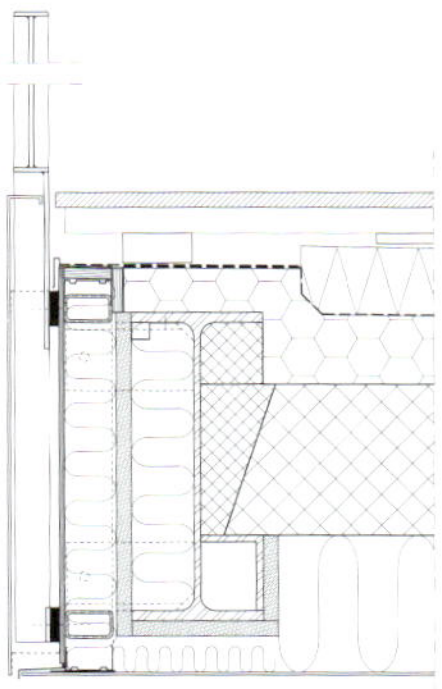

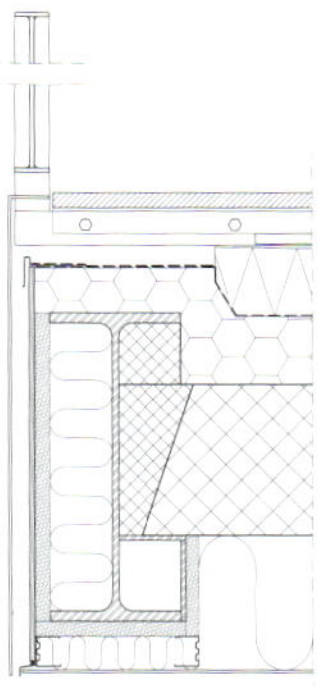

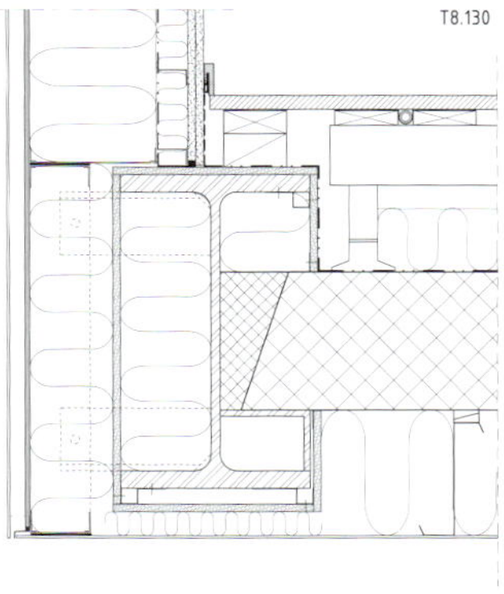

T8.130

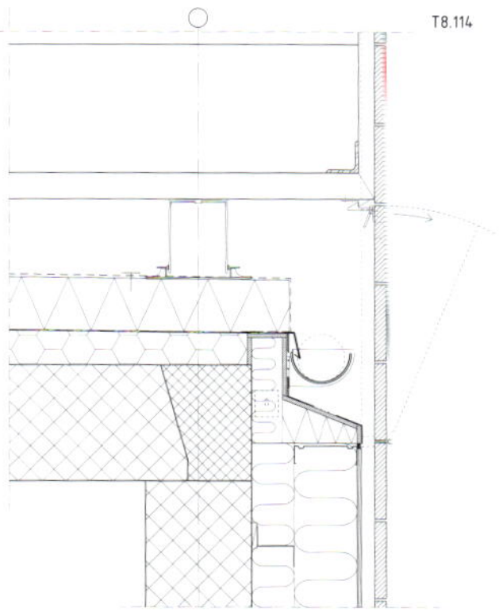

T8.114

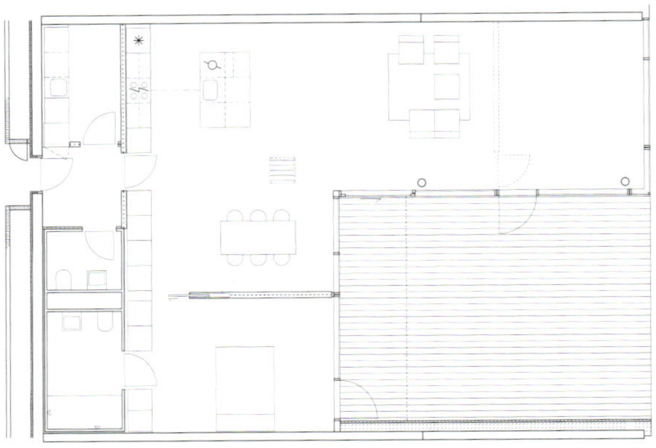

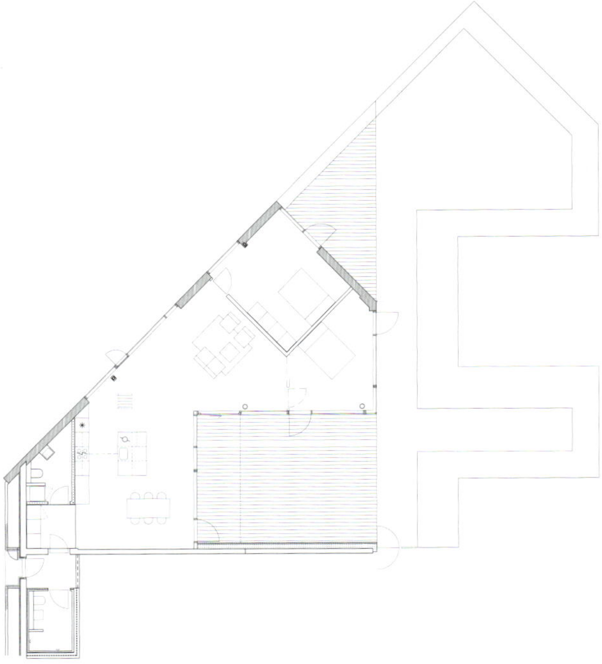

Architect: x architekten
Firm: x architekten
Location: Feilstraße, Linz, Upper Austria
Area: 1650m²
Materials: reinforced concrete construction
Photographer: Max Nirnberger

My Home is My Patio

Task:
There are leftover plots of land in every city which are difficult to build upon using conventional solutions. Intelligent solutions need to be developed for these challenging sites to seize opportunities for inner-city growth. The project for urban waste land offers innovative solutions which mitigate the need for the growth of city boundaries.
Concept:
A new type of house was designed for the plot in the Linzer Feilstraße which falls into a third category, after multi-storey buildings and terraced houses. A mixture of housing types are stacked up to give the building its structure. There are underground parking, business premises to liven up the street area with apartments above and two-storey courtyard houses on top. These two-storey courtyard houses offer a real alternative to a suburban detached house.
Implementation:
Extended wooden poles which help integrate the banisters and specifically placed areas of color accentuate the centrally located inner courtyard, with its arcades, as being the core of the development. Therefore, seven apartments with typical rooms can be found on the first floor. Apartments with patios and terraces can be found on two further levels on the floors above. These can be considered as small houses as each 92m² unit has its own intimate and completely hidden courtyard. The patios are opened up on the side towards the street to enable a connection to the surroundings. Roofs which are slightly sloped towards the courtyards and rounded off towards the back of the building give the apartments, with their ceiling height of 3.7m, a generous character.
The three courtyard houses with their round-arched roofs, give the impression that they are sitting upon the building structure below, creating the building's identity and its unique appearance from a distance. The patios face outwards and form an incision into the cubic structure giving the introverted and tightly packed building its structure and openness. With the building being situated at the corner of the Feilstraße this aspect is emphasized by a change in the material of the façade. The building presents itself with white plaster on the western façade and on the eastern side the roof is covered with trapezoidal sheet metal runs along the outer walls of the courtyard houses, along the façade and down to the premises.

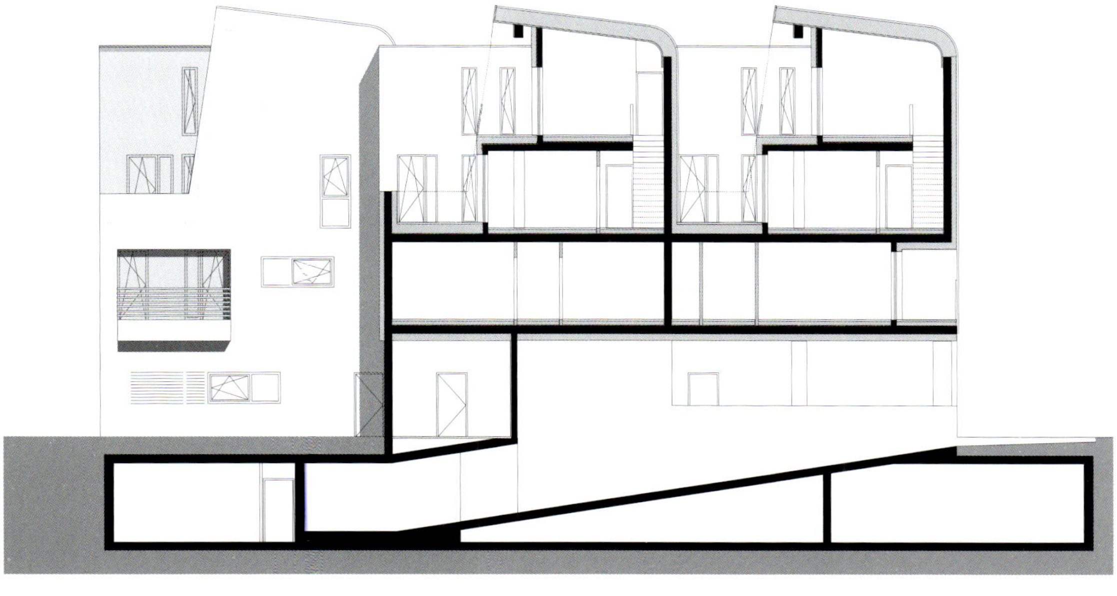

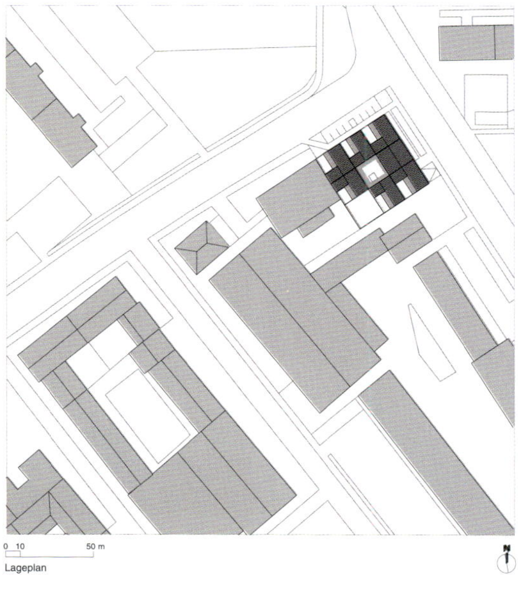
0 10 50 m
Lageplan

3.Obergeschoss

E

Architect: Francis-Jones Morehen Thorp
Firm: Francis-Jones Morehen Thorp
Documenting architects: Lend Lease
Consultants: Aecom, BBC Consulting Planners, Norman Disney & Young, Page Kirkland Group, Renzo Tonin & Associates, Robert Bird Group
Location: 4 Distillery Drive, Pyrmont, NSW, Sydney Australia
Area: 16506m^2
Photographer: John Gollings, FJMT/Andrew Chung

Sugar Dock, Jacksons Landing

Sugar Dock provides 132 residential units, associated communal open space and underground carparking. The design of the development seeks to complement and respond to the building forms and open spaces within the masterplan and its urban context characterised by its relationship to large scale elements.

Informed by a long trapezoidal masterplan envelope, the tower consists of a series of glass residential apartments enclosed and anchored by two precast shells combining the visual language of the new development with that of the old Pyrmont. Within the glass form, apartments are turned away from their primary boundary, and the adjacent tower, towards the view and optimal orientation. A podium e ement grounds the tower's masonry walls and reinforces the sandstone cliff-like Bowman street alignment. A generous communal and private landscaped space links the Sugar Dock to the next tower under development. Apartments along Bowman Street have been articulated to read as a series of houses inserted into the cliff face. They have an intimate character with their connection to the park on Bowman Street and are restricted in height allowing clearer views through site views from Distillery Hill and from the communal open space.

The material chosen for the shells is vertically articulated light coloured precast concrete providing a relationship with the prominent sandstone outcrops of the Distillery Hill development. The elevations have been developed to correspond with the more private functions within and to provide protection from afternoon sun.

To the north/east, a metal/glazed balustrade system has been developed to utilise daytime reflections to provide a surface to the building at the balcony line, suppressing the slab edge. Behind the balustrade is a series of fritted glass panels, arranged to create a greater sense of privacy and to provide a 'screen behind a screen' effect which operates both at the individual apartment scale and the scale of the whole elevation. The metal/glazed system provides a lightness to the buildings material character appropriate to its height and heightened environmental screening to the recessed balconies.

9518 8919

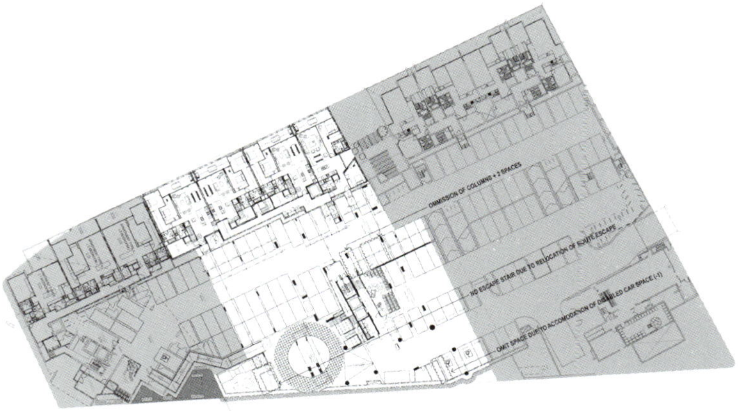

BOWMAN STREET
JOHN STREET

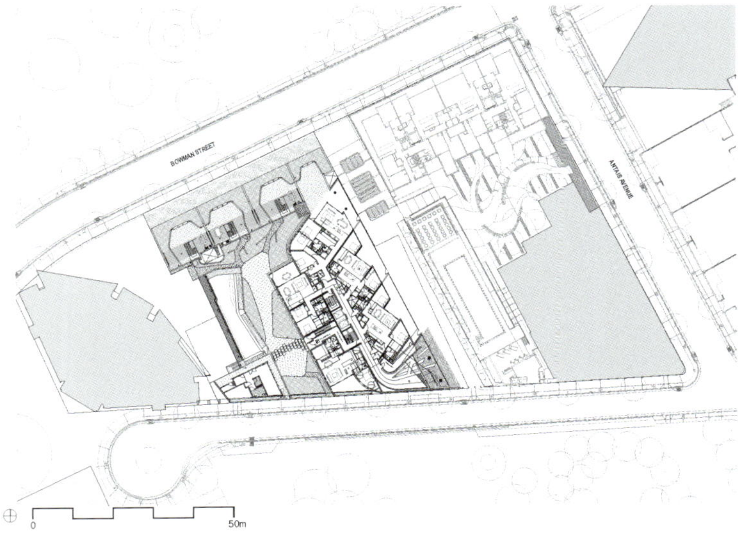
BOWMAN STREET
0
50m

Architect: Markus Kaasik, Andres Ojari, Ilmar Valdur, Merjer Müürisepp
Firm: 3+1 Architects
Location: Tallinn
Area: 1917m²
Photographer: Arne Maasik, Kaido Haagen

Koidula Apartment Building

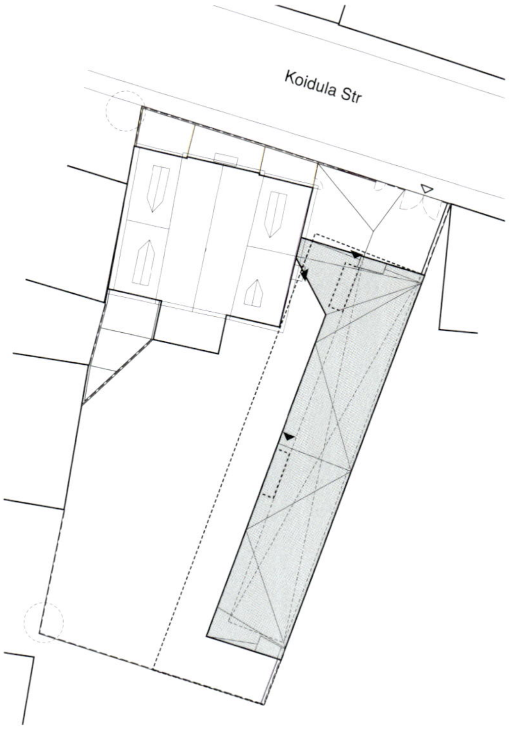

Building is located close to the Kadriorg park, one of the most important and historical green areas in Tallinn. Currently the socially and historically rich district is developing quickly into a prestigious living area. The residential buildings along Koidula St. date from different periods and it has caused the chaotic structure of the street.

The building is composed by the old renovated part and the new wing, which is in place of the old courtyard wing. Covered passageway connects the new wing to the existing building. The aim was to keep the courtyard and Le Balto office (Berlin) is designing a private garden there. It is possible to see the depth and beauty of the plot also from the street.

The room structure of the apartments in the older part is changed, creating new connections through the floors.

The new wing situates along the border of the plot and it opens only to the courtyard garden. The house could be divided into three linear sections. The first section opening only to the garden contains stairs to the apartments as well as terraces of all apartments. In the middle section there are living zone along with the bedrooms, the third section along the plot of the border contains so-called servant zone—kitchens, bathrooms, saunas and connections inside of the apartments.

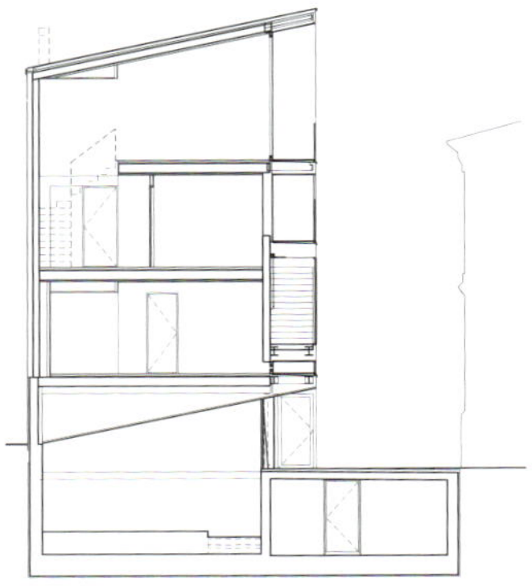

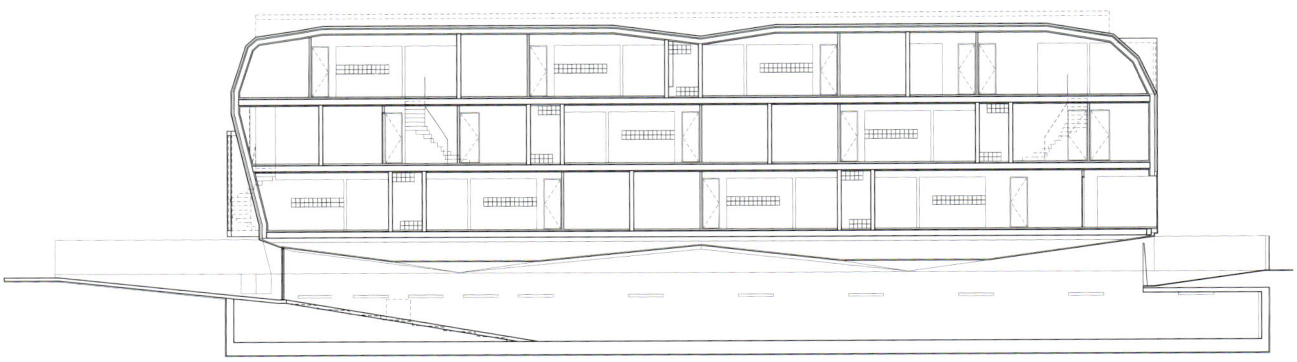

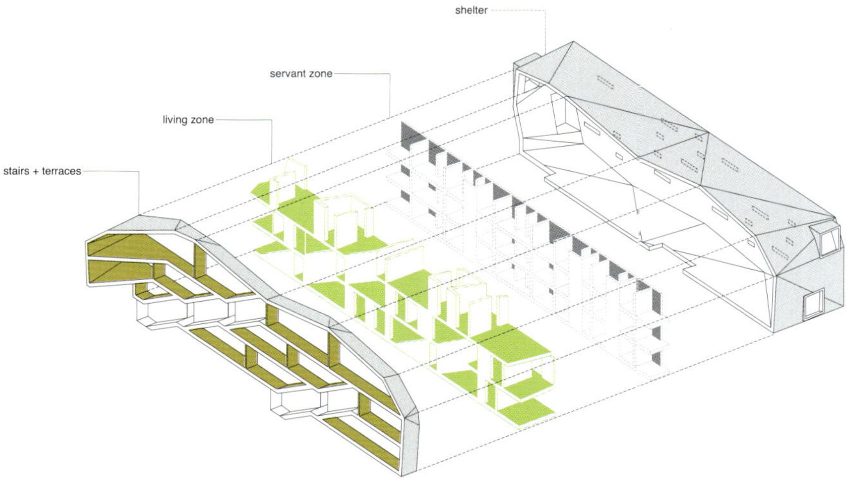

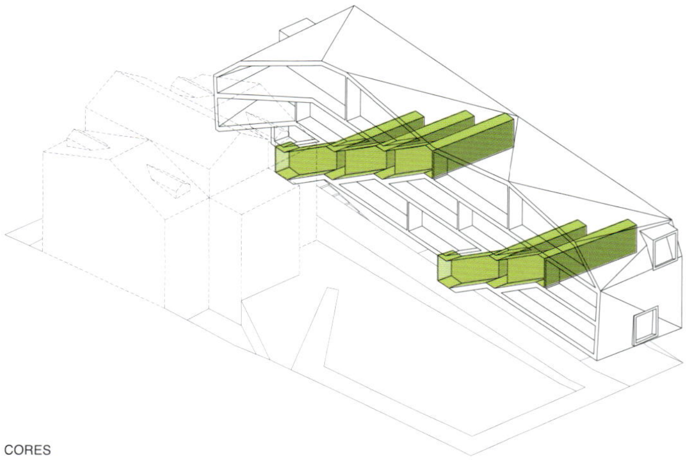

CORES

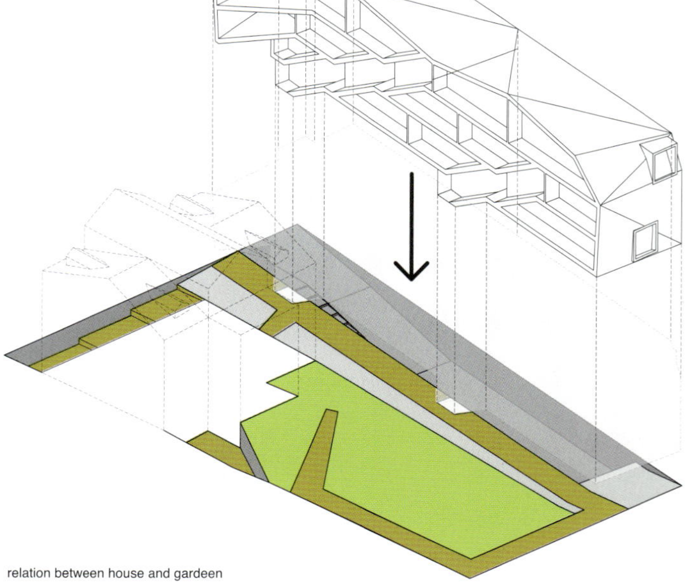

relation between house and gardeen

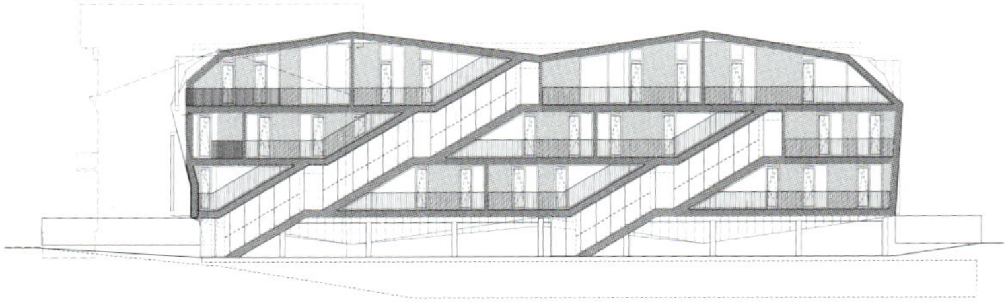

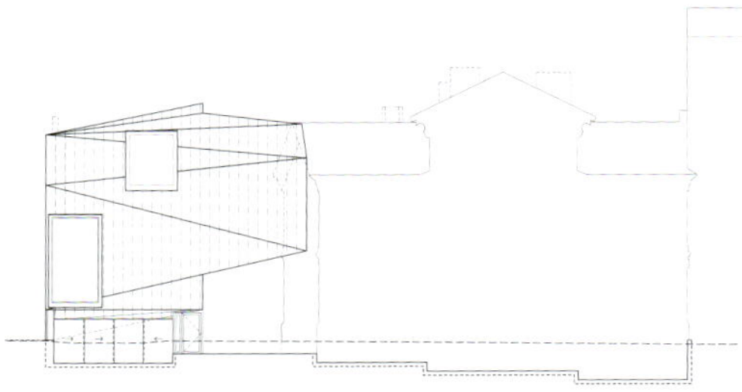

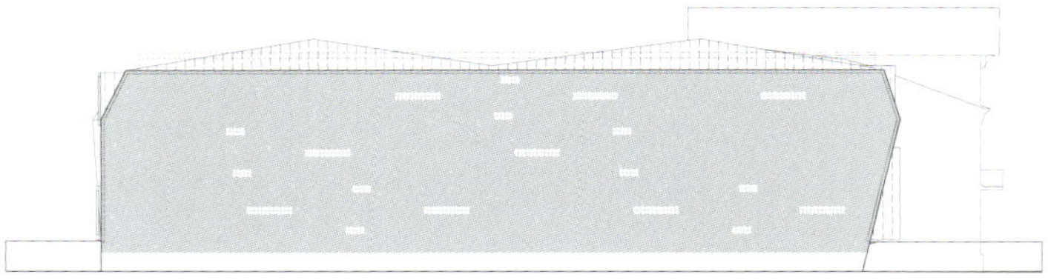

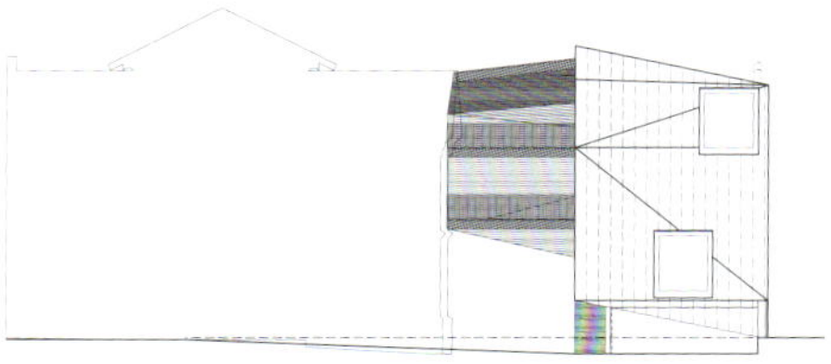

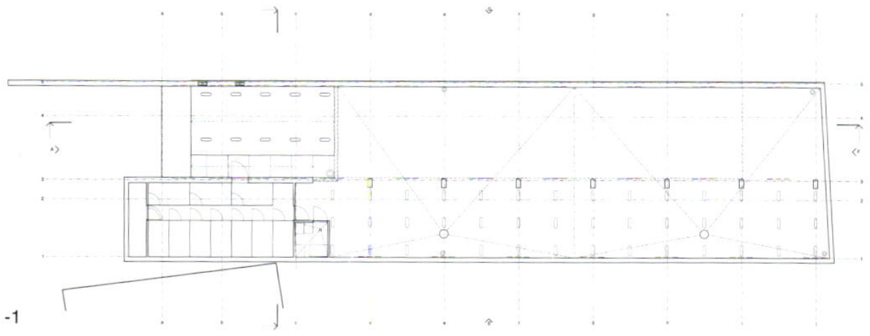

-1

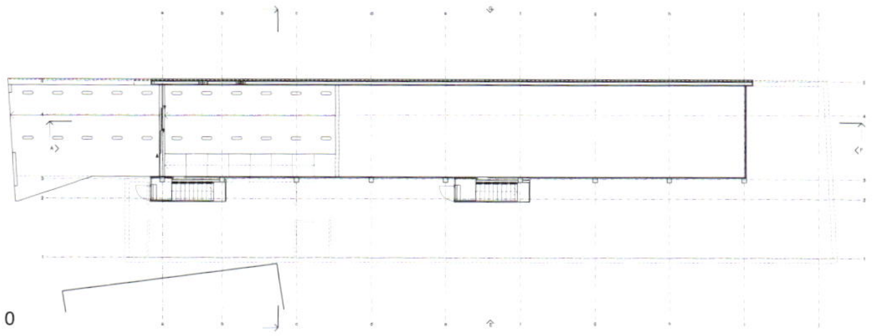

0

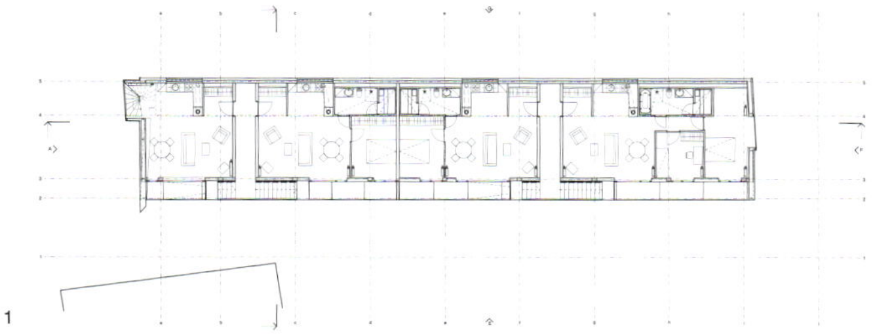

1

2

3

-1

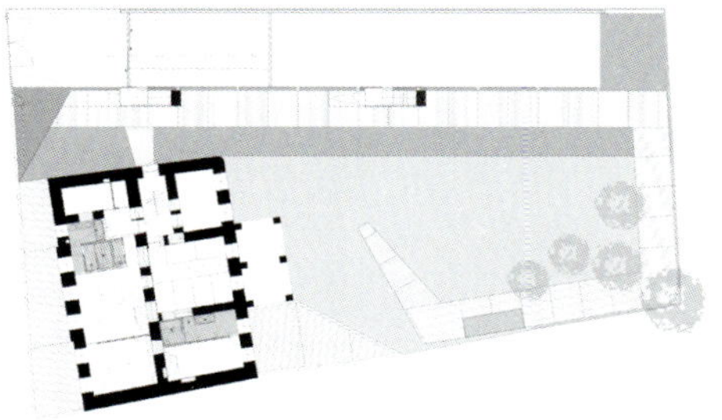

0

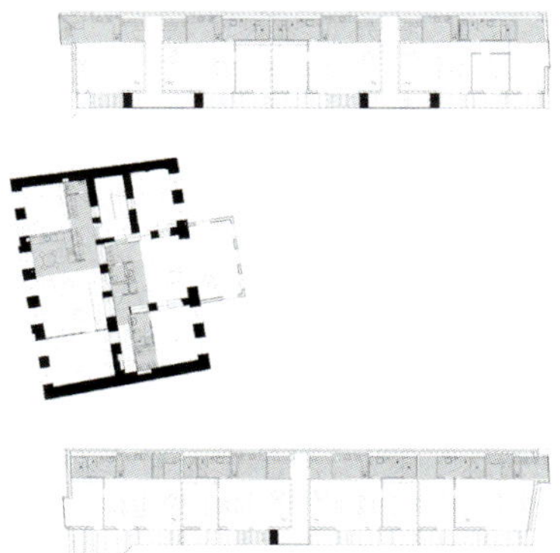

1

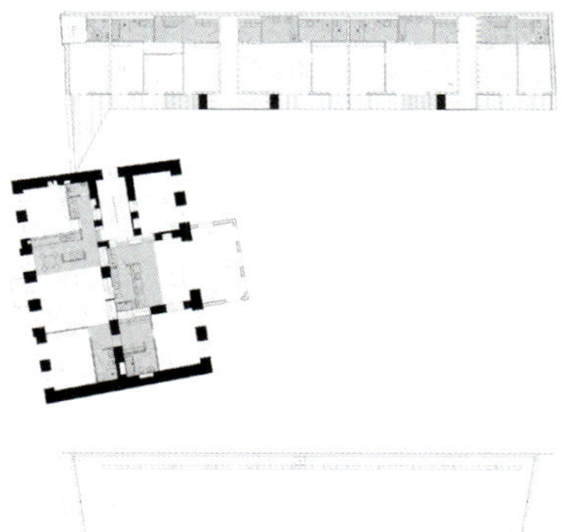

2

3

roof

Architect: Anders Strange, Anders Tyrrestrup and Torben Skovbjerg Larsen
Firm: AART Architects
Location: Njalsgade, Ørestad North, Copenhagen
Area: 7000m²
Photographer: Torben Eskerod

Bikuben Student Residence

The Bikuben Student Residence creates the settings for student life and extends the possibility that unity can arise from a broad social network.

The main shape of the building is quite clarified in its design and holds in it the fundamental idea of unity and equality. It has been the ambition to create a building with a strong and clear identity, with a direct connection to the contextual situation. The homes of the students and the common rooms are connected in a double spiral that surrounds an enlightenec courtyard. That offers direct contact and visibility between common and private space and at the same time, ensures the students' privacy in the home. The boundaries of every college home are expanded, basically filling all parts of the building volume, giving the social interface and every single student's latitude, maximum priority.

Apart from an often poor private economy, the biggest problem for the students in this day is loneliness and the lack of social relations. In this project the main structuralizing element is in fact, the possibility to enter the college community and the many different social relations that college life can offer. Local materials and pre-fabricated construction, façade and bathroom elements provide a both environmentally and economically sustainable structure.

Architecture and the sense of social responsibility intertwine and the building is captured by the context of the site, and changes its conditions.

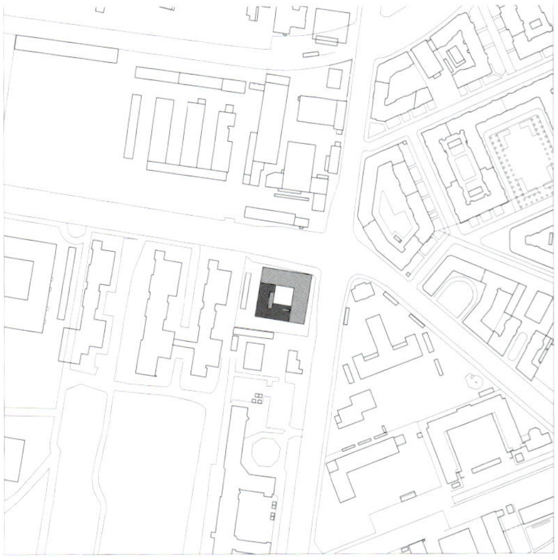

1 2 3 4 5 6 7 8 9 10 11
10m

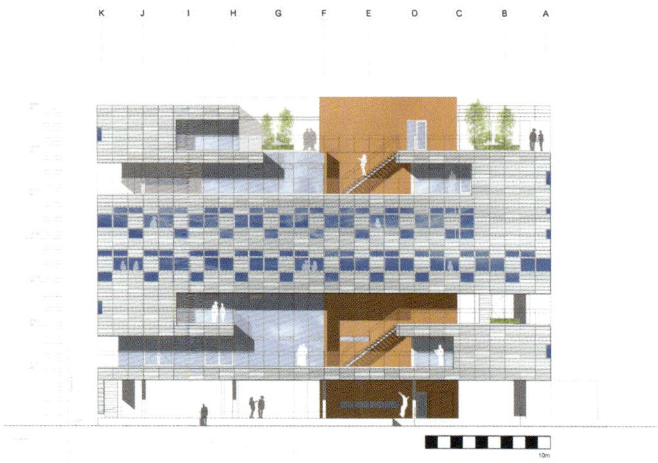
K J I H G F E D C B A
10m

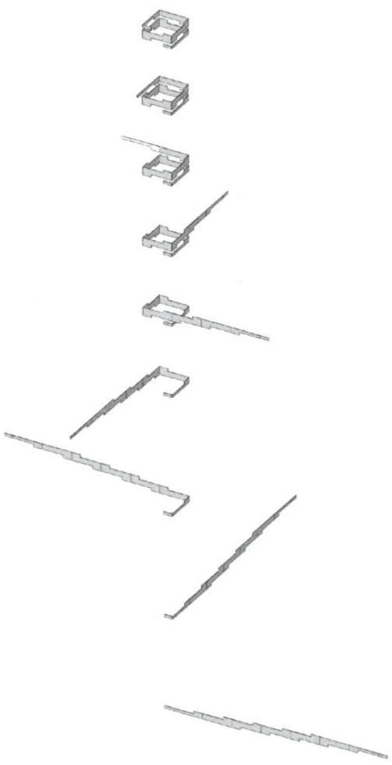

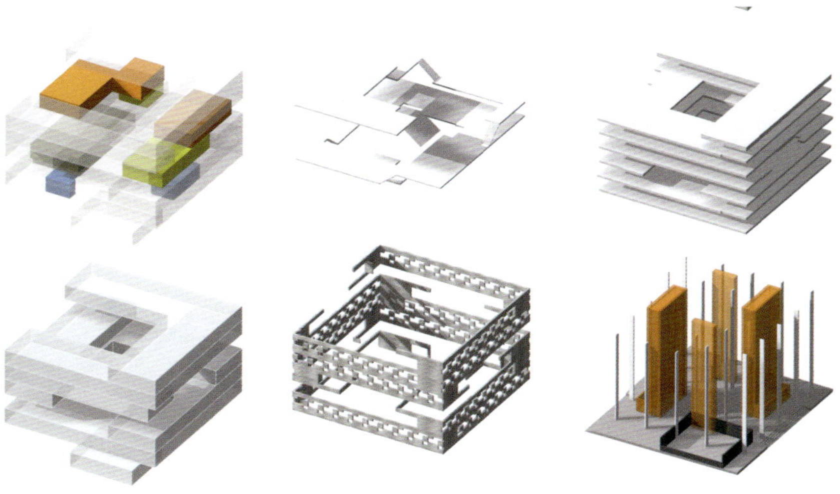

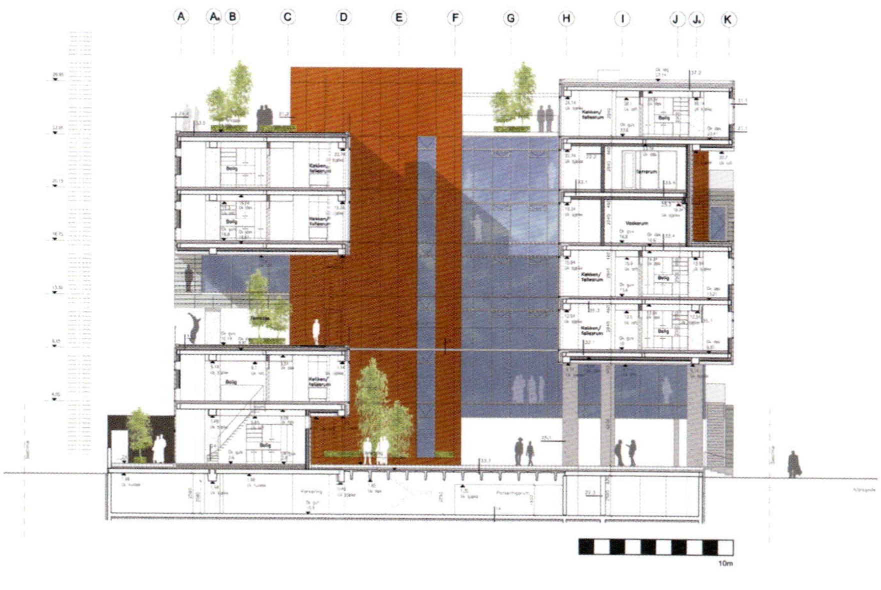

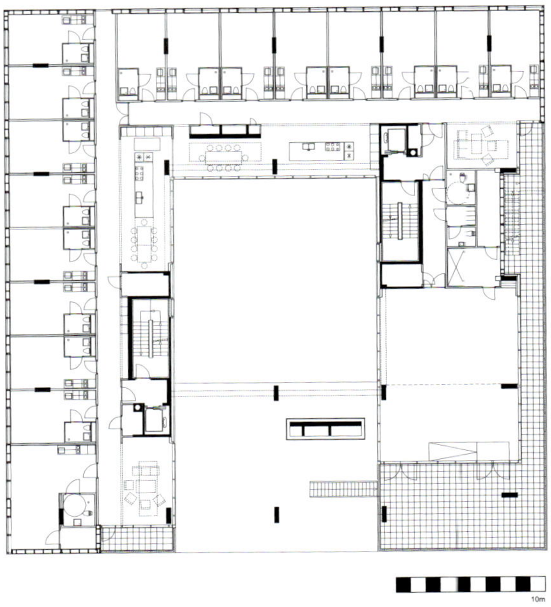
10m

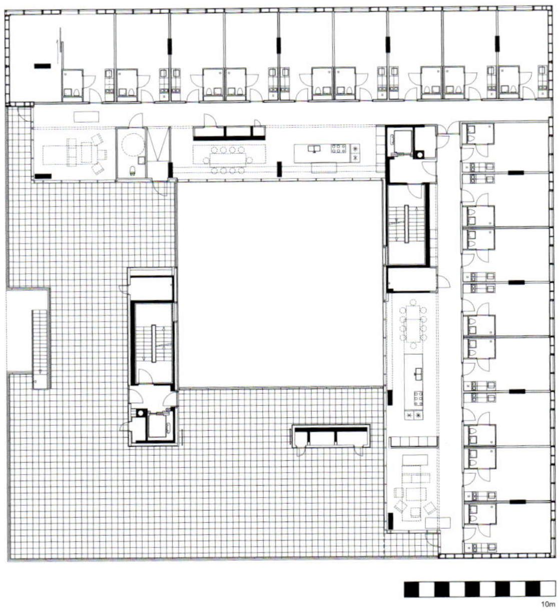
10m

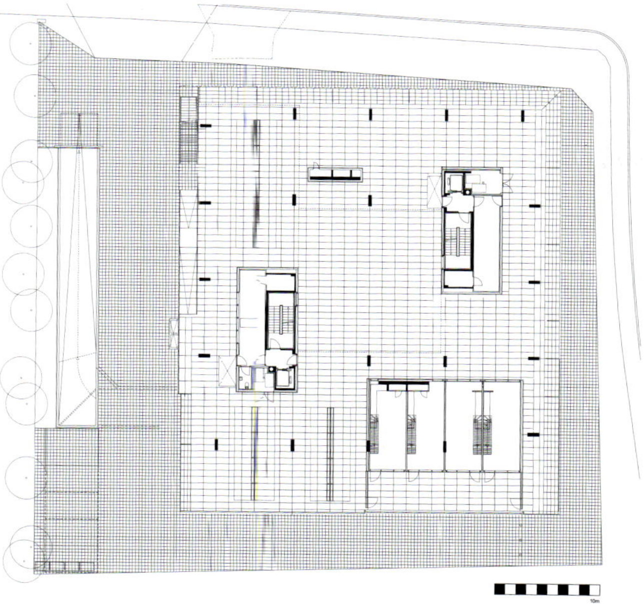

Architect: Stefano Boeri, Gianadrea Barreca, Giovanni La Varra
Firm: Stefano Boeri Architetti
Location: Seregno, Italy
Design team: Stefano Boeri, Gianandrea Barreca, Giovanni La Varra, Marco Erega, Andrea Caputo, Claudio Colla, Frederic Desmet, Marco Giorgio, Andrea Grippo, Susanna Loddo, Dino Polverino, Chiara Quinzii Patrizia Scrugli, Roberto Perego, Melanie Schmitt
Schemes and perspectives: Salottobuono
Structural engineer and Construction management: SCE s.n.c. (Ing. Stefano De Cerchio), Francesco Michelon
Area: 2700m²
Photographer: Paolo Rosselli

A House of Variable Geometry

The House of Variable Geometry is a project for affordable housing destined to give shelter to 38 families of different size and configuration. The project emphasizes the L-Shape of the site with a composition that strictly follows the lines of the given area yet transforms the building into a complex three dimensional articulation by alternating full volumes and voids over the given surface. The block is thus composed of patio-houses: voids generate and govern all different conditions of living, and give each dwelling an intimate relation to the outside. Centered on the concept of variability the project's starting matrix is a particular declination of the patio typology that guarantees that every apartment is allowed a private patio while these same open spaces create a configuration that permits the open spaces to be joined according to the needs of the inhabitants by eliminating one wall. The variability of the roofs, covered in heat-formed corrugated zinc sheets permits the presence of generous volumes that can also be exploited by the inhabitants through the realization of mezzanines and light structures according to the particular needs of each family.

PROSPETTO OVEST

SEZIONE C-C'

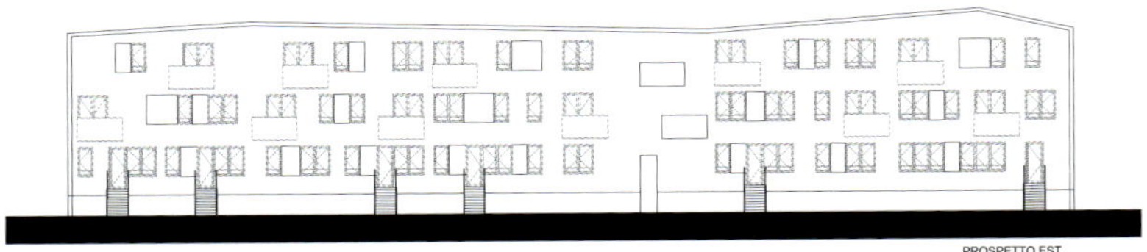
PROSPETTO EST

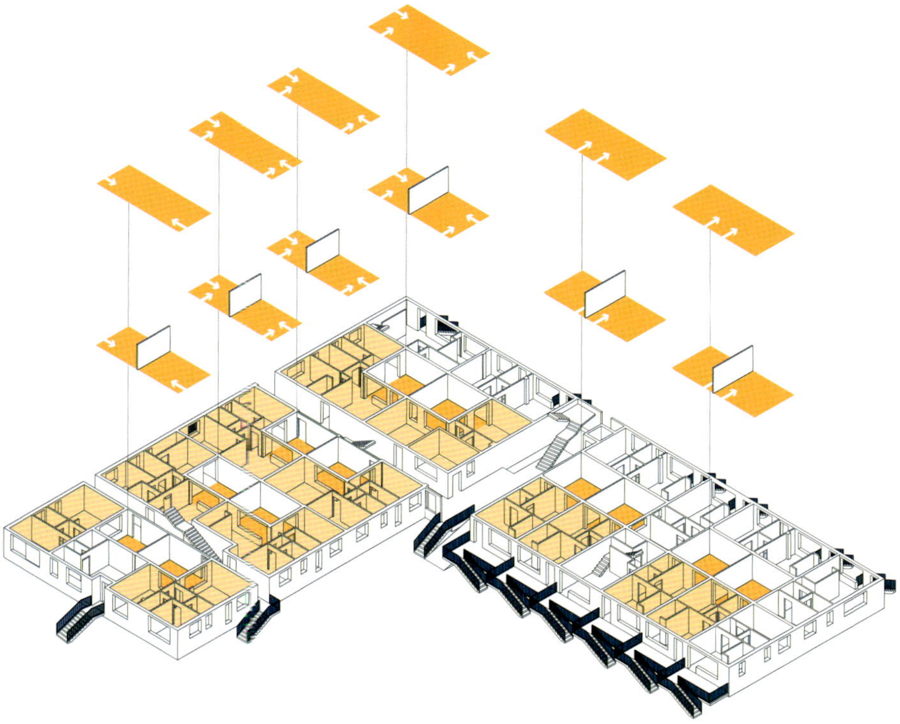

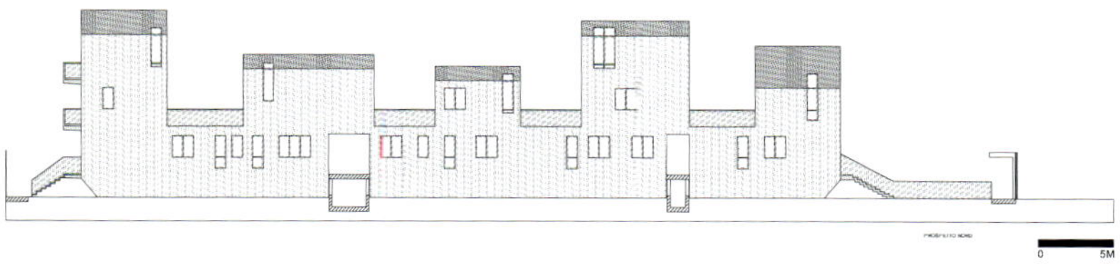
0 5M

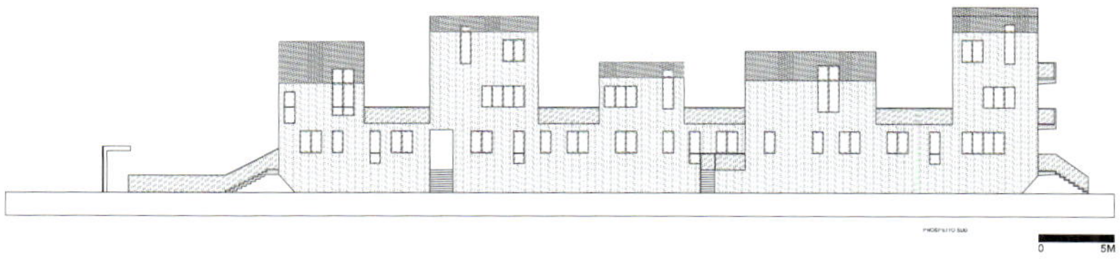
0 5M

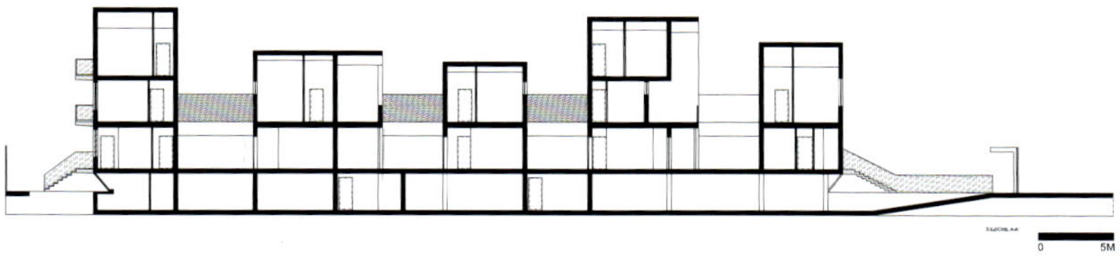
0 5M

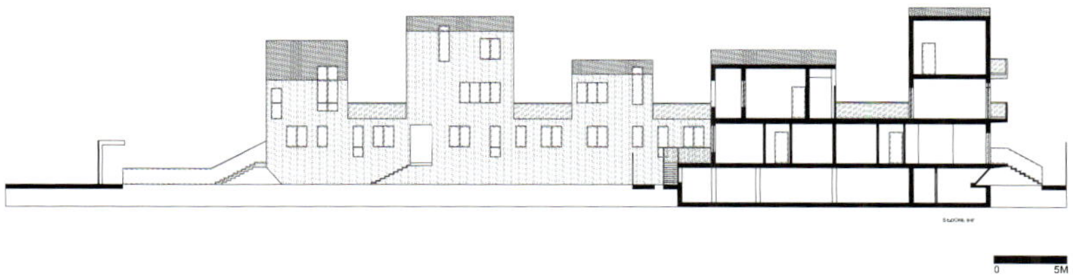
0 5M

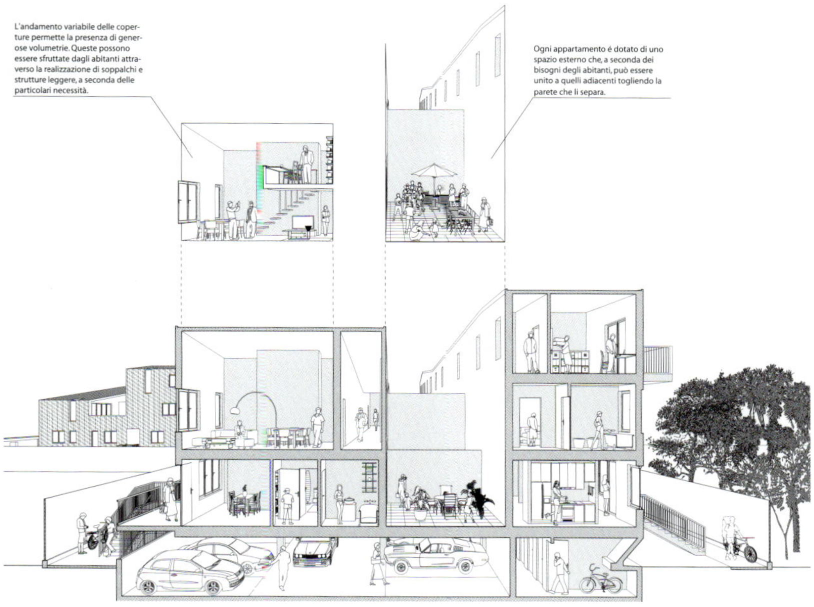
L'andamento variabile delle coperture permette la presenza di generose volumetrie. Queste possono essere sfruttate dagli abitanti attraverso la realizzazione di soppalchi e strutture leggere, a seconda delle particolari necessità.
Ogni appartamento é dotato di uno spazio esterno che, a seconda dei bisogni degli abitanti, può essere unito a quelli adiacenti togliendo la parete che li separa.

0 5M
PIANO PRIMO

0 5M
PIANO TERRA

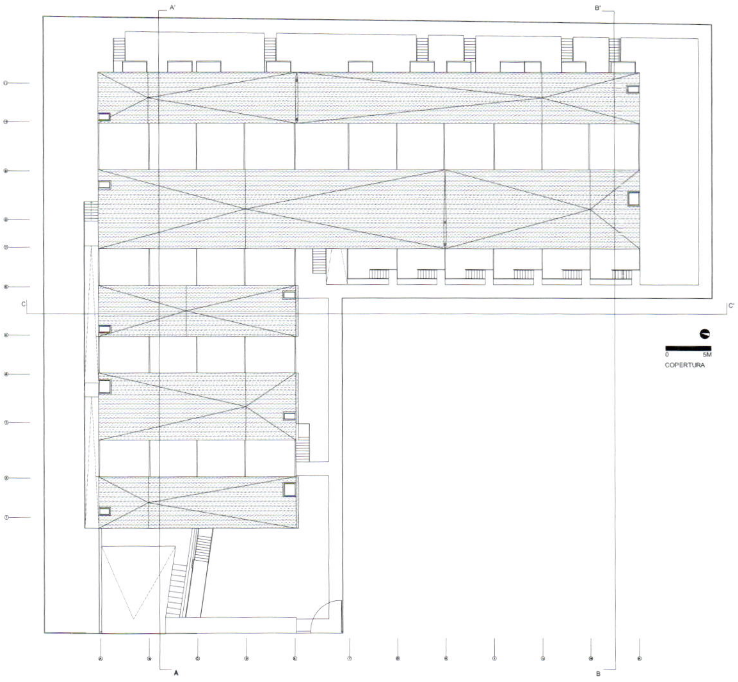
0 5M
COPERTURA

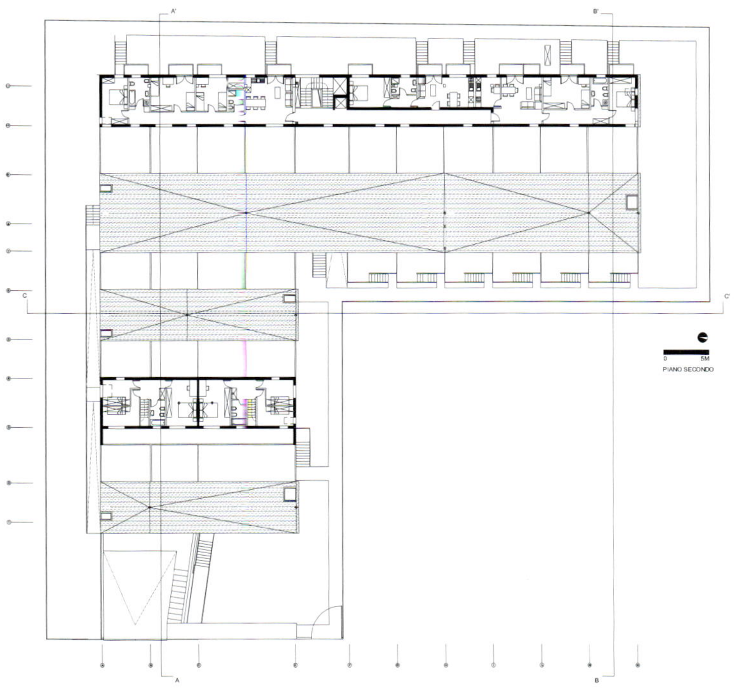
0 5M
PIANO SECONDO

Architect: Atelier Zündel, Cristea
Firm: A/ZC
Consultant: Oth Est
Area: 3500m²
Photographer: Yves Marchand, Romain Meffre

Retirement Home

This retirement home is located on the heights of Riedisheim (a suburb of Mulhouse), in a residential neighborhood.

The conjunction of geophysical and regulatory constraints rendered this plot of land difficult to build upon: its elongated form, its gradient exceeding 10%, and the very restrictive POS regulations applicable to it were a priori incompatible with the size and demands of the project.

These conditions led us to develop the entire project within a construction adopted to the form and constraints of the land. The building unfolds in length from east to west, while its height, made of up of four trays slightly shifting away from one another, follows the natural slope. Its height never exceeds the maximum authorized 10m (the equivalent of three floors) at any point level with the land.

The building was also positioned near the plot's northern dividing boundary, which allowed us a POS (H=L), hence a progressive spacing in relation to this boundary, as it goes along and the land descends. This proximity allowed us to free up the southern side so as to build the living spaces of each unit with a terrace attached to each.

The roofs, visible from the principal access road located at the top of the plot, are all planted, which helped dampen the impact of construction in the neighborhood.

We divided the length of the building into little modules the size of several rooms, in order to better integrate the length within the residential environment. These interruptions bring about the entry of a generous amount of natural light into the interior circulatory paths of all floors.

The flour floors are composed of: two large units on the ground floor and upstairs, the Alzheimer unit at garden level, and the service premises at the level of the courtyard.

In spite of the shifting between one floor and another, we concentrated all the service spaces of each unit one atop another, around the vertical circulatory paths, which creates a shortening in the distances the staff must travel.

The building complex is covered in wood, whose reassuring touch is present at the center of the lives of residents.

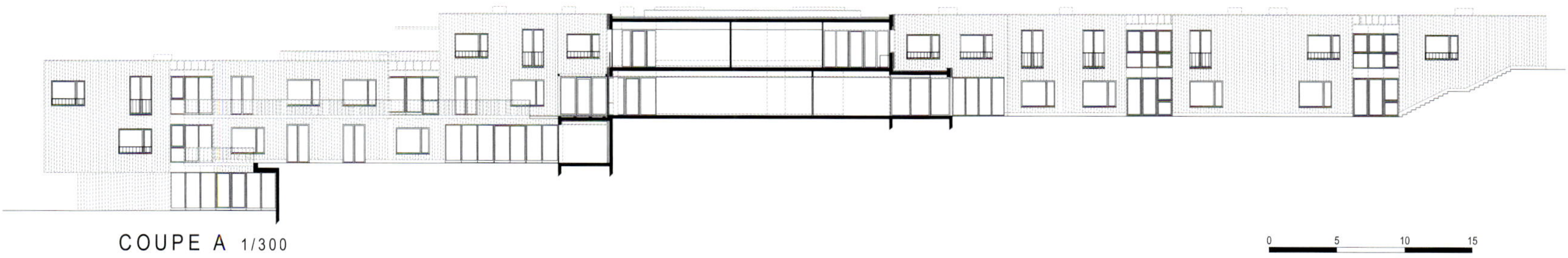

COUPE A 1/300

PLAN MASSE 1/1000

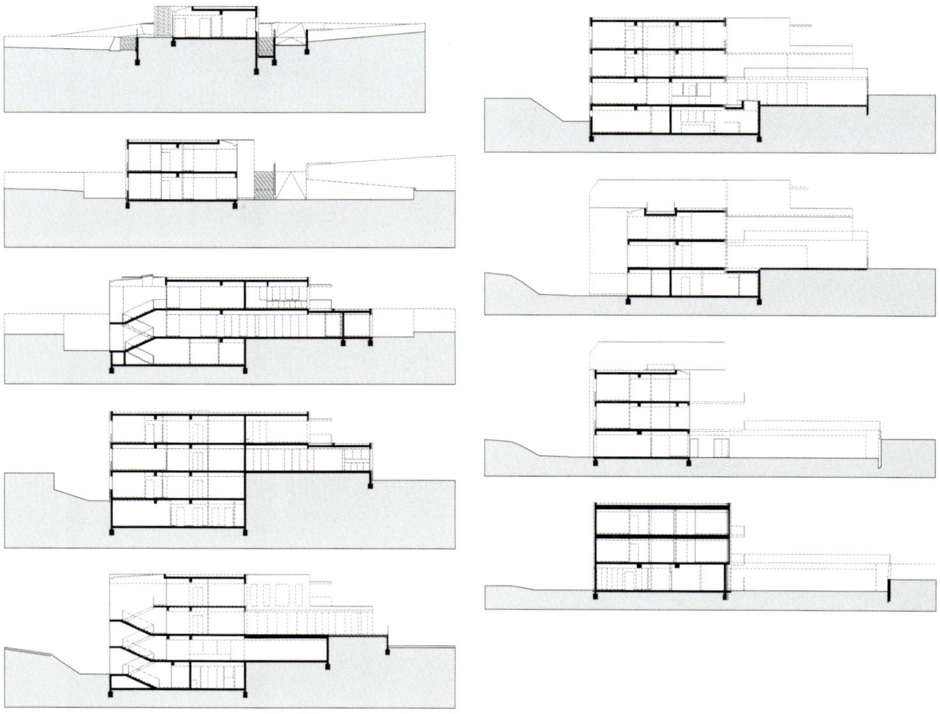

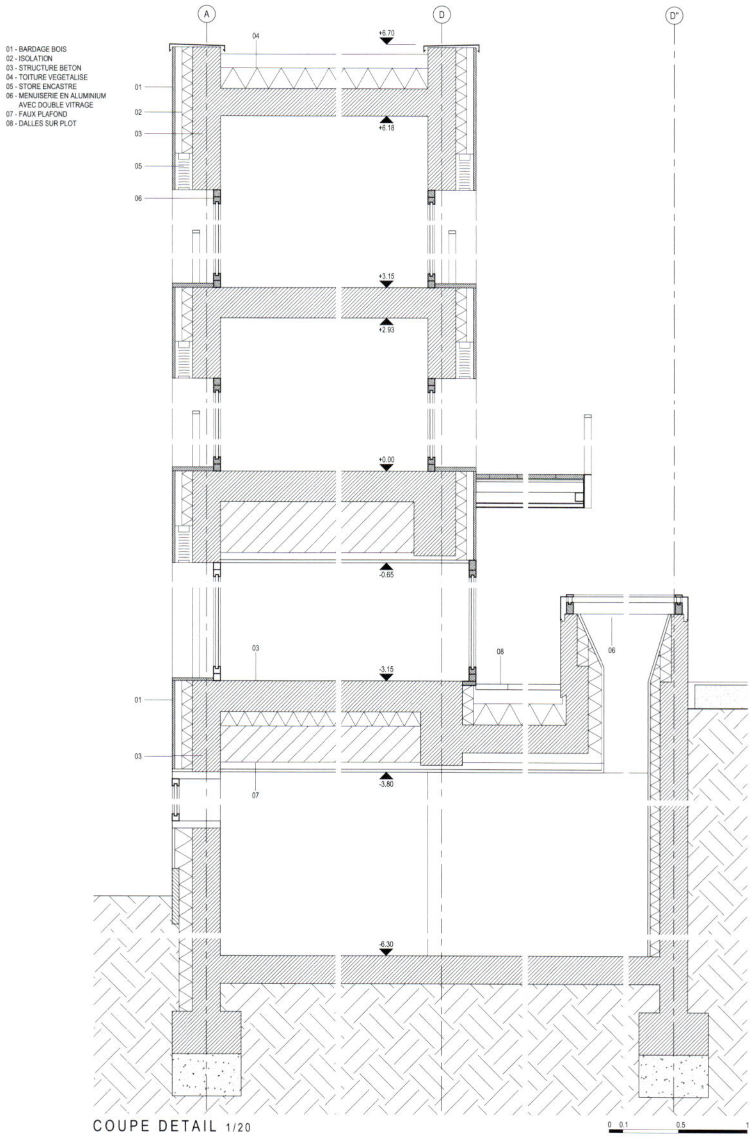

COUPE DETAIL 1/20

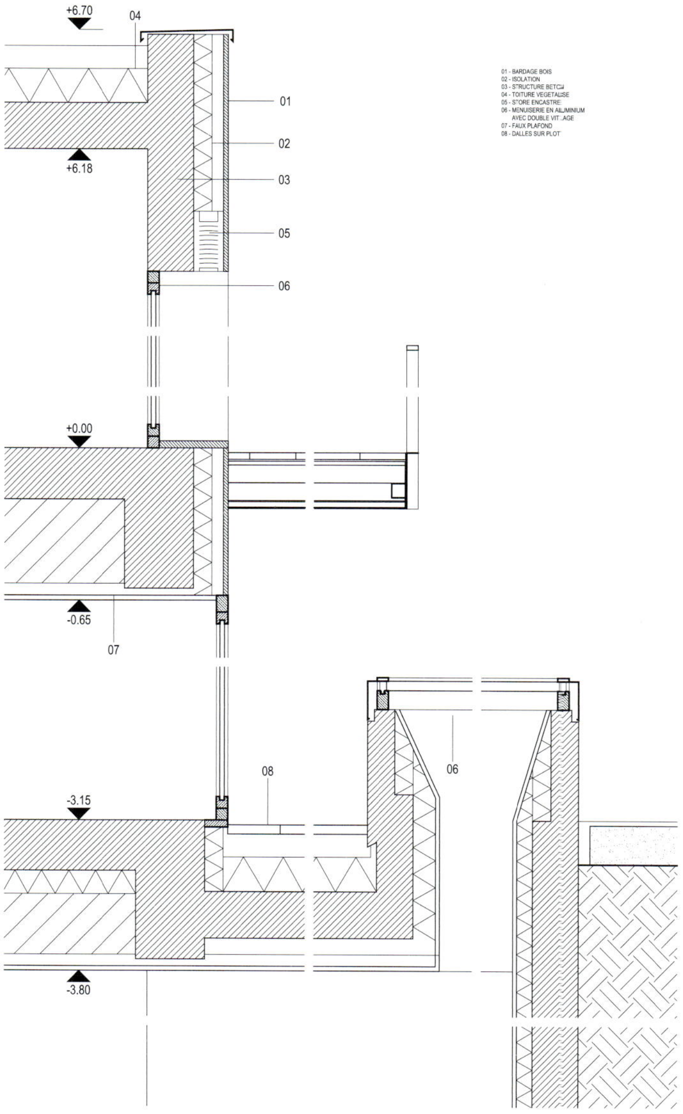

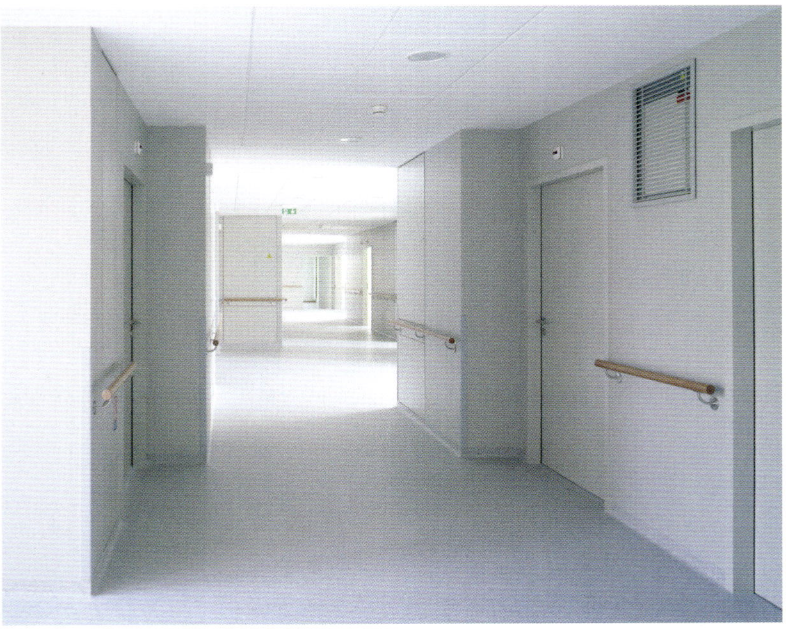

0
5
10
15

0
5
10
15

Architect: Lawrence Scarpa, FAIA, Brooks+Scarpa
Firm: Brooks+Scarpa
Project team: Lawrence Scarpa, Stephanie Ericson, ngela Brooks, ilke Clemens, Joshua Howell, Ching Luk, Charles Austin, Gwynne Pugh, Lawrence Scarpa
Engineering: BPA, Cobalt Engineering
Environmental consultant: Albert Bicol, Cobalt, Greg Reitz, ReThink
Façade consultant: CR Laurence, Inc.
General contractor: JT Builders
Landscape: FormLA
Location: 751 N. Fairfax Blvd., Los Angeles, California
Area: 2973m²
Photographer: John Edward Linden, Tara Wujcik

Cherokee Mixed-Use Lofts

The main architectural feature of this project is the building's owner-controlled operable double façade system. By allowing the occupant to adjust, at will, the operable screens of the building façade, the facade is virtually redesigned "live" from within the space, reflecting the occupants of the building within, in real time. The screens also enhance the existing streetscape and promote a lively pedestrian environment. By visually breaking up the façade into smaller articulated moving elements, the building appears to move with the passing cars and people.

The perforated anodized aluminum panels of the building creates an ever changing screen that sparkles in the sun and glows at night, while simultaneously providing shade to cool the building, reducing noise, enhancing privacy, and still allowing for spectacular views, great natural light and ventilation from ocean breezes which pass through its millions of perforations even when all panels are closed. The material reappears as a strategic arrangement of screens on the east, west and south-facing walls, lending a subtle rhythm to the exterior circulation. South-facing screen walls filter direct sunlight that lends unexpected visual depth while creating a sense of security for the occupants. Enhancing the structure's geometric texture, the irregular array of moveable openings variably extrudes from the building's surface. Its unique architectural form and integrated function create a high-performing building that is an expression of the people who live there and the environmental and cultural context in which it is built.

The planning and design of Cherokee Lofts emerged from close consideration and employment of passive design strategies. These strategies include: locating and orienting the building to control solar cooling loads; shaping and orienting the building for exposure to prevailing winds; shaping the building to induce buoyancy for natural ventilation; designing windows to maximize day lighting; shading south facing windows and minimizing west-facing glazing; designing windows to maximize natural ventilation; utilizing low flow fixtures and storm water management; shaping and planning the interior to enhance daylight and natural air flow distribution. These passive strategies alone make this building more than 40% more efficient.

I've got
new fries.
They're free this Fryday.
Yup, Friday.
CLEARCHANNEL
DIGITAL

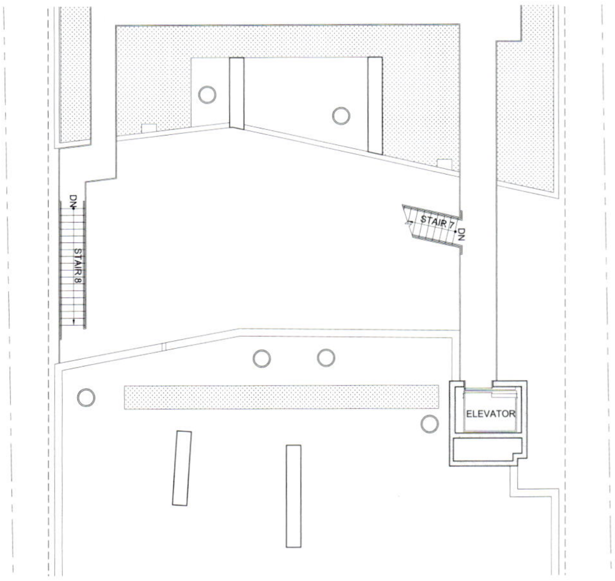
DN
STAIR 8
STAIR 7
DN
ELEVATOR

11
12

ROGERS
organi

BLACK LAKE
ON SWEAT
ON SWEAT
ON SWEAT
SCREAMING
THEY SAY
HES SICK
HES OBSCENE
NO MORE
SICK SICK!!

Architect: Maartje Lammers, Boris Zeisser, Olav Bruin, Anja Verdonk, Albert-Jan Vermeulen, Floor Kokke, Valentine van Dam
Firm: 24H architecture
Structural engineer: CAE structural engineers, Rotterdam
Location: Nieuw Leyden
Area: 4000m²
Photographer: 24H Architecture

Musician's Housing, Hoogvliet

For the Co-housing project that was developed by WIMBY (welcome in my backyard) as part of the International Building Exhibition Hoogvliet, a new urban concept was developed. The concept gives not only expression to the thought that a musician is an individual with a communal passion, but also expresses a form for new collective space within the contemporary city.

Especially for musicians, 38 houses were recently completed that all have their own music room. Most of the music rooms are placed in a collective mountain. The individual music rooms all have their own light pyramid that protrudes through the grass mountain. The houses are arranged in various 'stamps' which are repeated multiple times. On the edge of the 'stamps', individual zinc covered music rooms are added like parasites to a tree. These parasites accentuate the main urban routes through the site.

This project is build with an extreme low energy consumption avarage (EPC 0.46), do to the urban heating system that reclycles energy from the nayboring Rotterdam harbour. The wood facades are made of hydro thermal treated wood from local European forests.

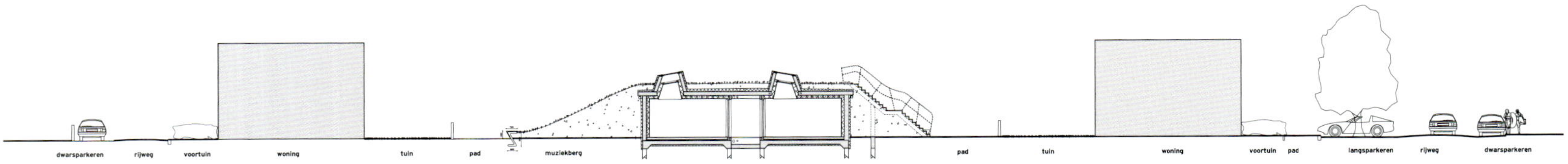

30

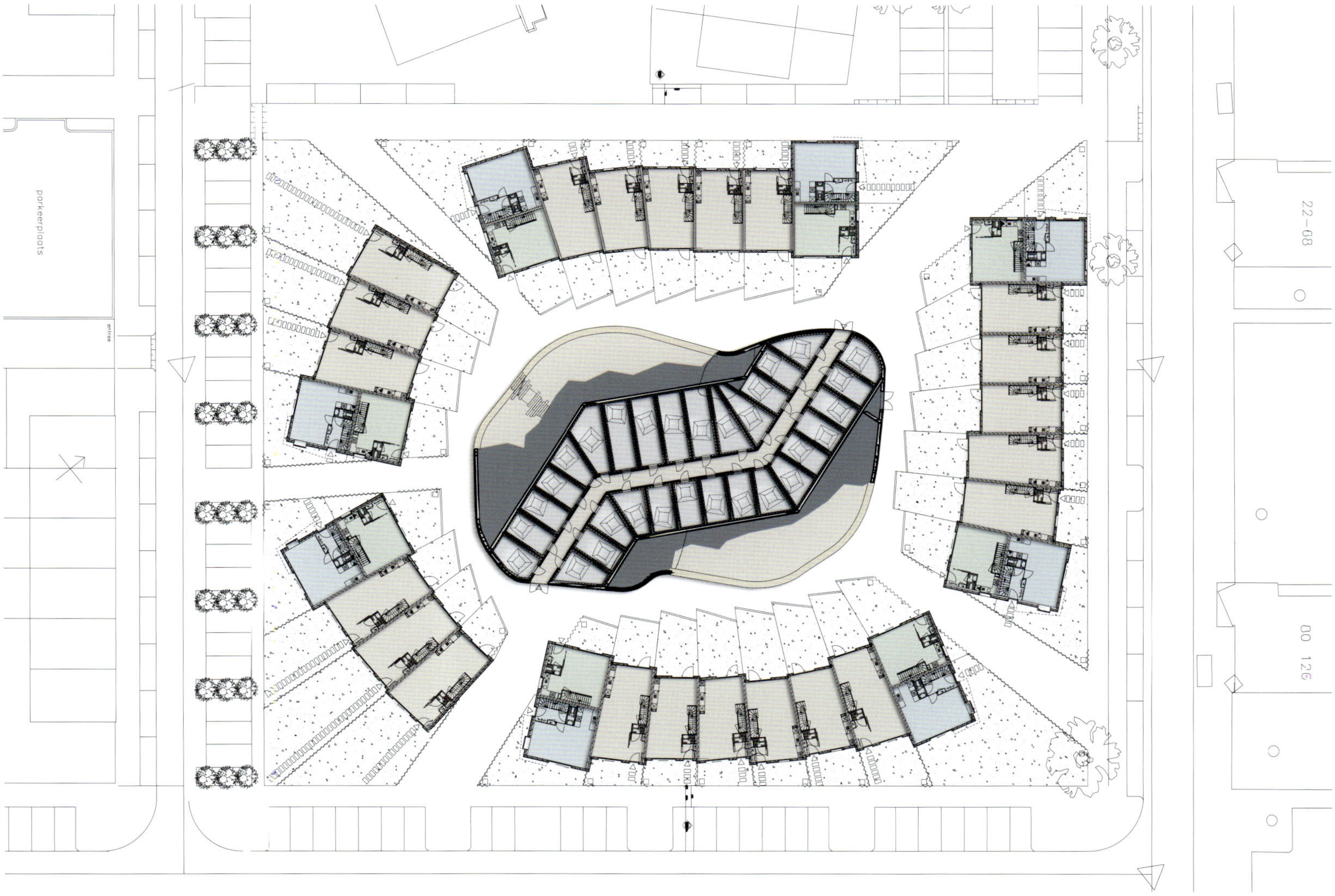
parkeerplaats
entree

Architect: AquiliAlberg
Firm: AquiliAlberg
Design team: Ergian Alberg, Laura Aquili, Alessio Pucci, Alexandros Franzolini, Christian Ronchi
Structural engineer: Studio Polis
Location: Cremona, Italy
Area: 1000m²

Shift Housing

The scenographic effect is technically obtained by means of a simple trick: the basement of the main elevation that goes out from the ground of 1.2 meters, rotate clockwise, meanwhile the first floor of the main elevation rotate in a counterclockwise direction. Two volumes are generated with two opposite directions, which create a X shape facade.

The view through the natural park guides the main orientation of the building and the opening of the lodges. To complete the residential use, the fruition of the green areas is studied deeply. Lodges that open themselves but private spaces become finally a filter between the intimacy dimension of the habitations and the external environment. Then, the first level has been resided up of about one meter from the ground, to increase the level of privacy for its apartments. This vertical movement adds space to create openings which bring natural light to the large parking at the basement level.

Every apartment will have an open space: a double garden at the first floor and a large lodge at next levels. So an innovative compromise between public and private space is defined, in which people could see themselves part of a small community but simultaneously safe in their private flats.

Finally the result, with the big lodges one span roof, is a strong dialogue and a seductive exchange between volumes and voids, lightness and dynamism.

The roof is designed flat to facilitate (and drive to) the installation of photovoltaic systems. In fact, from an energetic point of view, AquiliAlberg office, thinking to sensible habitations in continuous evolution, predisposed the building for the best integration of a photovoltaic system, which being designed previously it will be perfectly integrated in the architectonic project.

Coherently with mutations of the contemporary way of living, the exchange between human being and environment do not ignore the strong identity of the project, necessary component to increase the sense of belongings of inhabitants to a innovative model of sustainable development.

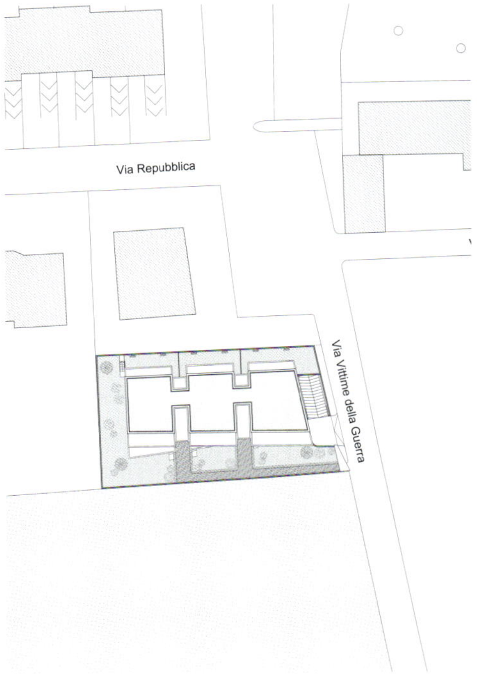

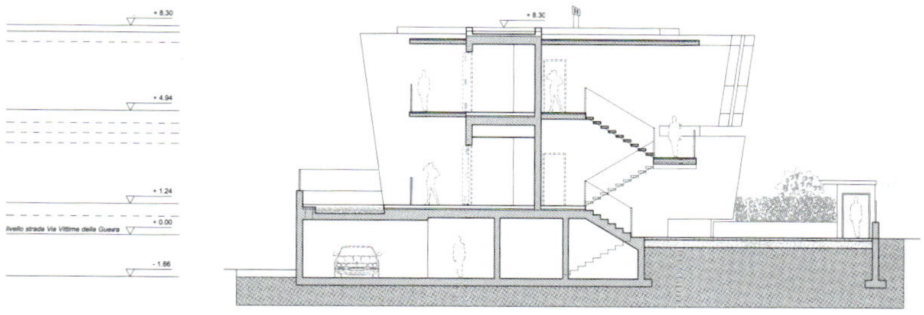
+ 8.30
+ 4.94
+ 1.24
+ 0.00
livello strada Via Vittime della Guerra
- 1.66

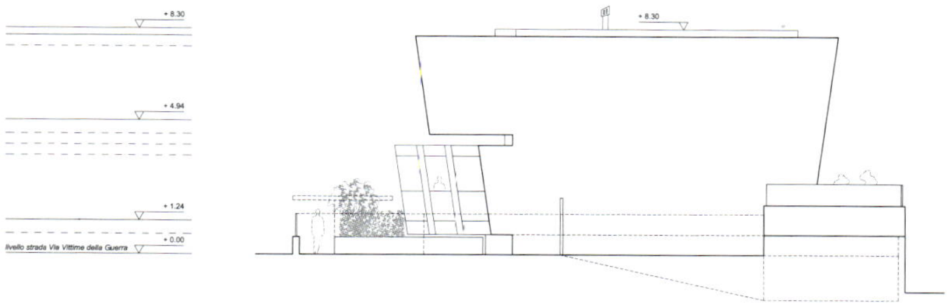
+ 8.30
+ 4.94
+ 1.24
+ 0.00
livello strada Via Vittime della Guerra

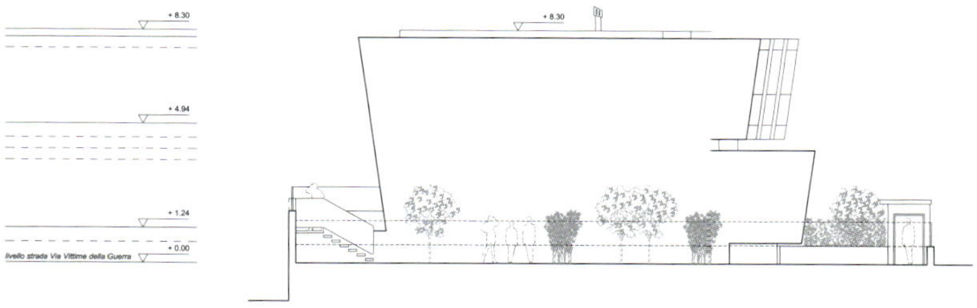
+ 8.30
+ 4.94
+ 1.24
+ 0.00
livello strada Via Vittime della Guerra

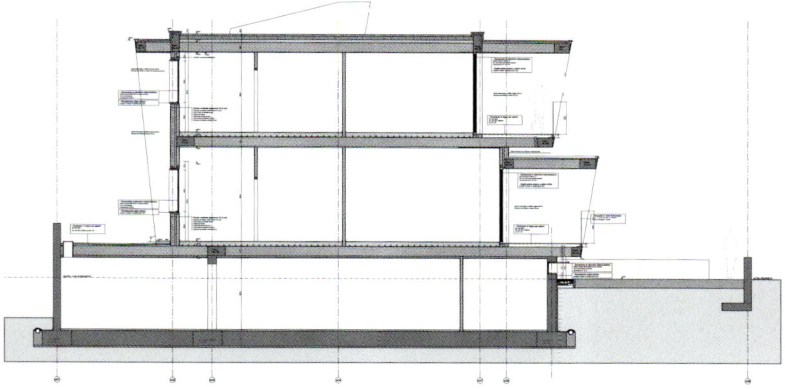

+ 8.30
+ 4.94
+ 1.24
+ 0.00
livello strada Via Vittime della Guerra
- 1.66

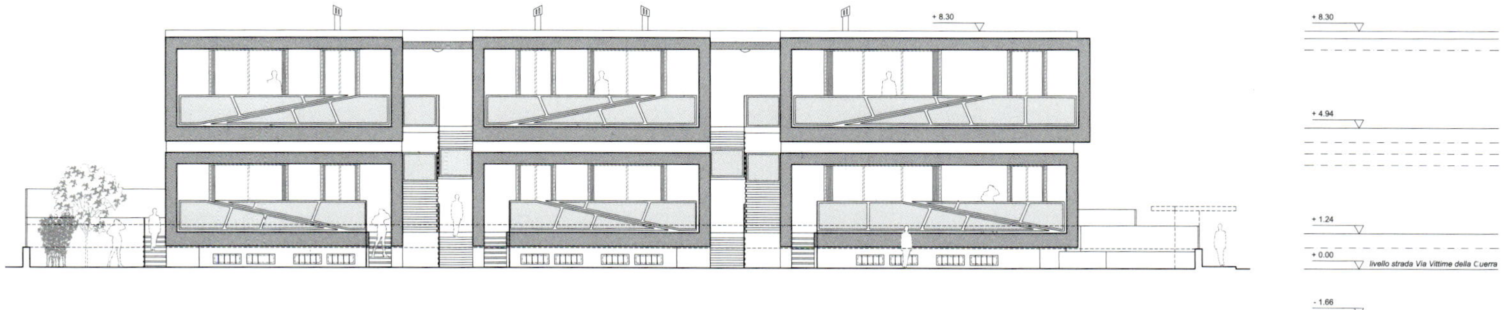
+ 8.30
+ 4.94
+ 1.24
+ 0.00
livello strada Via Vittime della Guerra
- 1.66

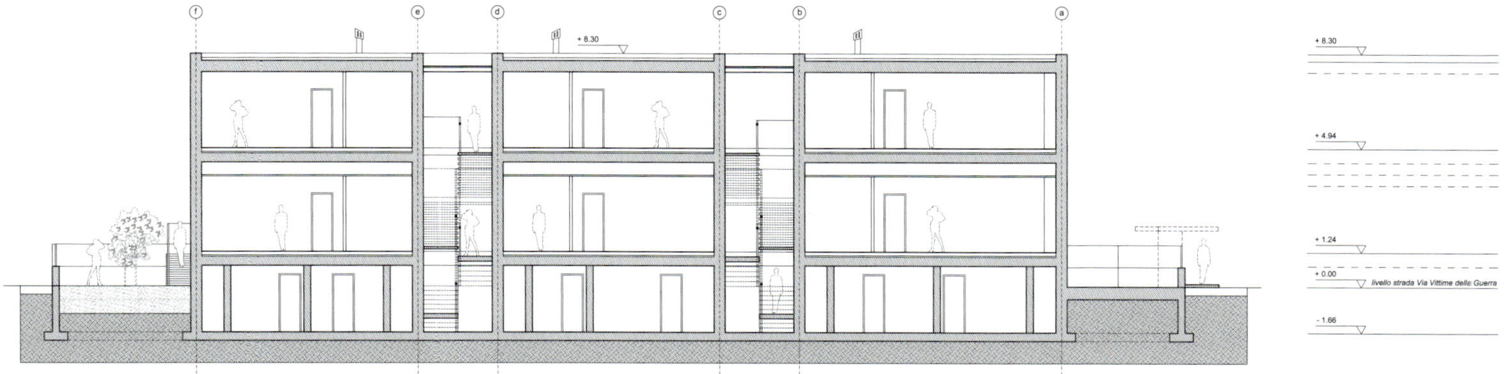
f
e
d
c
b
a
+ 8.30
+ 4.94
+ 1.24
+ 0.00
livello strada Via Vittime della Guerra
- 1.66

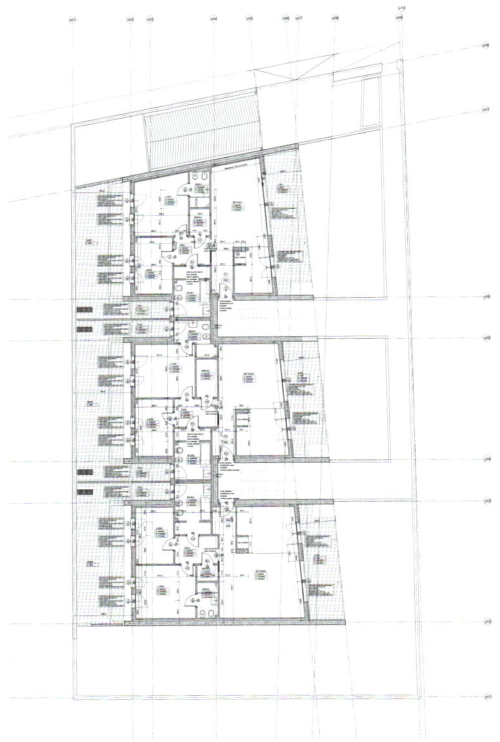

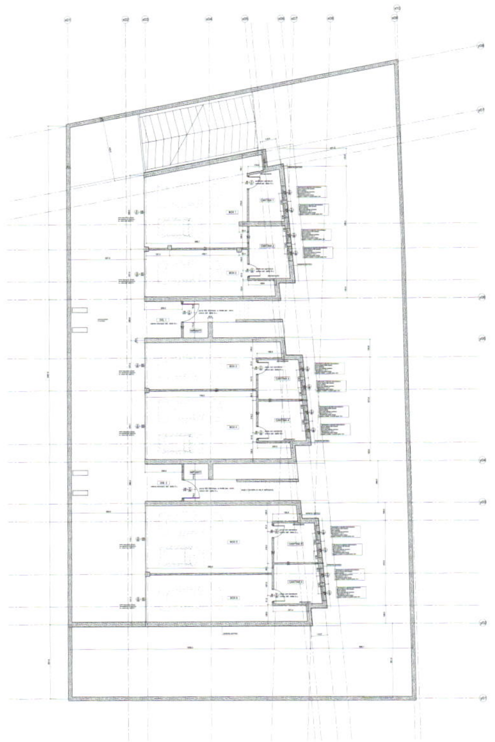

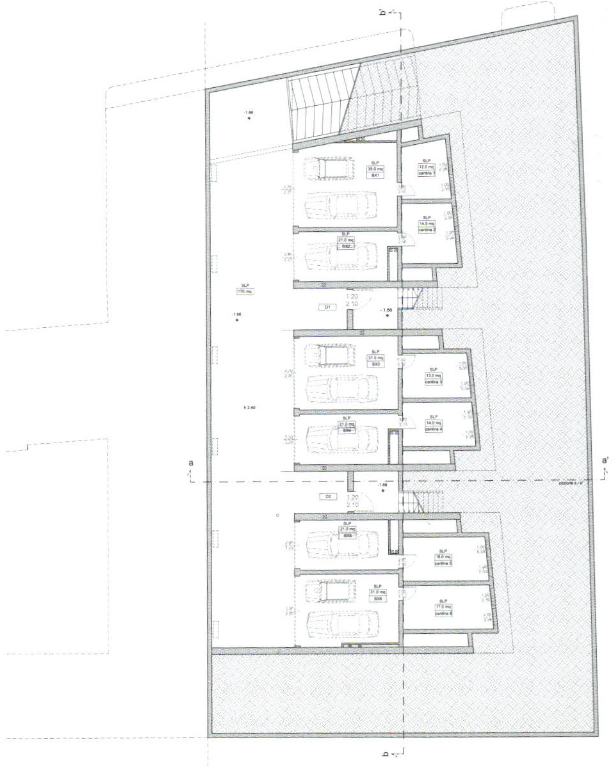

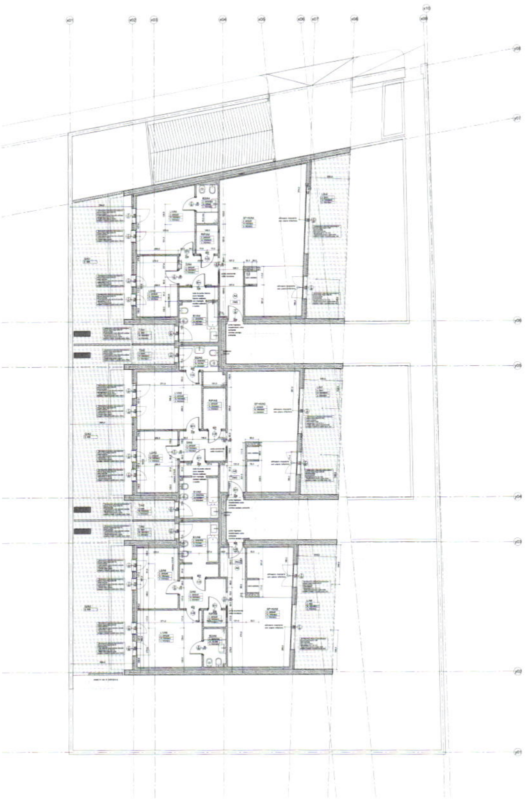

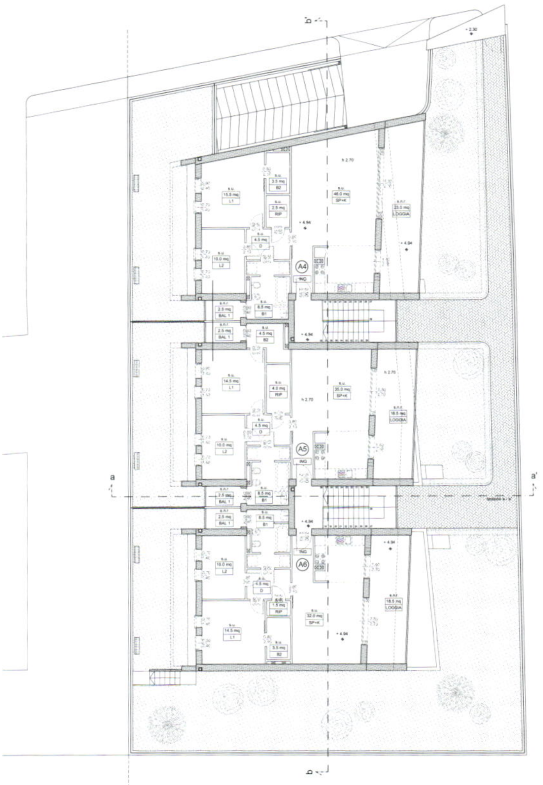

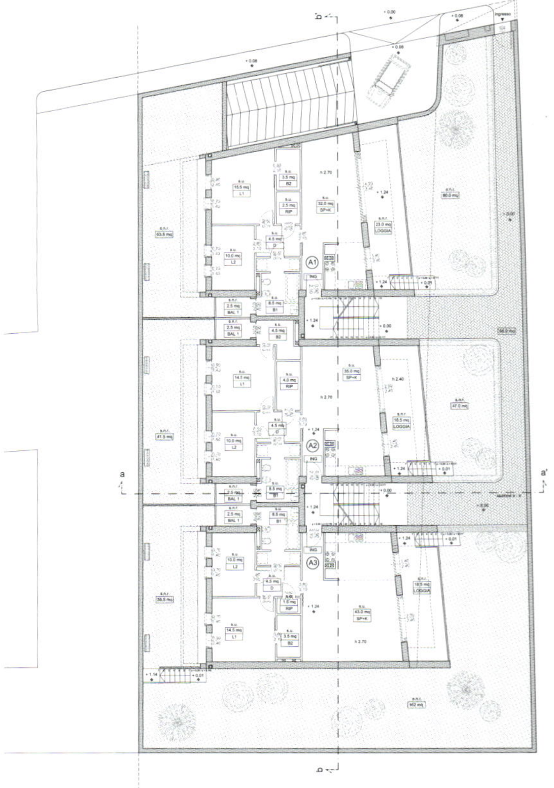

Architect: BFLS
Firm: BFLS
Acoustic consultants and fire engineer: WSP Group
Environmental advisor: URS Corporation Ltd
Construction advisor: Brookfield Construction UK Ltd
CDM coordinator: Construction Project Services UK Ltd
Planning supervisor: Construction Project Services UK Ltd
Planning: Consultant DP9
Development manager: Brookfield Developments UK Ltd
Townscape consultant: The Richard Coleman Consultancy
Aviation consultant: Donald Butler Associates Ltd
Wind engineer: RWDI-Anemos Ltd
Wind turbine: Designer/Manufacturer Norwin AS
Location: London, UK
Area: 28428m^2

Strata SE1

Strata SE1 is designed for sustainable urban living. Features include:

- Bespoke high performing façade with an air permeability leakage rate that is 50% better than current building regulations.
- Whole house ventilation system with heat recovery to each apartment (SAP Appendix & compliant).
- Low energy lighting in all landlord areas and 40% of the lighting in each apartment.
- Absence and presence lighting control system to all landlord areas.
- Car park lighting control provided with motion detection that reduces lighting by 50% when there is no movement.
- 96% recycling of all waste material generated during the construction stage.
- Designed to connect 'day one' to the planned Elephant & Castle Multi Utility Services Company (MUSCO) district system which will provide heat and electricity from Biomass CHP energy centers, grey water to all toilets for all affordable units within Strata SE1 and a connection to the site wide fibre network, providing high speed data access to all residents.

Independent analysis by Integrated Environmental Solutions Limited (IES) has compared CO^2 emissions from Strata SE1 against the Building Regulations 2006 Part L2 targets. The result demonstrates that Strata SE1 (with the connection to the MUSCo system in place) will achieve a predicted 73.5% reduction in CO^2 emissions when compared against the Building Regulations benchmark.

This independent analysis also concludes that Strata SE1 is predicted to achieve 2050 CO^2 target emissions. Strata SE1's sustainable design means that energy costs per apartment will be substantially less than the typical UK housing average significantly reducing residents and tenants occupational costs.

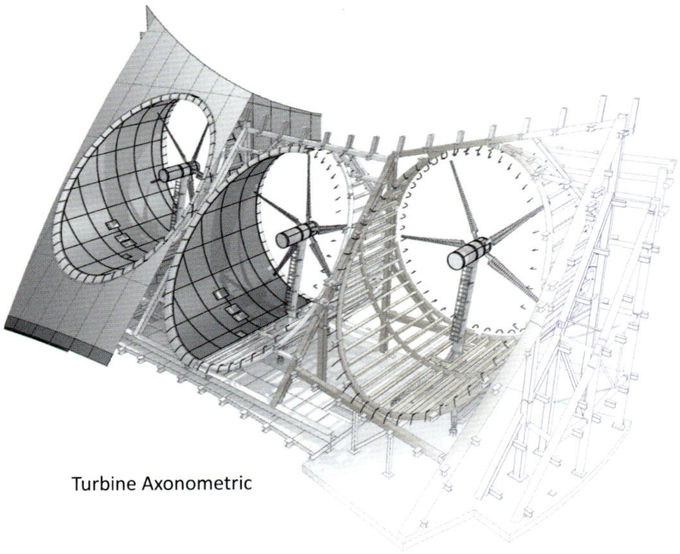

Turbine Axonometric

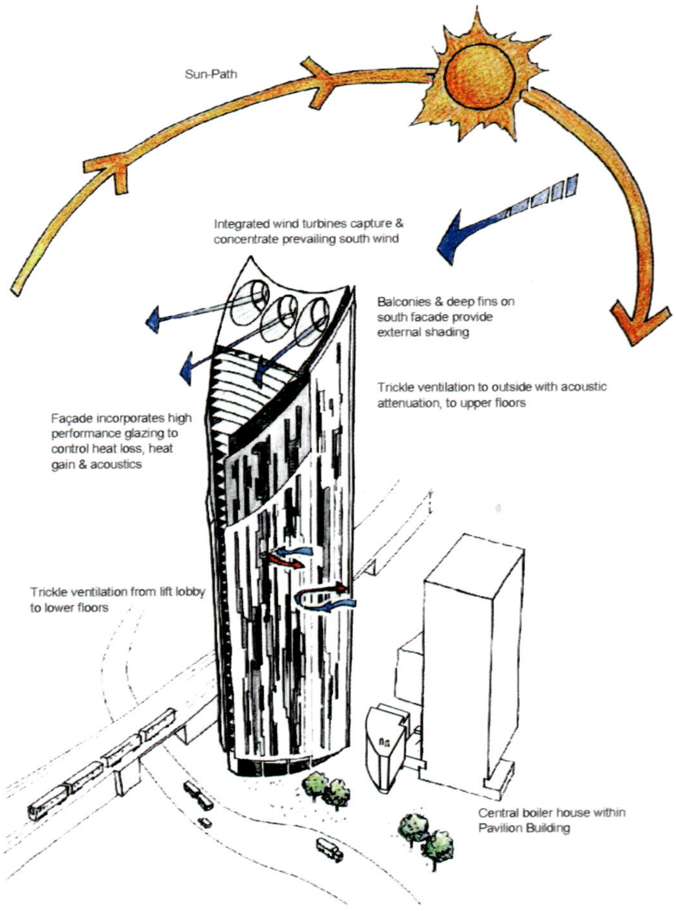

Environmential Study

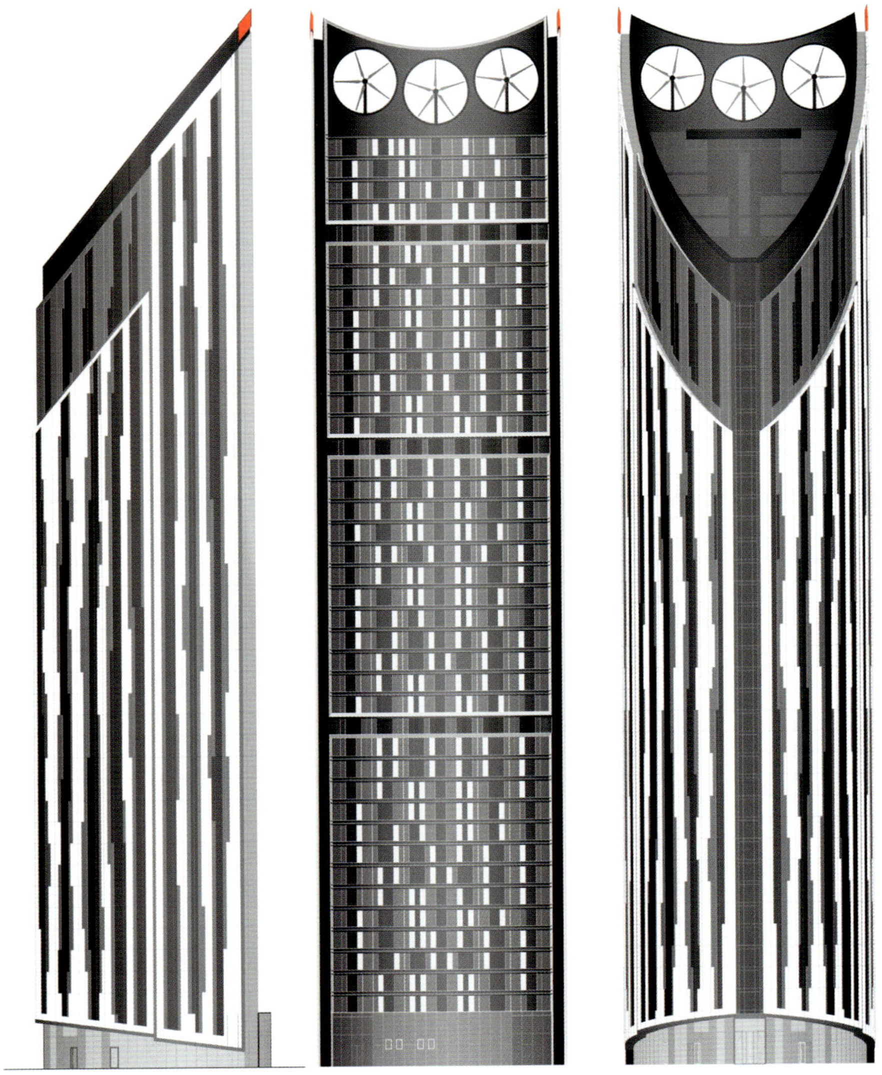

West, South+North Facades

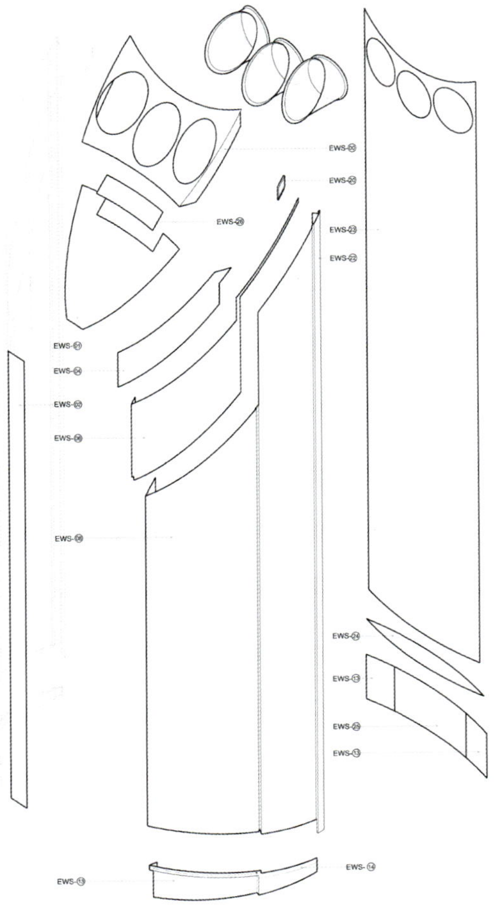

Kit of Parts

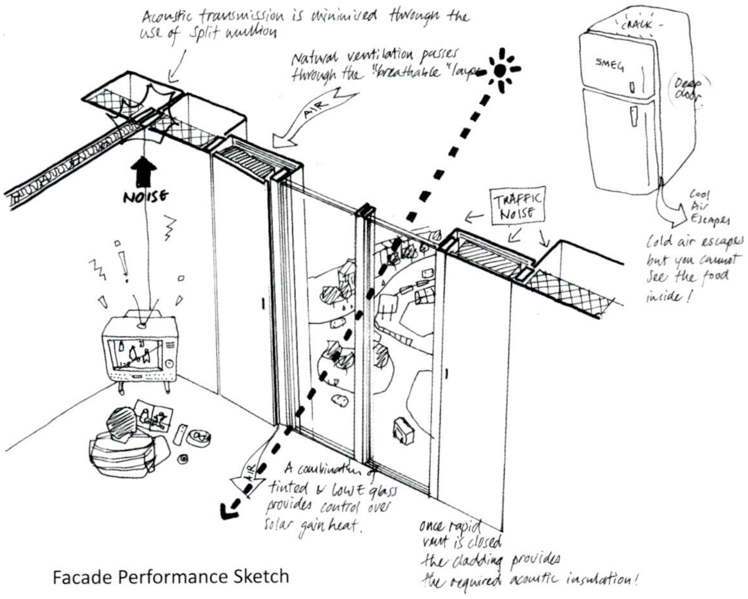

Facade Performance Sketch

Initial Concept Sketch

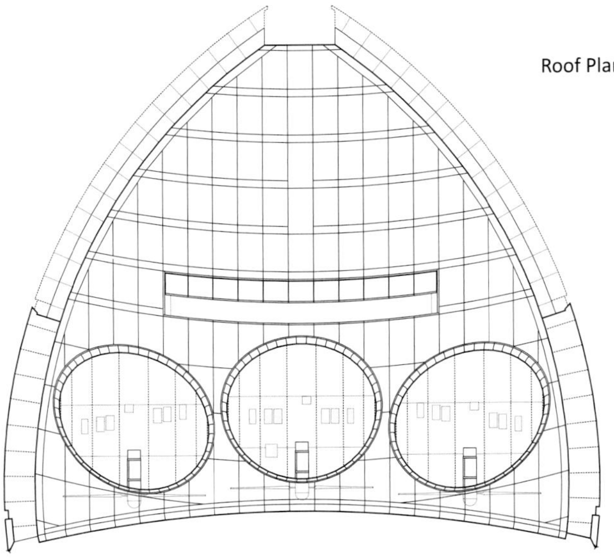

Roof Plan

Ground Floor Plan

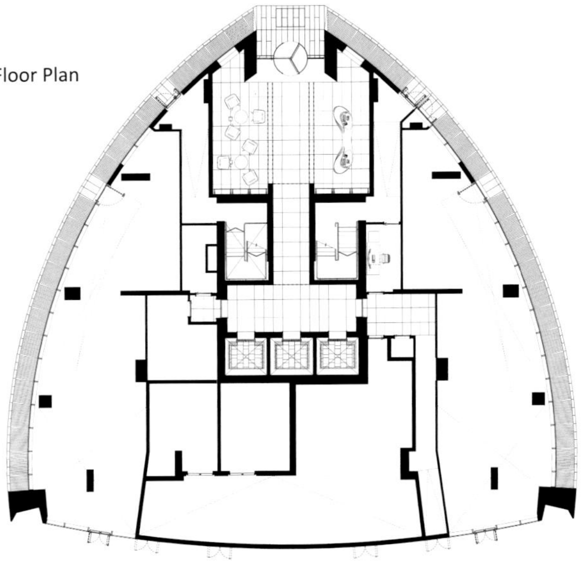

10th Floor Plan

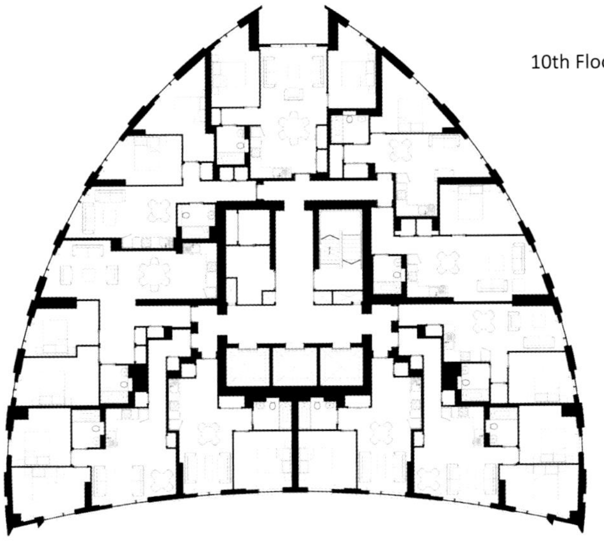

20th Floor Plan

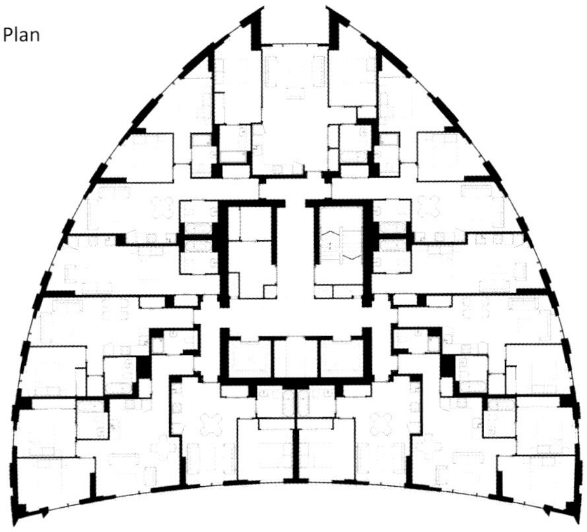

29th Floor Plan

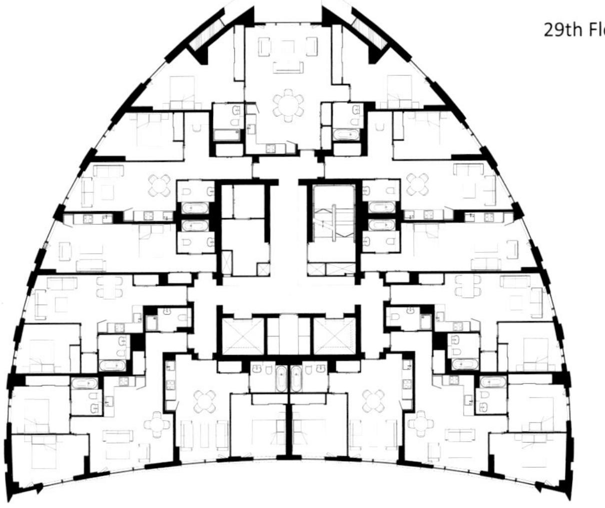

37th Floor

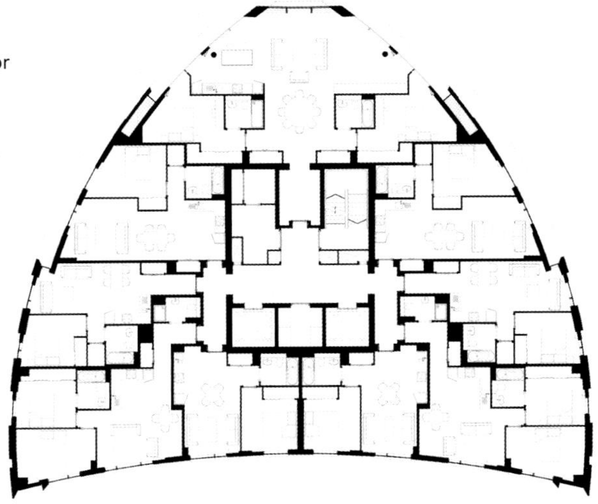

39th Floor Plan

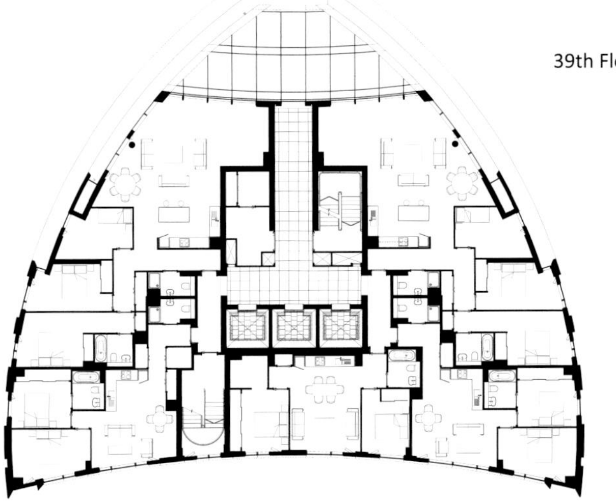

Architect: Bert Dirrix
Firm: diederendirrix
Location: The Netherlands
Area: 5320m²

Marienburg ENGELS

This dual block was the first of four housing complexes to be realized in this park like area, as part of a development plan by the architectural firm Lubbers. The complex consists of forty houses for rent or sale and three penthouses, with underground parking and storage facilities. The size of the complex, which is considerable for the village, has been minimized by the terraced layout and the graduated exterior space. This reduces the massiveness of the architecture without robbing it of its strong expressiveness. Loggias, terraces, receding and overhanging balconies ensure that the apartments are optimally oriented towards their green surroundings.

The two blocks, which are at an angle to each other, each align themselves differently with the environment.

This angle has created a shared inner courtyard between the two volumes that connects to a terrace in the park through pivoted gates.

Generous, circling galleries, and a staircase and lift that have been designed as separate elements grant access to the apartments. Rented and owner-occupied apartments are mixed randomly, and various income brackets live next to each other, just as in a regular village.

MARIENBURG situatie 1:2000

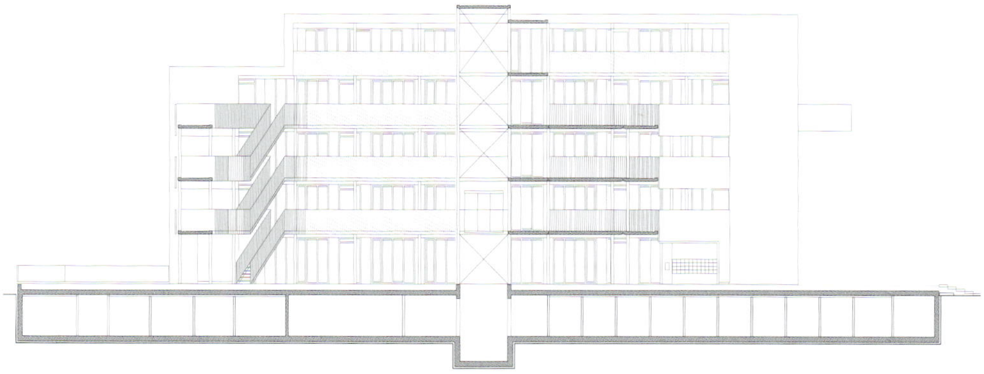

MARIENBURG doorsnede 2 1:200

MARIENBURG doorsnede 1 1:200

MARIENBURG gevel zuid 1:200

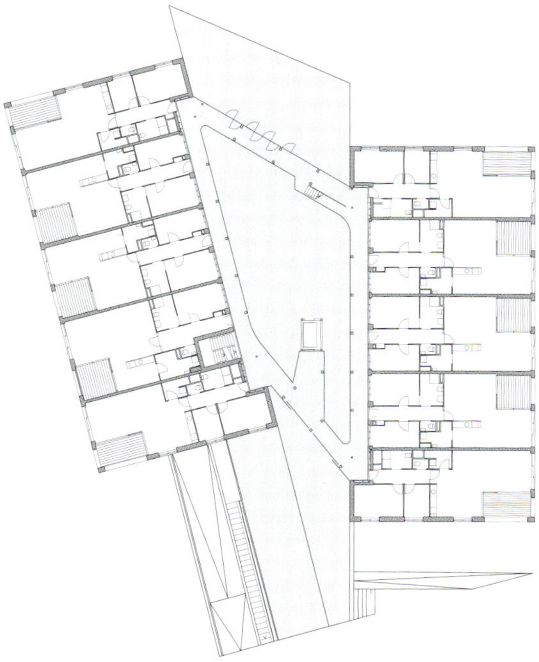

MARIENBURG plattegrond begane grond 1:500

Architect: LOVE architecture and urbanism
Firm: LOVE architecture and urbanism
Location: Hamburg, Germany
Area: 6110m²
Photographer: Anke Müllerklein

Baufeld 10

LOVE's special challenge: The Baufeld 10 project was developed in a joint building venture. This means that the various future residents worked together to create a real community for the new building. Within this model, individualists connected with each other with the goal of building THEIR communal house. The building typology had to meet this expectation. This is why the building houses many different building typologies with all kinds of furnishing standards: from very large apartments (up to approx. 225m²) to smaller units (approx. 50m²), which feature entirely different designs – from one-storey apartments to maisonettes that stretch across four storeys.

For Baufeld 10, individualisation was the top priority. Each of the 28 new residents can now enjoy his or her unique lifestyle within his or her apartment – whether horizontal or vertical, and whether in a small or a big apartment. These different visions were blended into one building to everyone's satisfaction.

Viewed from the outside, the residential building presents a gleaming white structure formed of slightly bevelled cubes and with generously proportioned, slightly bevelled window openings. The configuration of these window openings matches the layout of the apartments behind them. Each apartment has a balcony and/or bay which protrude from the building. Within the rigid row of buildings, this configuration provides maximum views in an "exciting direction" – namely, the harbour.

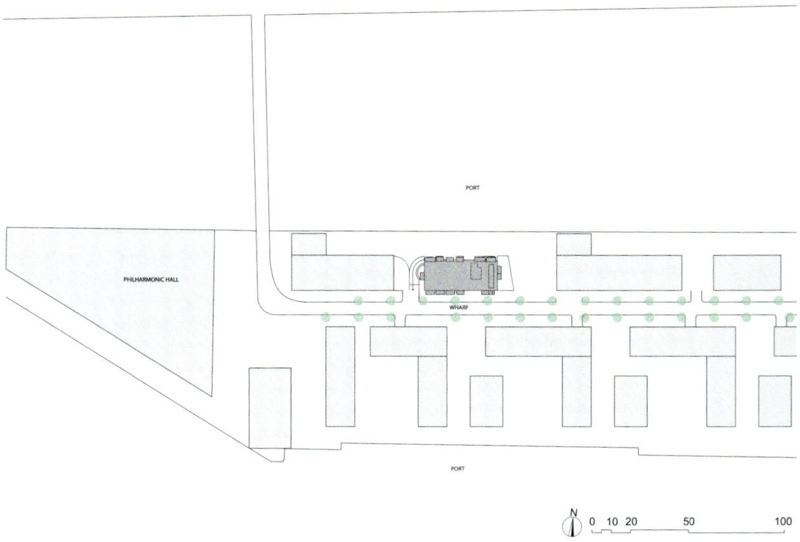

SITEPLAN

ELEVATION SOUTH

ELEVATION EAST

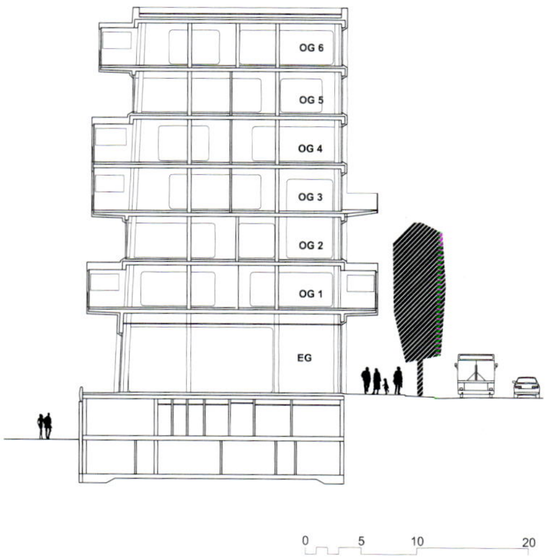

SECTION

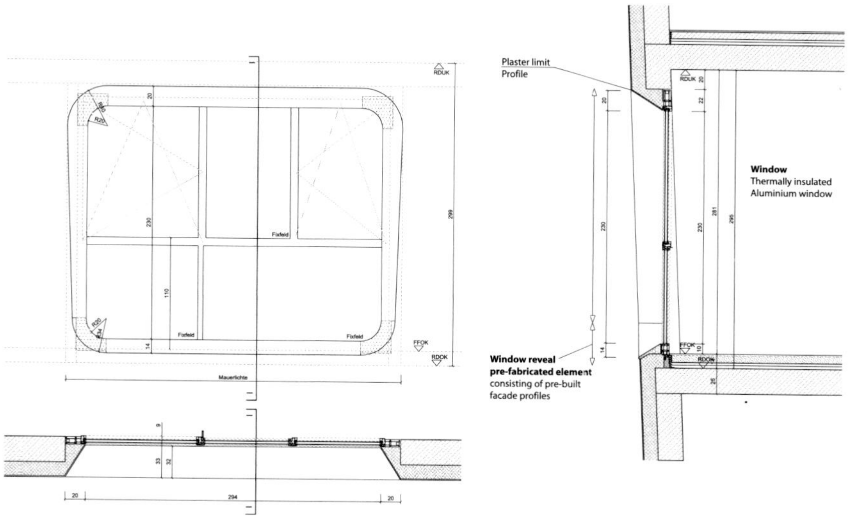

DETAIL SECTION

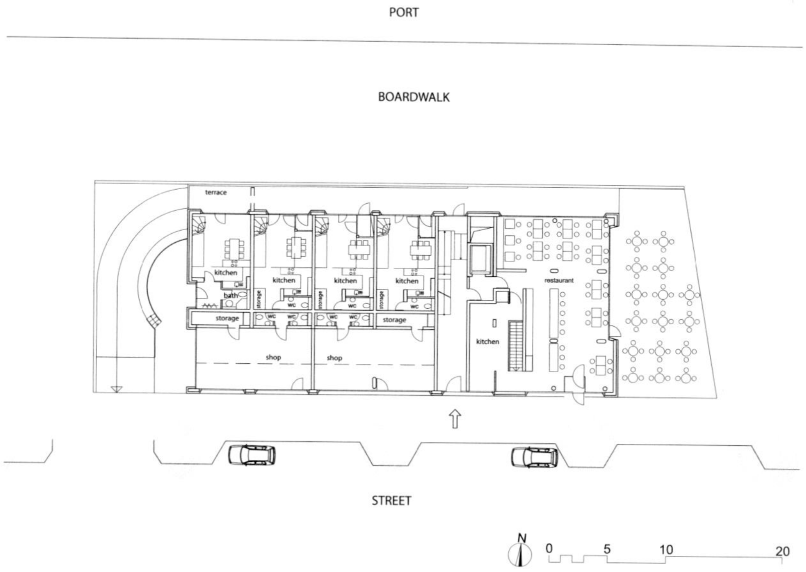

GROUND FLOOR

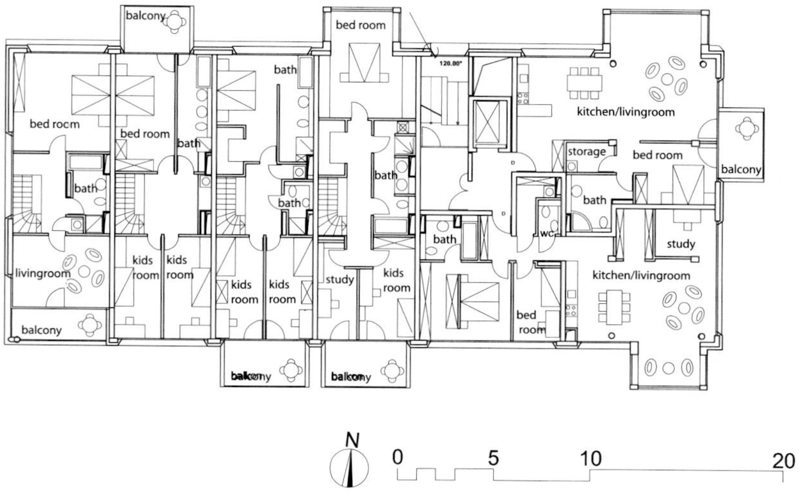

1st FLOOR

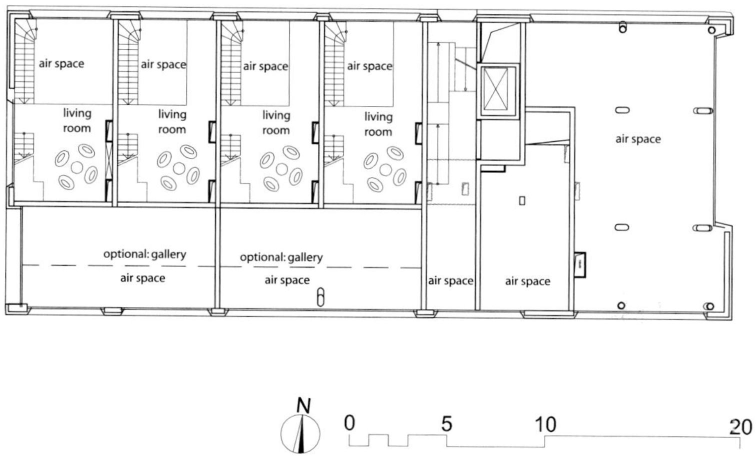

INTERMEDIATE STOREY

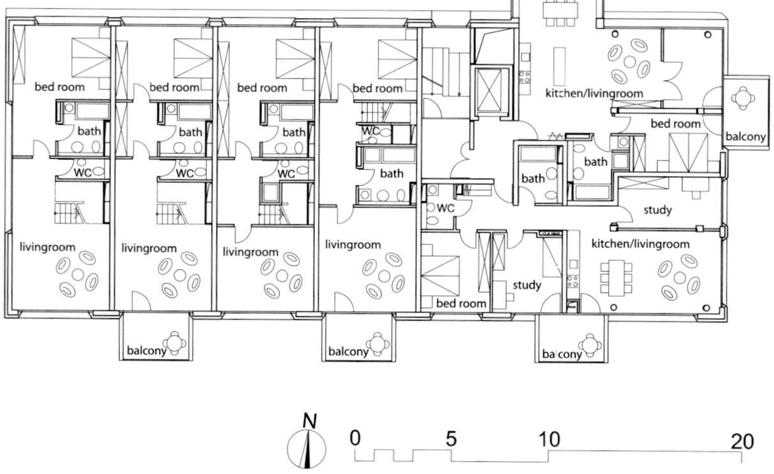

2nd FLOOR

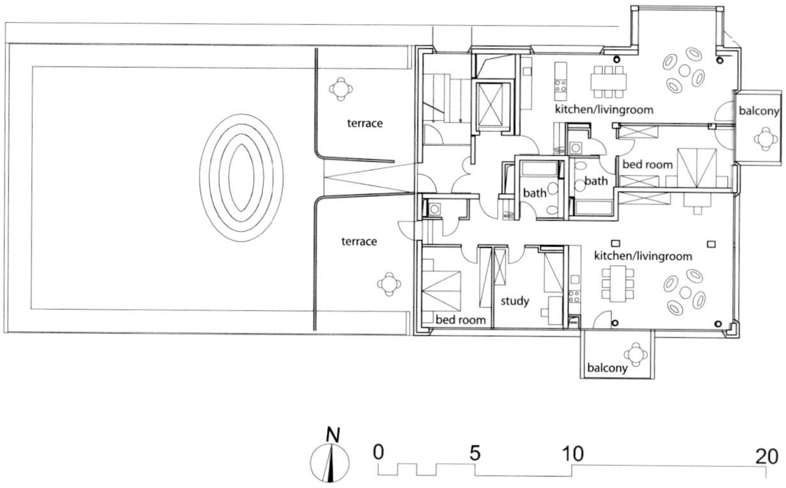

6th FLOOR

Architect: Caramel Architekten
Firm: Caramel Architekten
Landscape architecture: Idealice - Alice Größinger
General planing: Stumpf Develepment Gmbh
Location: Vienna, Austria
Area: 2705m²
Photographer: Hertha Hurnaus

Krautgarten

Although the krautgarten project is a four-story building, each individual living unit was designed to be as self-contained as possible. Each apartment, therefore, has its own floor plan and the various loggias and balconies allow the outside observer to recognize each unit as a distinct "house". Private yards, roof gardens, loggias, and balconies provide each apartment with various levels of direct contact to the green surroundings – like a house surrounded by grass and trees. The individual internal structure of the building gives the outer shell a playful character as a positive atmosphere in a functional unit "for living". This is emphasized by the haptic impression of the very course plaster of the building exterior and the cheerful choice of colors.
From an urban development perspective the site is located in a very heterogeneous environment. The corner of the building facing the public traffic area assimilates the main directions of the adjacent buildings. A green zone is produced along the street as a buffer between public and private space.
The main target group for this housing development with "green" surroundings at the periphery of the city comprises young families and the like. More than half of the apartments have their own gardens; the rest have terraces or loggias.
In addition, the apartments – mainly terrace-house-style duplex unit -– feature a high degree of individuality; the differences in the apartments can also be seen in the exterior design and the façade. The loggias indicate the living rooms; the smaller windows arranged in varying patterns mark the bedrooms.
Located virtually at the "heart" of the building, at the geometrical center of the ground floor, adjacent to the stairwell is a multipurpose zone and children's play area with a sandbox and monkey bars.
The common zone consists of a wooden surface including a large sunken area for playing with sand and a vulcanized rubber play area with a hill. In addition, there is a long multipurpose element good for both sitting and lying on. Despite its small size, the area is intended as a space where residents can meet, talk, and play together.

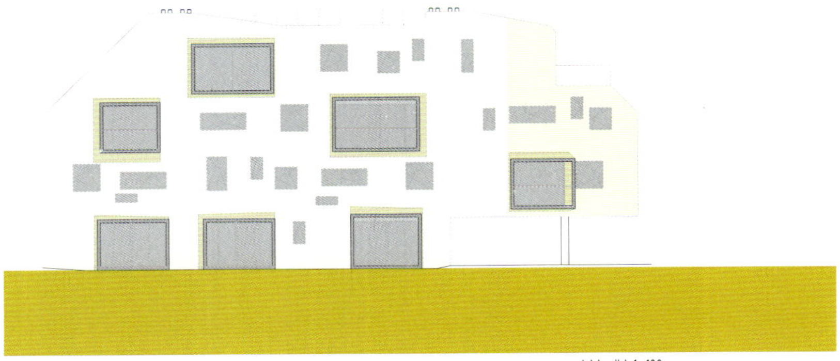
ansicht süd 1_100

südost

blick strassenseitig

west

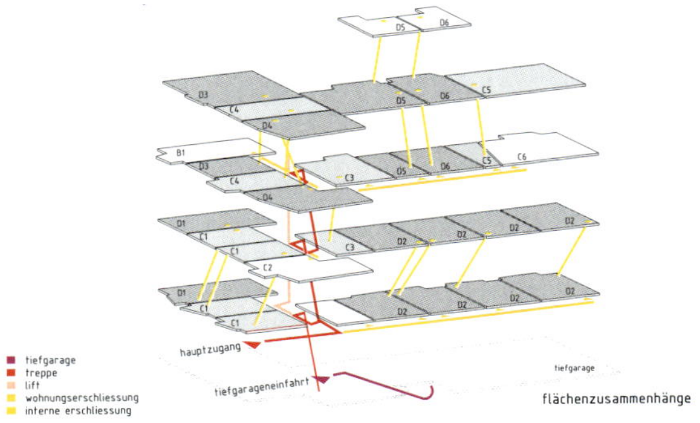

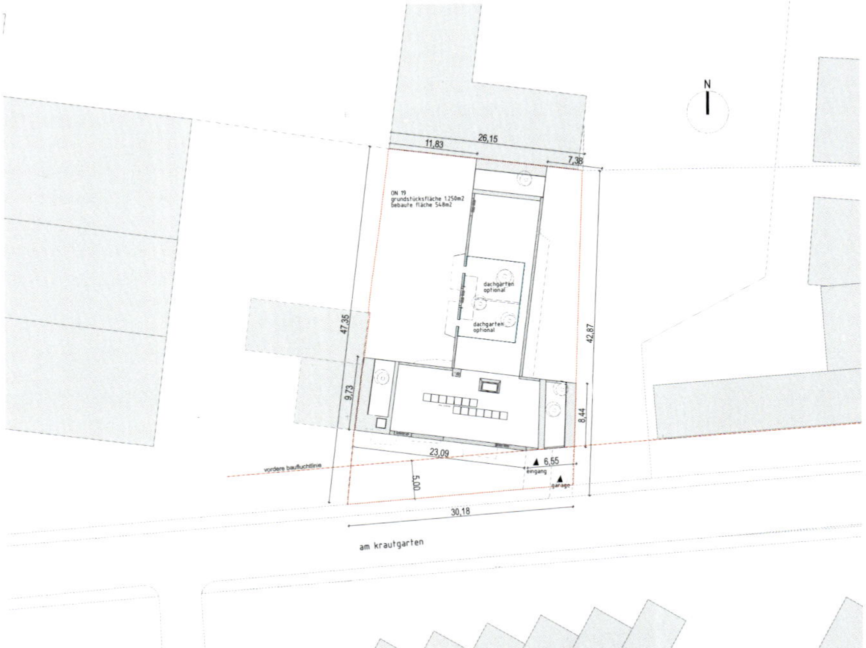

lageplan 1_250

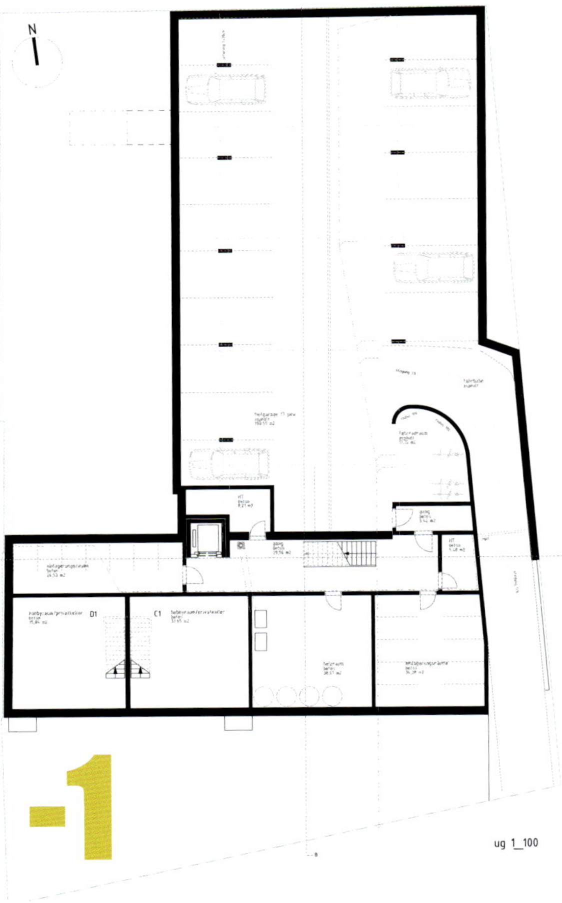

ug 1_100

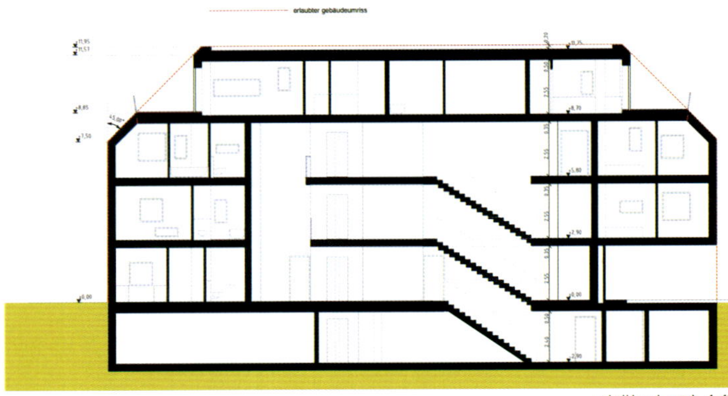

schnitt ost-west 1_100

ansicht süd 1_100

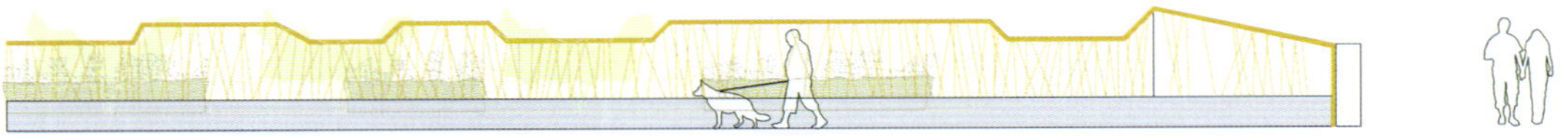

zaunelement zur strasse 1_50

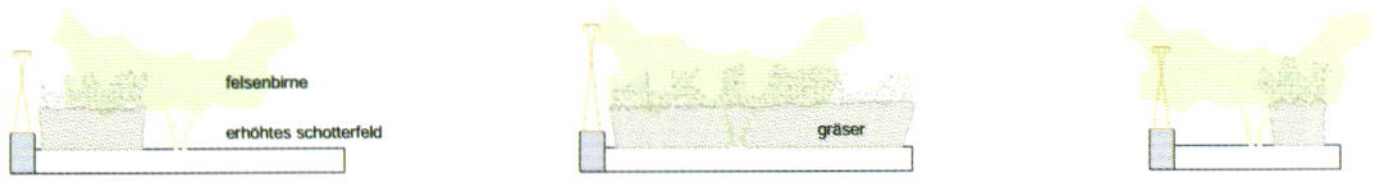

schnitt zaunelement und grünpuffer zur strasse 1_50

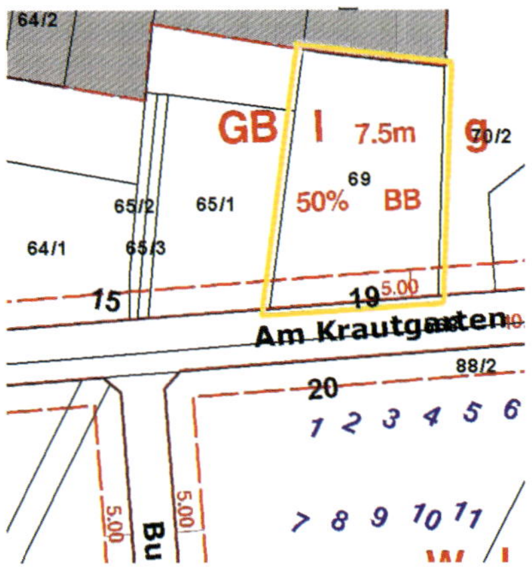

auszug plandokument flächenwidmung 1_500

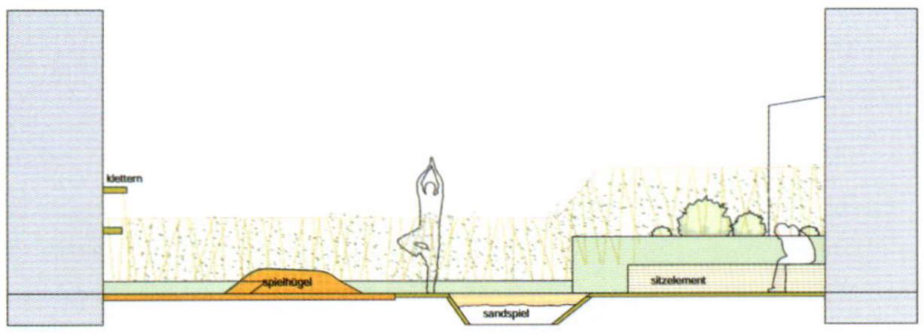

abtrennung kinderspielplatz - privatgärten 1_50

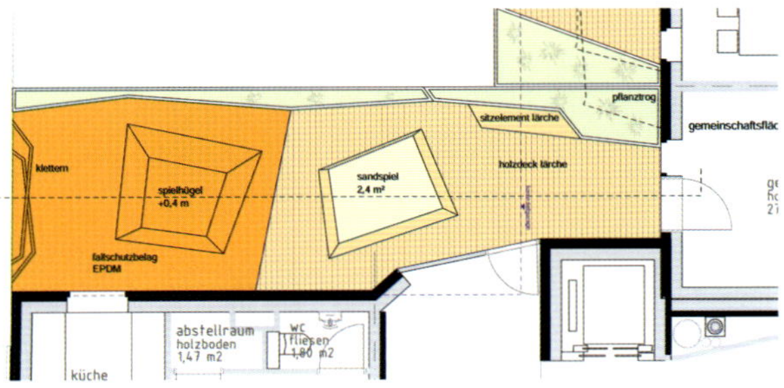

kinderspielplatz bei gemeinschaftsraum 1_50

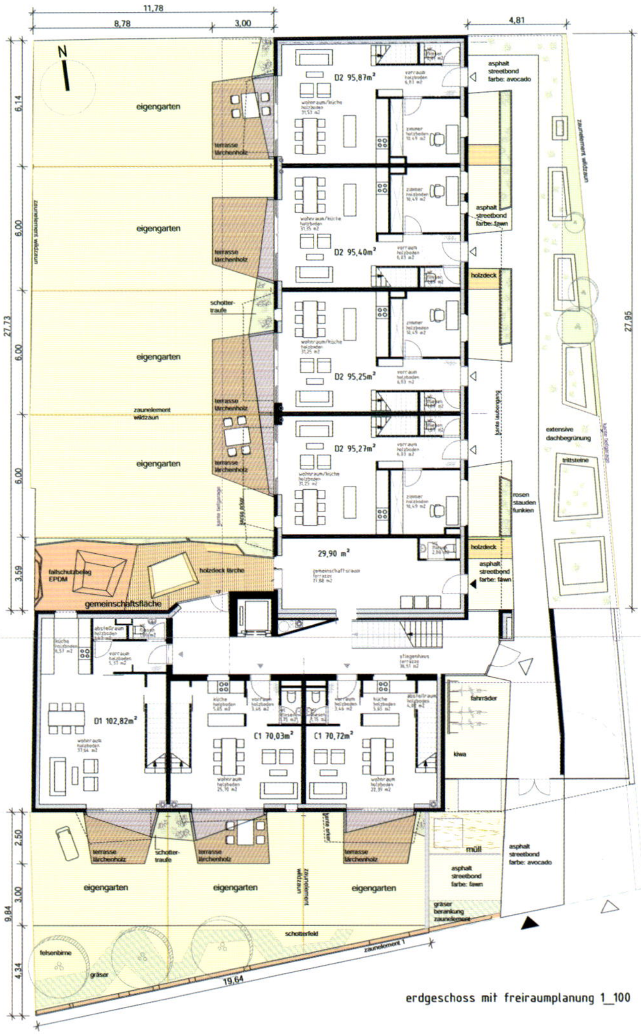

erdgeschoss mit freiraumplanung 1_100

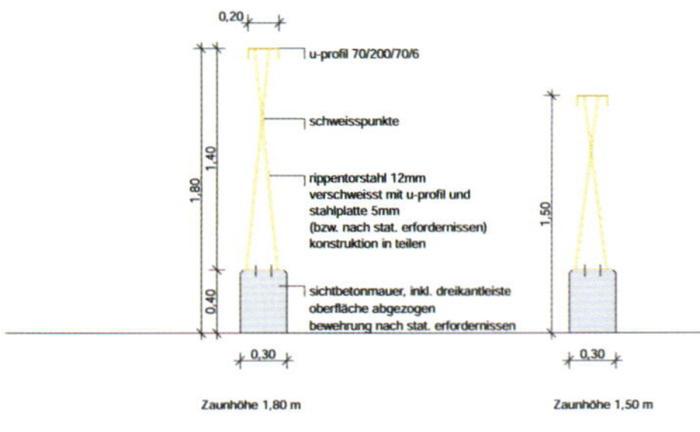

detailausbildung zaunelement

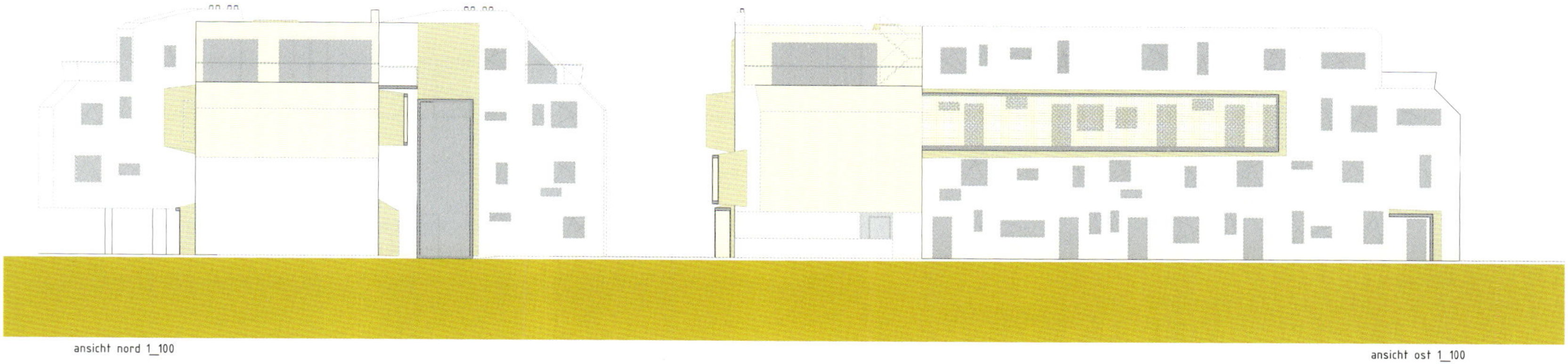

ansicht nord 1_100

ansicht ost 1_100

1

1. obergeschoss 1_100

2

2. obergeschoss 1_100

zaunelement zur strasse 1_100

ost

schnitt nord_süd 1_100

3

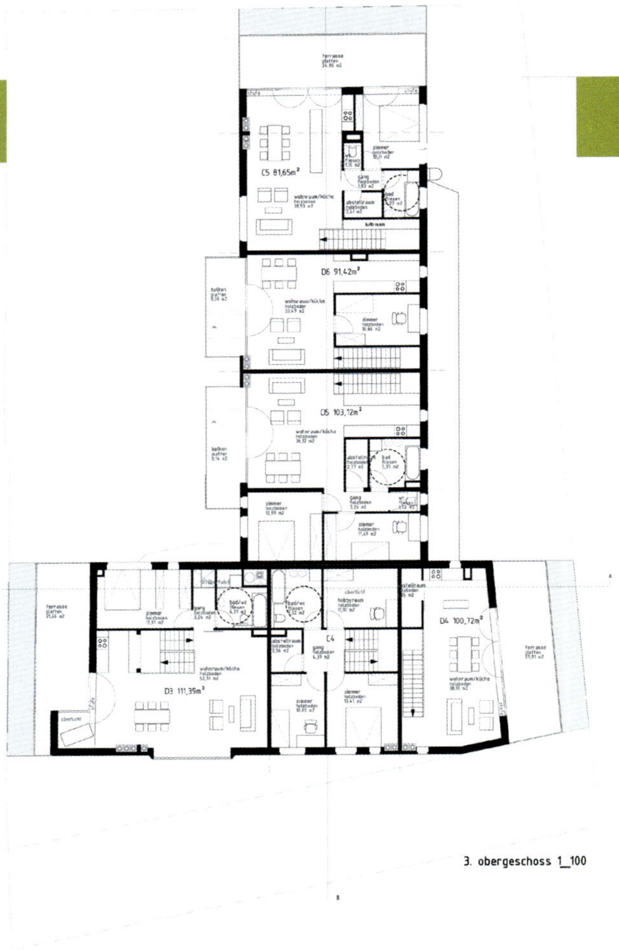

3. obergeschoss 1_100

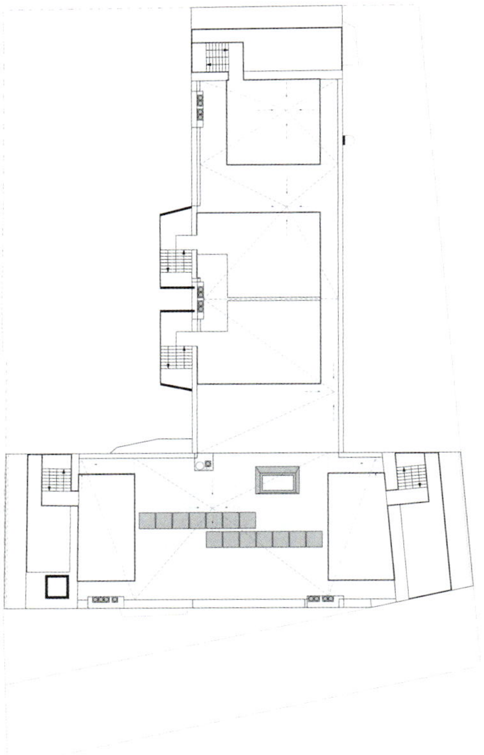

Architects: José Miguel Roldán, Mercè Berengué
Firm: Roldán+Berengué, arqts.
Collaborators: Zana Bosnic, Irma Arribas, Vicenç Sanz, Alfredo Garza, Núria Monfort
Constructors: Construcciones Caler, S.A.
Structure: Bernuz-Fernández Arquitectes, S.L.
Installation: Manel Comas i Javier Mateos
Technical architect: Enric Vijande
Location: Montcada i Reixac, Barcelona
Area: 3314m²

Masrampinyo

There was a 1989 book, Sigurd Lewerents 1885-1975. *The Dilemma of Classicism* published by the "Architectural Association Press".
The first two hands made book written by Peter and Alison Smithson, called *The Silent Architects*. From that "quiet society" with members such as Pikionis, Melkinov, Eames Architects, among others, now what would had happen if the article had been written after 1989.
Quiet architects, quiet buildings, sandpaper.
Is the sandpaper roof evocative as an echo of textures of the churches mortars? Or, is it speculative because expects venture from that quite architecture that finishes being rough and unlabeled? Is it difficult to swallow? Or it is made of sandpaper because the book wants to say that Lewerentz is among the architects who tryes to soften modern architecture and remove the hasty and brutal aspects of the newborn prototypes to achieve matt objects?
People believe that Smithson discoveres this guideline in the way of these architects and portrayed with them another way in the contemporary architecture.
All this come into it: in that scratchy roof finished giving him topic to the first scale model for the contest of small housing buildings planned for young people, that now has been finished in here. The project barely is a cubic volume cut in two pieces, to the right and to the left hand, following the law of the aisles of each floor, some big winfows that put into groups the smaller ones and a folded facede over the ground floor. A mask-dress made for a program with 42 minimum cells.
To group and break up to adjust its scale and to manipulate the size of the building until make it seem more intimate of what it is. A house of houses situated in the corner of a plot located on the outskirts and that always finish perceiving from foreshortened figures that pivot on the corners. Bronze coloured bricks, metallic black frames that mantaing floating this skin and show the white interiors of the niches.

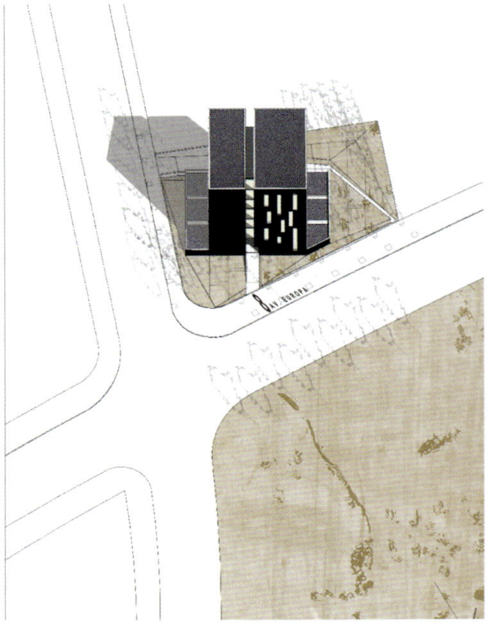

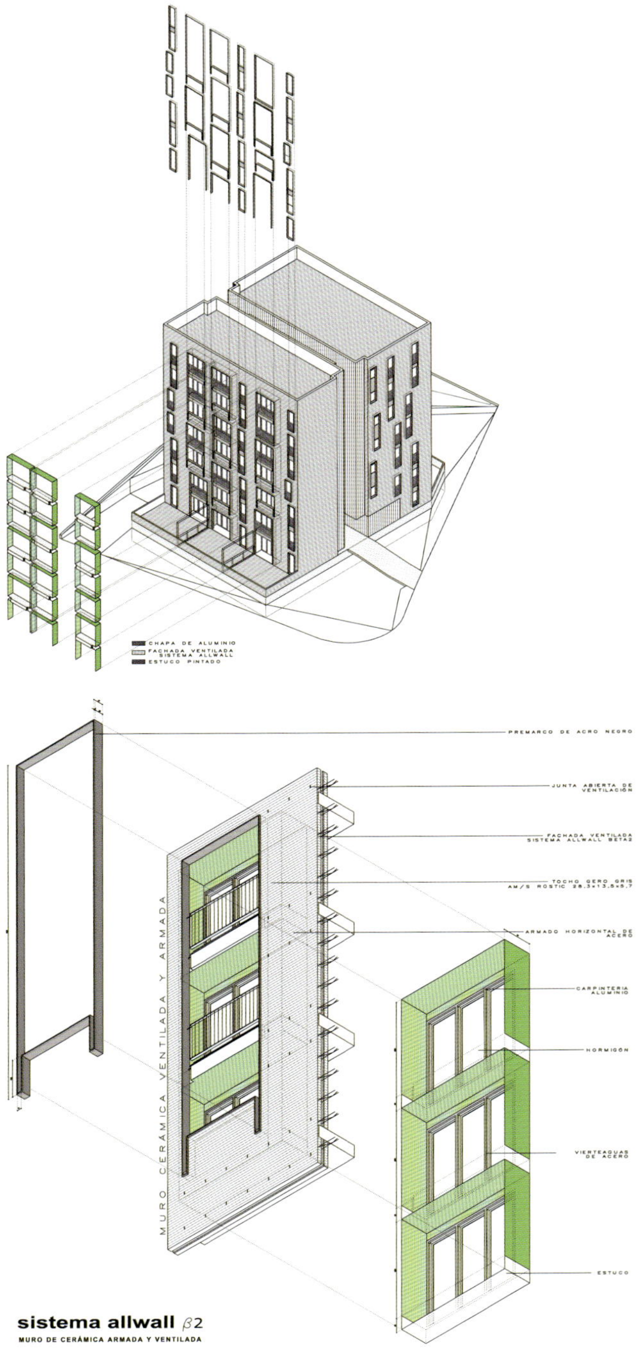

sistema allwall $\beta 2$

MURO DE CERÁMICA ARMADA Y VENTILADA

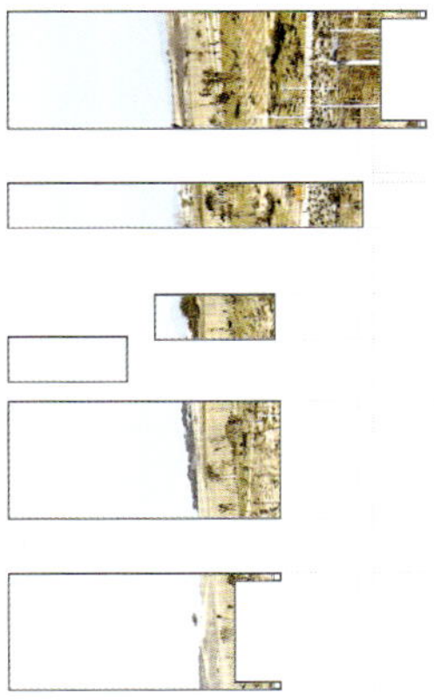

pD detalle

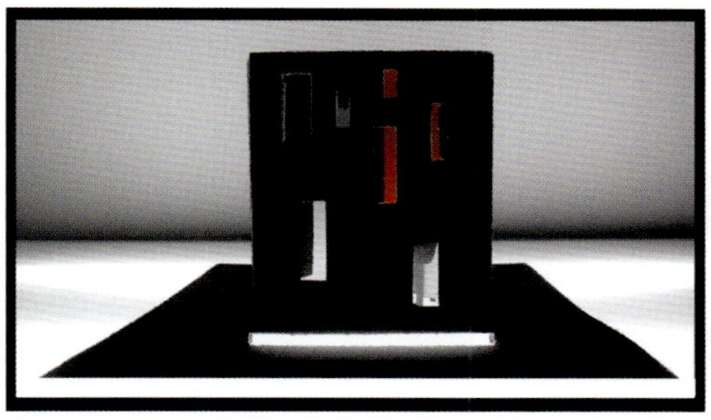

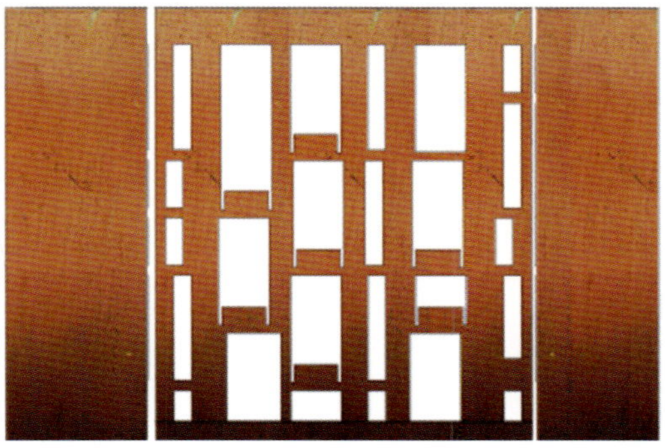

p B
BAJA

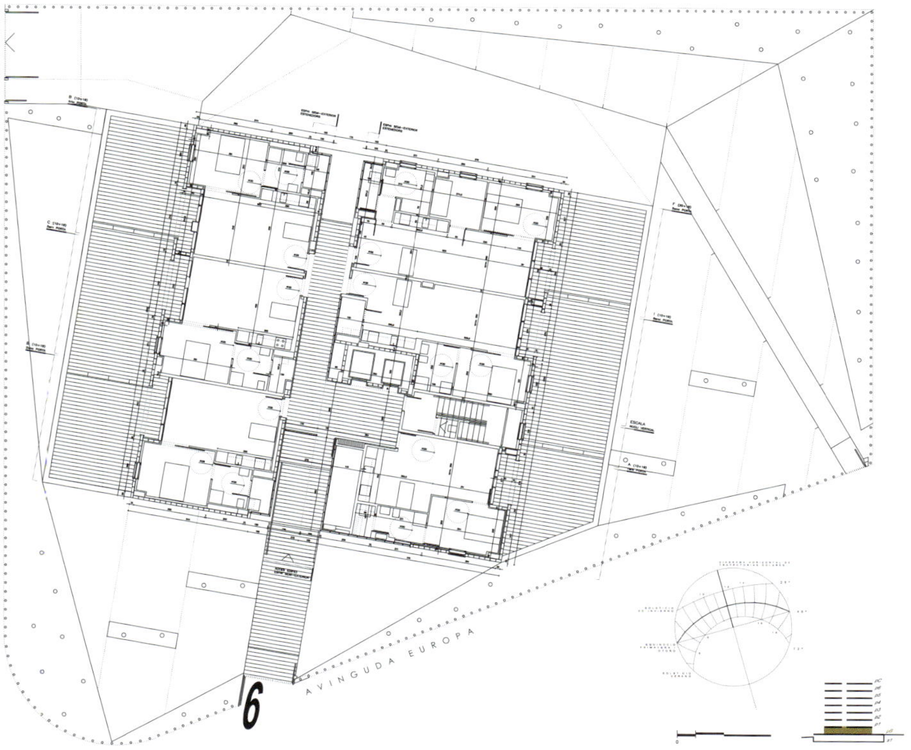

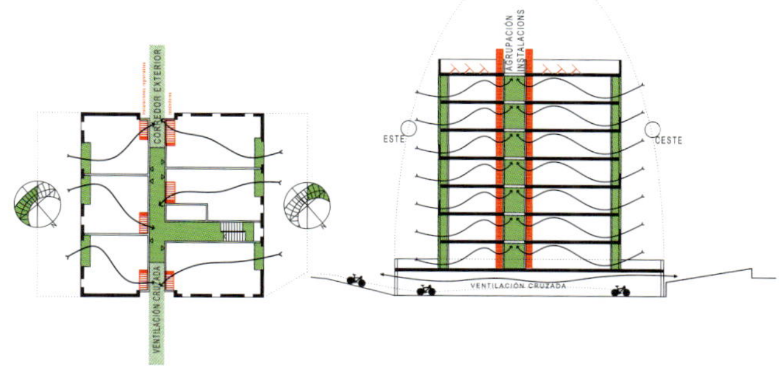

s 1
SÓTANO

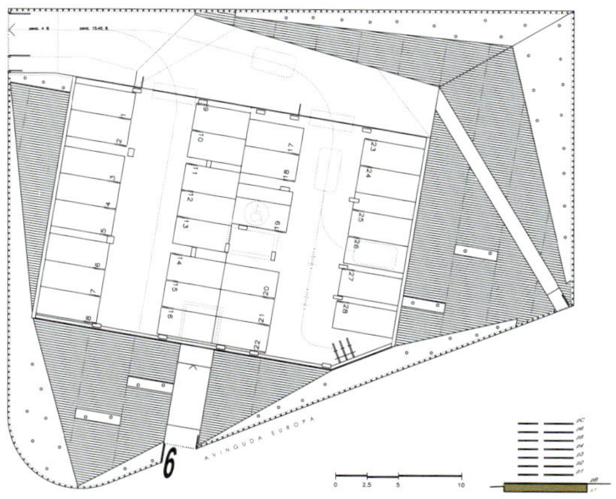

p T
TIPO

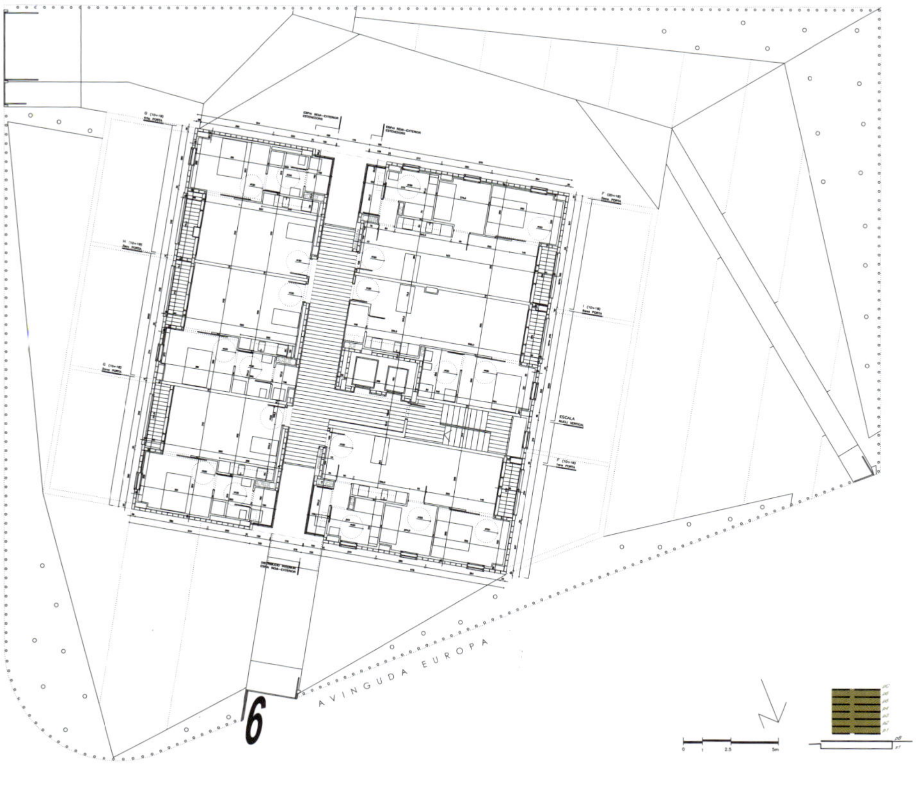

f N
NORTE

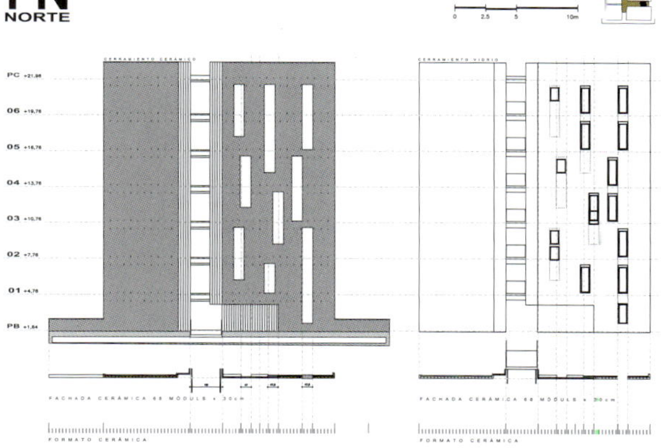

f O
OEST

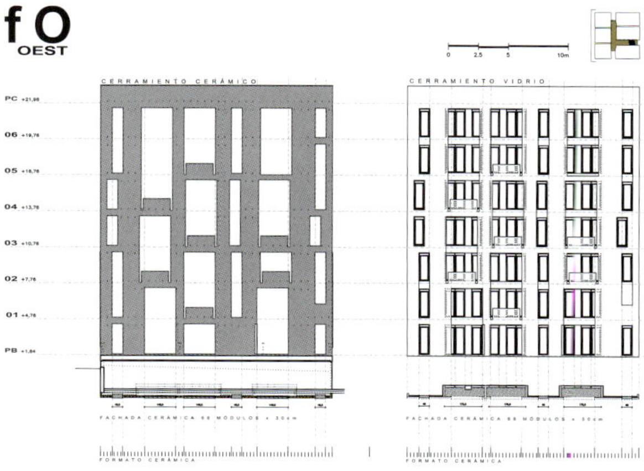

Architect: Igor Franić
Firm: Studio za arhitekturu d.o.o.
Collaborators: etc.
Contractor: Zagrebgradnja d.o.o.
Structure: Tehnicke konstrukcije d.o.o.
Mechanical installations: SM – inzenjering d.o.o.
Electrical installations: Elektro-ekspert d.o.o.
Sewage and water supply: SM – inzenjering d.o.o.
Location: Zagreb
Area: 19927m²

Business and Residential Building in Nova Cesta

The verge of the very city center encourages free defining of the content and the intensity of activities. The proximity of the railway, a sports center with a large parking lot, and a block structure with residential verticals provide the entire area with pronouncedly urban character. The new hybrid structure is complex and multilayered, but unfortunately limited in height by the brief. The interweaving of amenities is realized in the spatial interplay with the street, back-yard structure of city roofs, and indented verticality of residential floors, which stretch the directions and views over and between the existing composition, highlighting the voids.

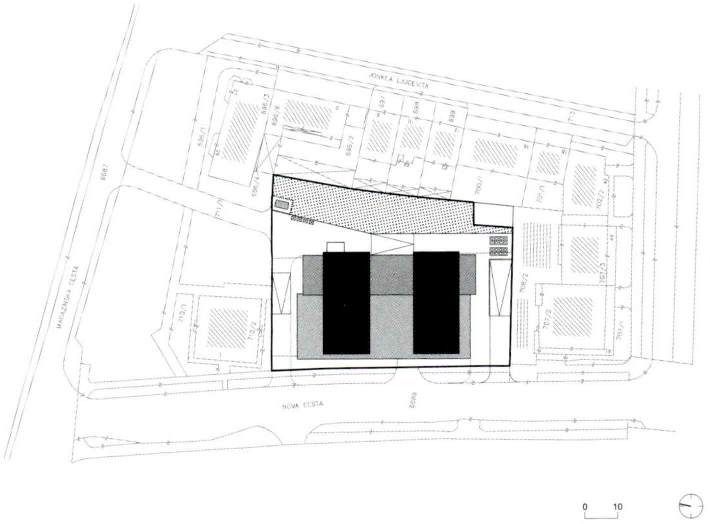

ZG 3175 ML
ZG 5093 DK
ZG 3808 DH

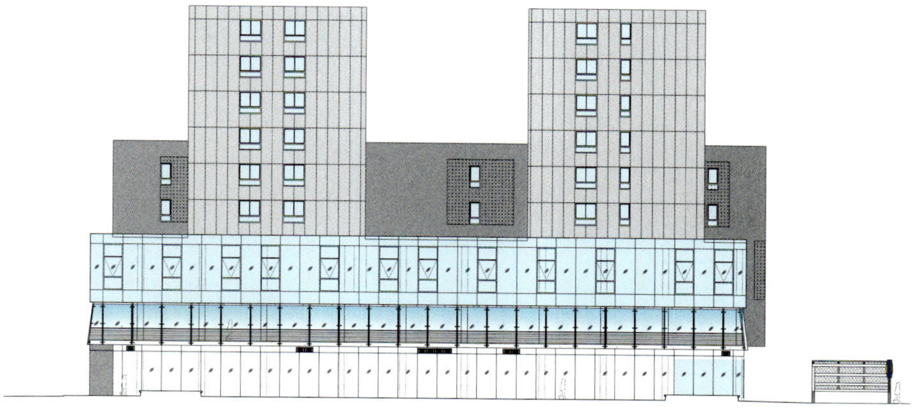
West elevation

North elevation

East elevation

North elevation

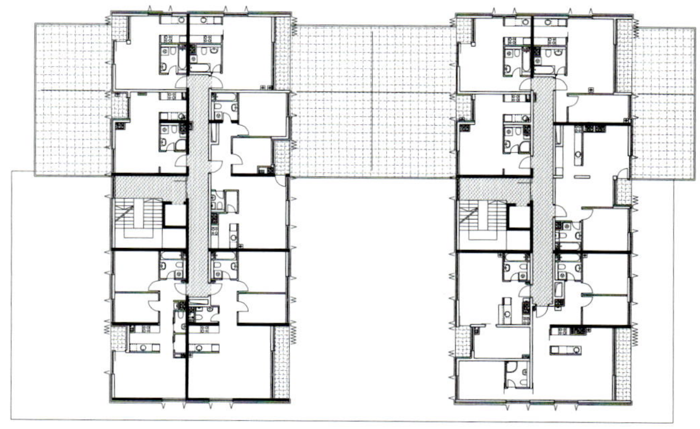
tipical floor plan

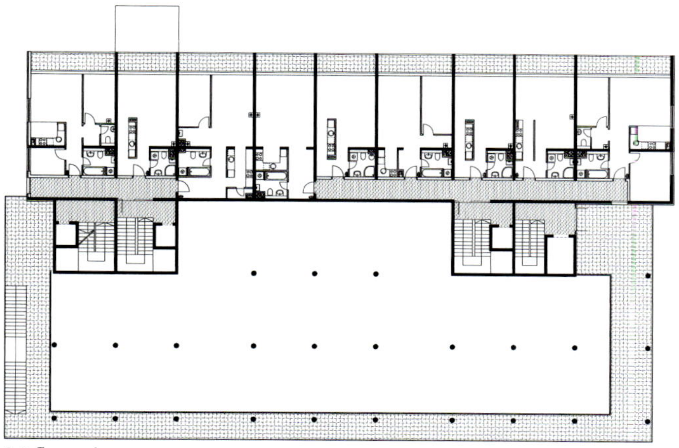
1st floor plan

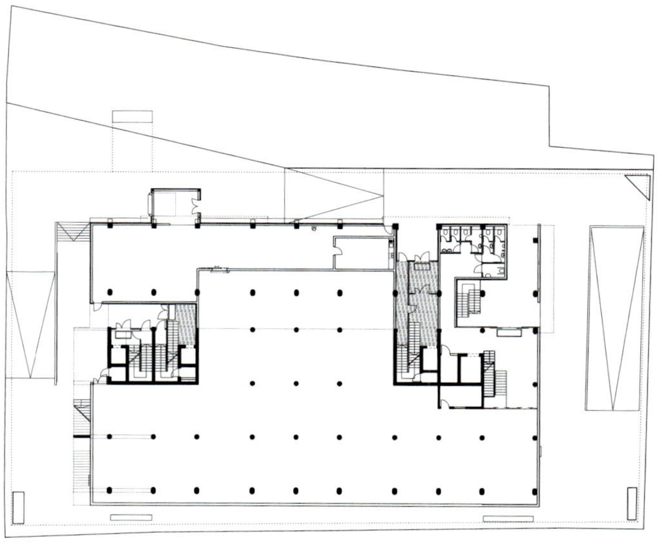
Ground floor plan

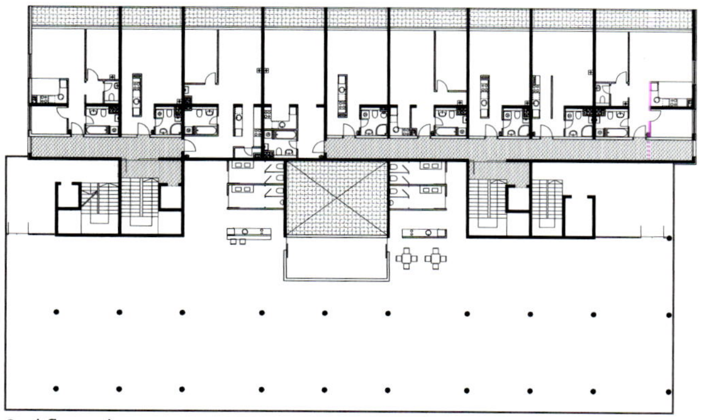
2nd floor plan

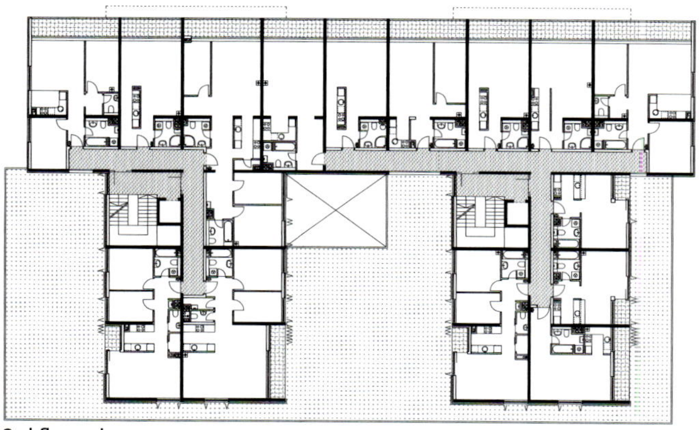
3rd floor plan

Architects: Jan Pallado, Aleksander Skupin
Firm: Biuro Projektów Architektonicznych
Location: Katowice, Poland
Area: 1074m²
Photographer: S.Polański

Debowa Housing Estate

Dębowa social housing estate is located in the west part of Katowice, in the district Dąb, near the Voivodship Park of Culture and Recreation in Chorzów. The development consists of two buildings; one residential & commercial, complementary to the west frontage of the Dębowa Street and the other for residential use only, located further back within the plot.
The residential & commercial building is attached to an early 20th century corner town house and honors its building line, but also accommodates itself to the curve of Dębowa Street by creating shape reversals accordingly. The building is raised on the southern side, providing a visual counterbalance to the high attic of the old corner town house. The residential building situated further back within the plot offers similar reversals as the front one that helps to exhibit a historical villa enclosing an urban courtyard from the south.
The housing estate consists of 85 flats, mainly of one bedroom flats ranging from 42m² to 46m² as well as some studio and two bedroom flats. Large three bedroom flats have been located in the raised part of building and it is designed for family with numerous children.
A limited budget determined use of simple means in shaping the urban courtyard, dominated by green car parks and formation of new buildings with colorful parts of walls and uninhibited windows—counterbalanced to the monotony of repeatable flats.
This composition is supplemented by floral and paper bird motifs on the front facade.

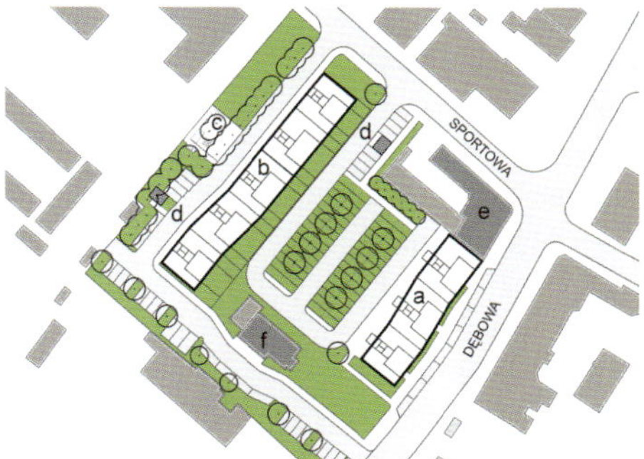

Site plan:

a - residential commercial building, **b** - residential house, **c** - playground, **d** - refuse storage, **e** - 20th century corner town house, **f** - historical villa

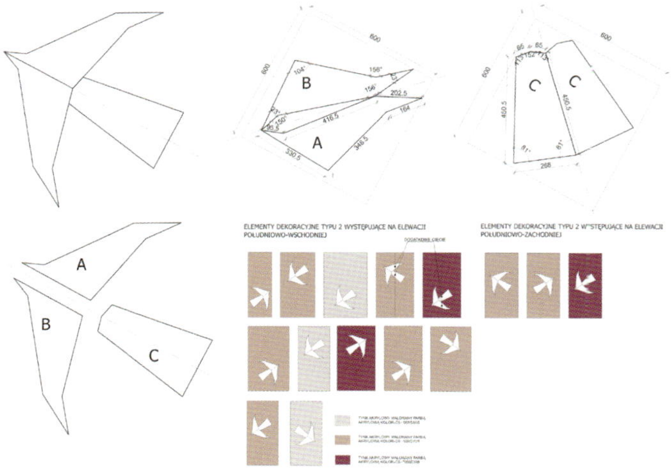

DĘBOWA STREET FASADE

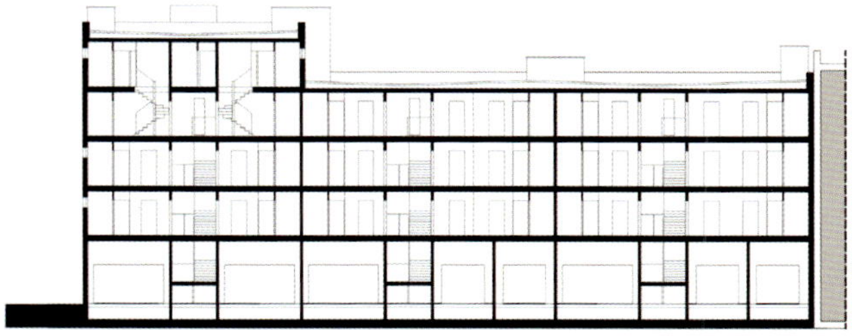

SECTION a-a

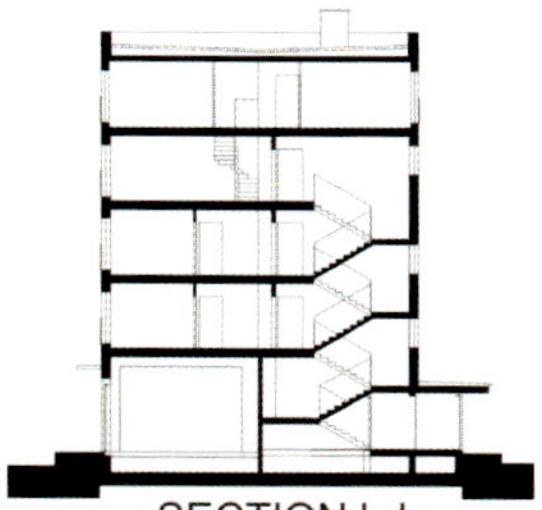

SECTION b-b

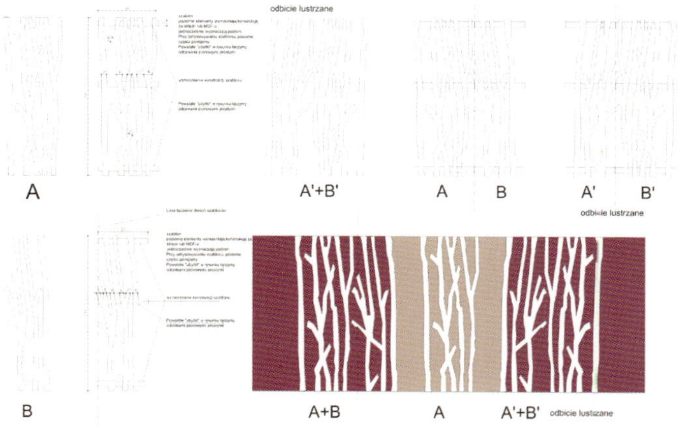

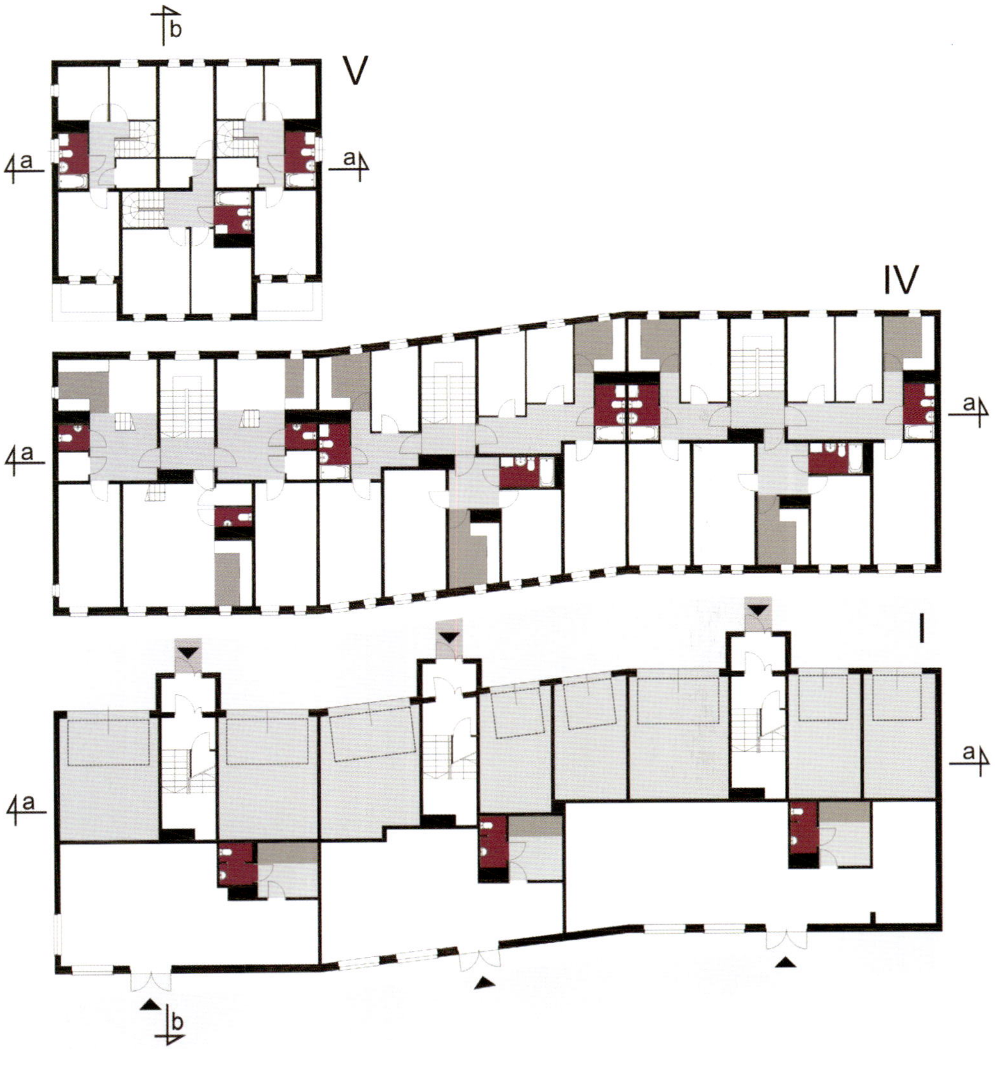
b
V
a
a
IV
a
a
I
a
a
b

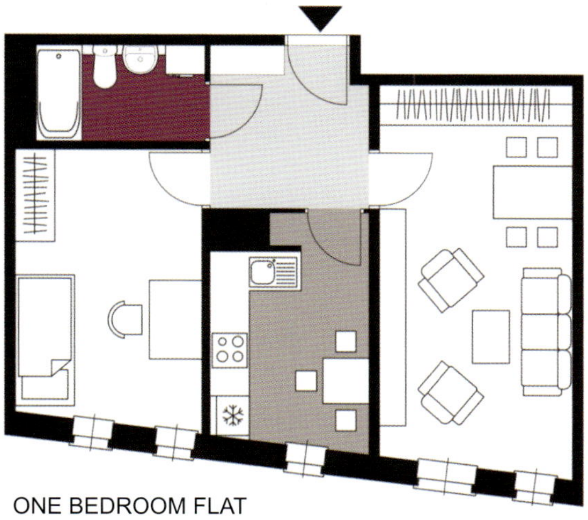

ONE BEDROOM FLAT

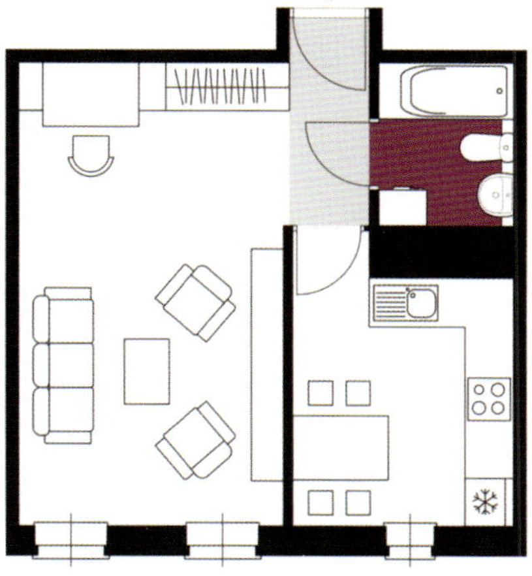

STUDIO

THREE BEDROOM FLAT

Architect: Leen Borst, Mark Snitker
Firm: Snitker Borst Architecten
Contractor: Du Prie Bouw en Ontwikkeling
Structural Engineer: Pieters bouwtechniek Haarlem
Location: Roomburg, Leiden, the Netherlands
Photographer: Roos Aldershoff, Dennis Sies

48 Single-Family Houses

In Roomburg SNITKER BORST ARCHITECTEN intended to make a repetitive housing project with a larger ceiling height and a spatial section. These qualities are not often seen in this segment of the housing market. The office does research into the possibilities to create larger ceiling heights, especially in apartment buildings. The research consists of the analysis of realized projects and the development of new spatial models.
The project is part of the new residential area Roomburg in Leiden, the Netherlands. The houses are situated along the Octavialaan, which is the northern border of the subarea 'Landscape'. This is the last stage in the development of Roomburg. The façade wall is 400 meters long and it is divided into three blocks.
The project in Roomburg contains 48 single-family houses with a floor space that varies between 158m² and 175m². The project consists of five housing types with alternating terraces. A subtle linking of these five housing types results in a vivid façade with rhythm, relief and sculptural expression.
The cubist architecture from the twenties and thirties of Dutch architects like J.J.P. Oud, Jan Wils and J.B. van Loghem was a reference for the project. The project has a small relief in the plinth and a more expressive relief in the upper structure. The plinth and the upper structure are made in two kinds of brickwork. They are separated by a zone of windows and prefabricated concrete elements on the first floor. The façade recedes from the building line and the top floor terraces provide for extra space and light in the street.
The houses have a split-level floor on the first floor. The split-level generates a living room with a ceiling height of 3.4 meters. Also the rooms on the first floor at the street side have larger ceiling heights. The split level floor creates a spatial staircase that makes the daylight fall into the heart of the house.

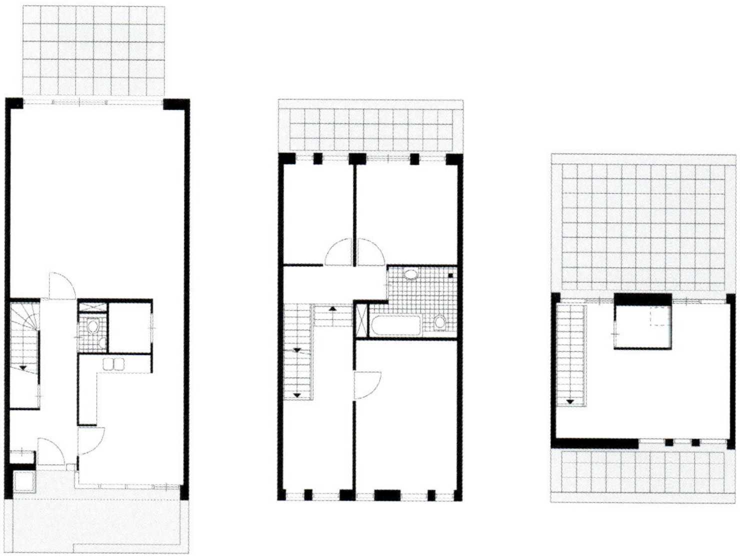

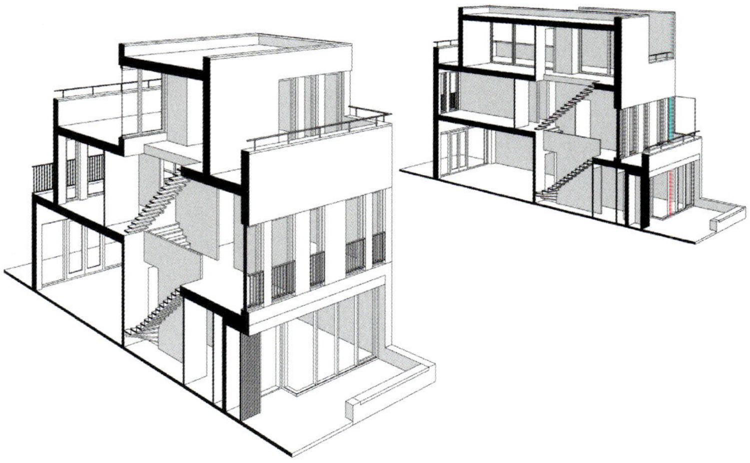

92
94
96

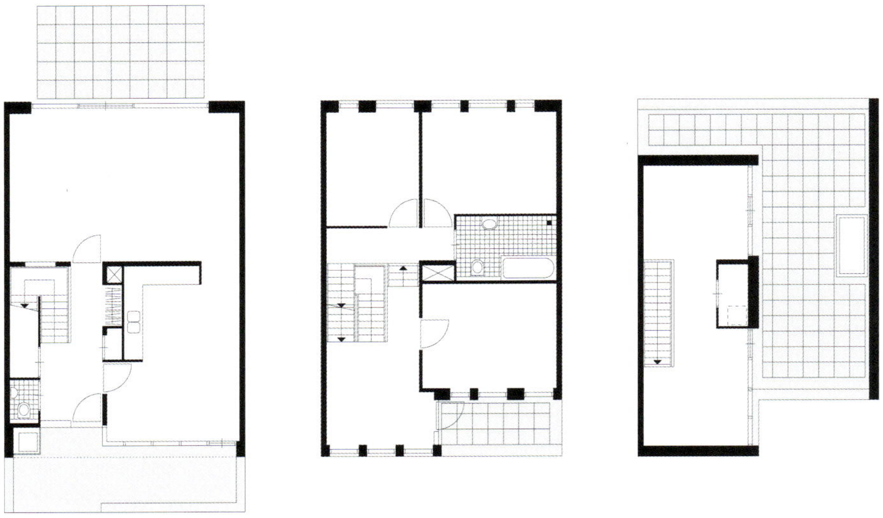

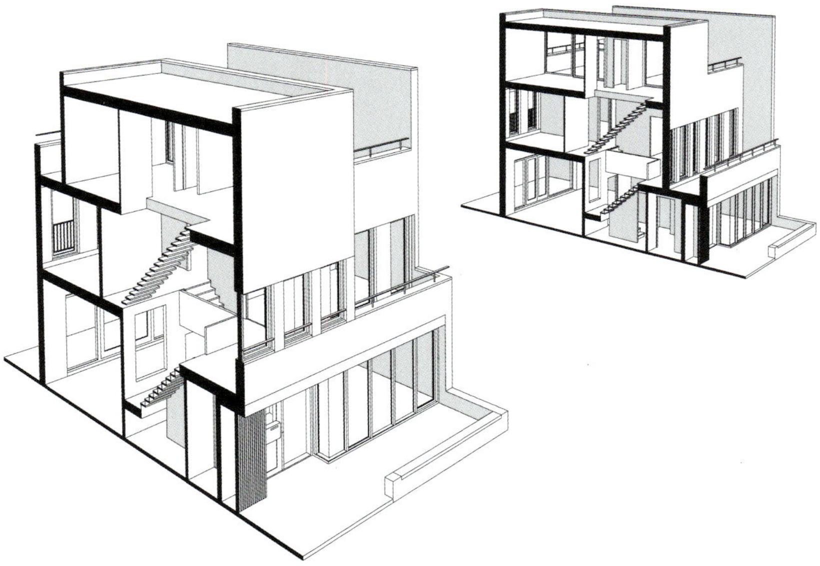

Architect: Biuro Projektów Lewicki Latak
Firm: Biuro Projektów Lewicki Latak
Principal designer: Kazimierz Latak, Piotr Lewicki
Design team: Begona Herrera Rodriguez, Piotr Knez, Mateusz Manecki, Agnieszka Zajaczkowska, Zbigniew Piotrowicz, Bartosz Pionka, Anna Rytter, Widold Opalinski, Kinga Raczak, Hannelore Elias
Location: Wroclaw, Poland
Area: 20461m²

Corte Verona

In a close vicinity there are examples of using that material both while constructing buildings of the same function and in creating the so-called small architecture objects: walls, benches and fences. An open-work brick wall in the neighbourhood became a direct inspiration of shaping the structure of the new architecture by the authors. Bricks were rescaled to a size of one flat and the gaps between one another became loggias.

The building was given a well-known, well-functioning form of an urban block with a division of the space into: public (outside), private (flats) and semi-public (the courtyard).

The ground floor houses services, available from the outside (from the side of the streets), above there are flats, available from the courtyard through entrance halls and below the ground surface there is a garage, ensuring a high ratio of parking space.

Small flats (45m²) consisting of a living room with a kitchen, a bedroom, a bathroom and a small cloakroom together with a relatively large loggia of ca. 8m² are prevailing in the building. Thanks to the loggias the authors could give more light to relatively long flat tracts by using corner windows.

The structure, of reinforced concrete, was designed as a set of transverse vertical walls, which, while separating particular flats, isolate them acoustically and horizontal slabs which are elements of the floors.

All installations are run in shafts in longitudinal walls, isolated acoustically from the side of the flats and the corridor by two independent layers of the wall.

The façade, in the ground floor part there is silicate plastering together with aluminium doors and windows. Above there are ceramic tiles, placed interchangeably on each side of the quadrangle together with wooden windows and glass balustrades.

The space of the courtyard, used only by pedestrians, was treated with special attention, planning its composition including green areas, paths, ramps and stairs supplemented by a sand pit, swings, see-saws and benches.

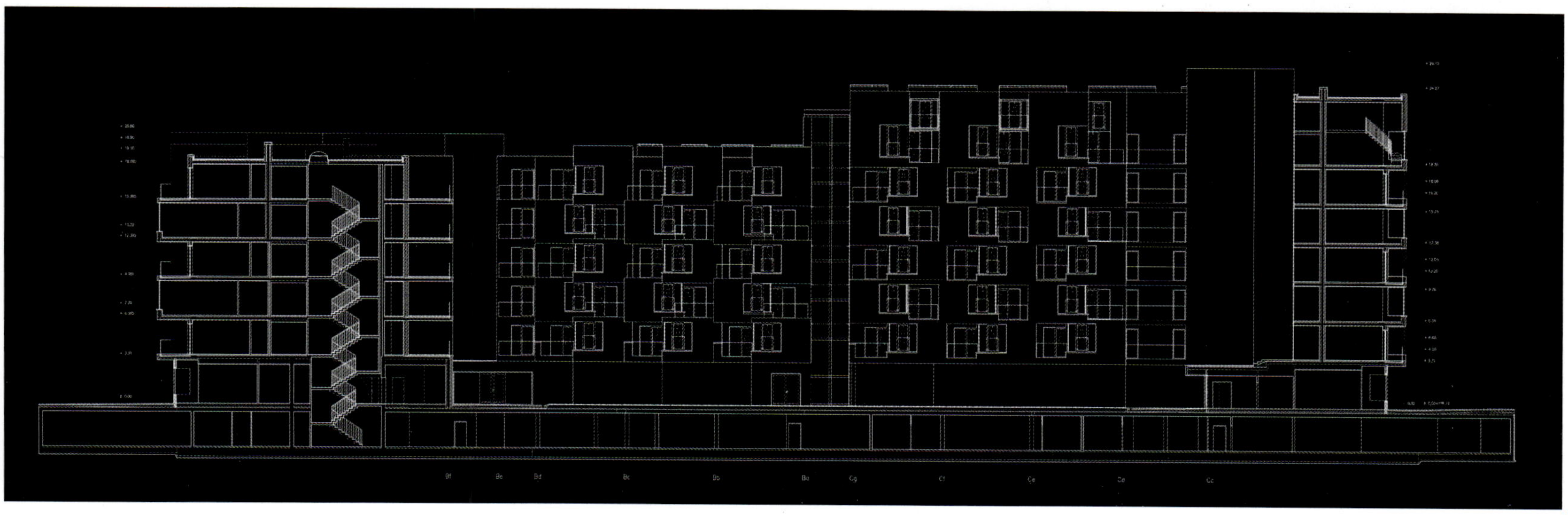

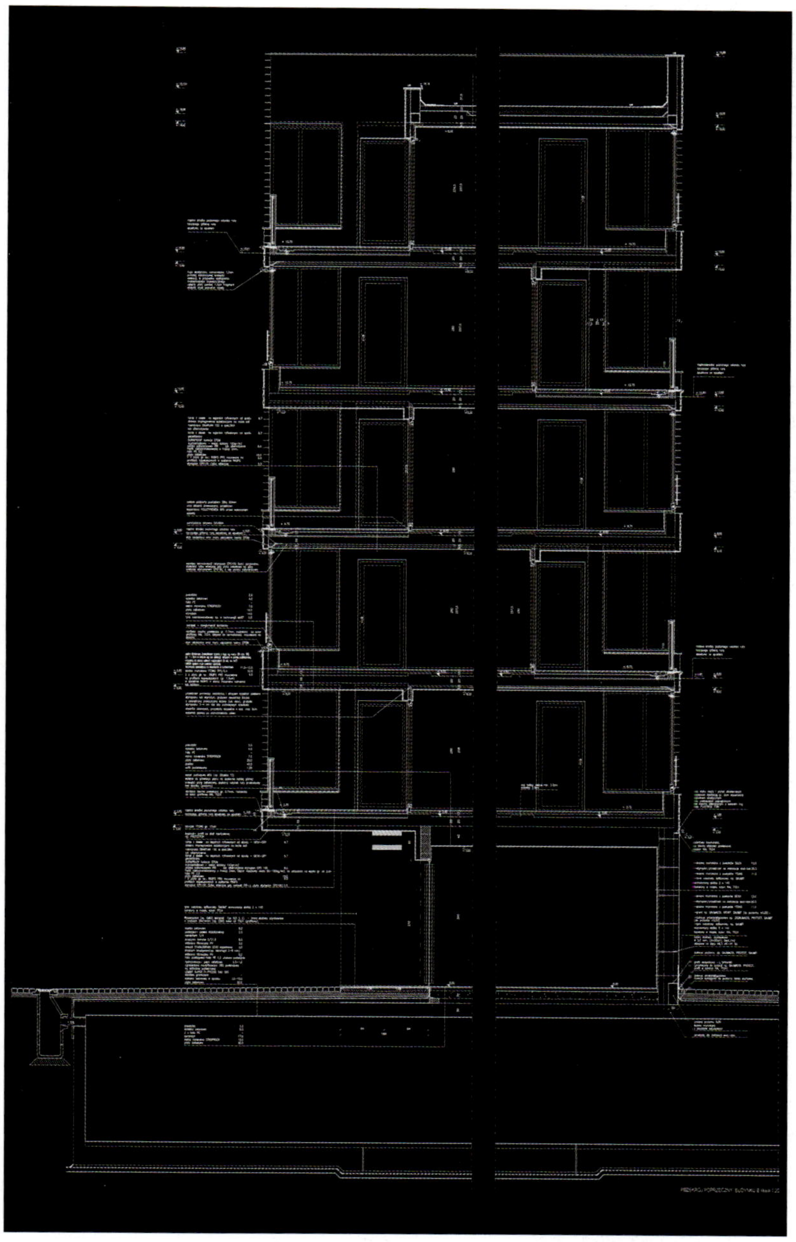

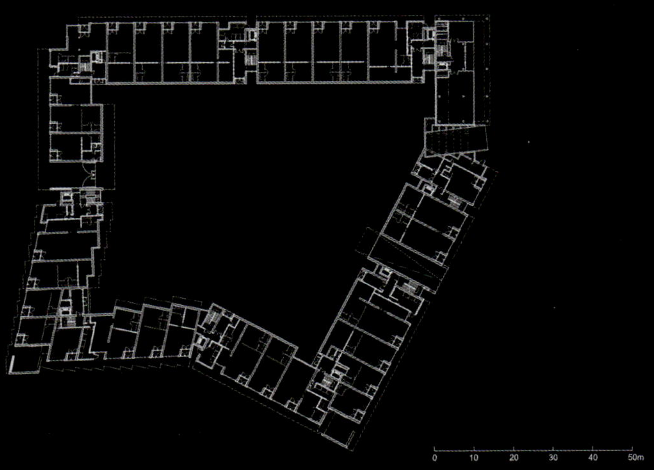

RZUT PARTERU

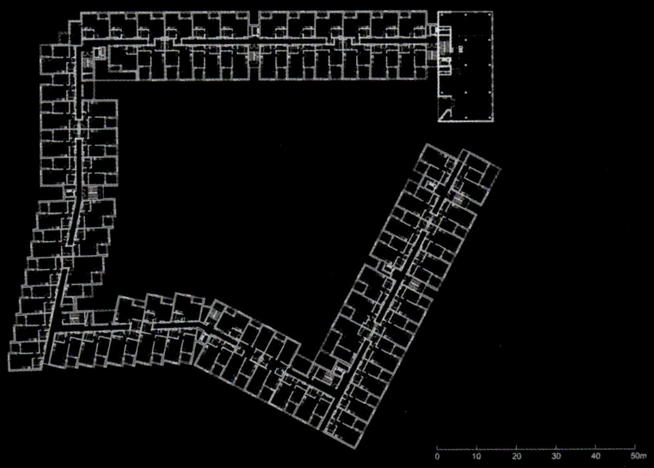

RZUT PIĘTRA: 1, 3, 5

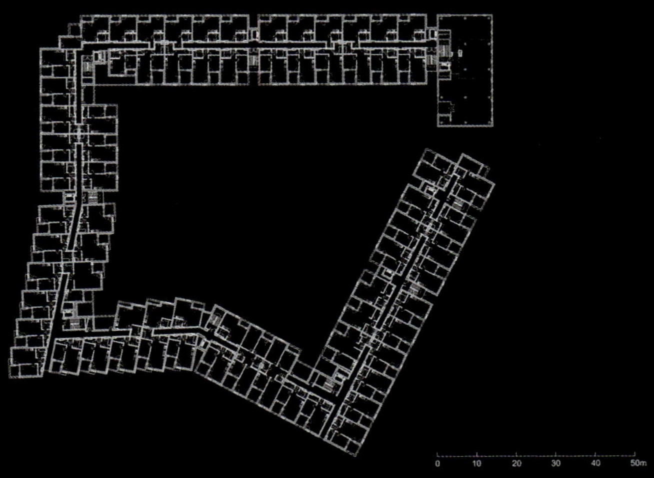

RZUT PIĘTRA: 2, 4

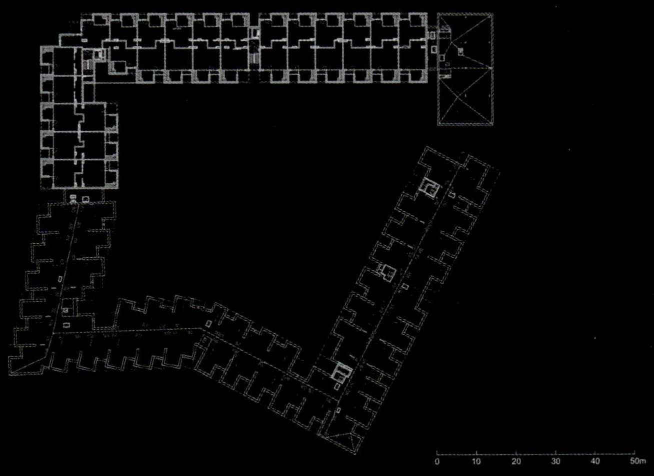

RZUT ANTRESOLI

Architects: Vojteh Ravnikar, Robert Potokar, Marjan Starič, Petra Slukan
Firm: Ravnikar Potokar architectural office
Competition designers: Vojteh Ravnikar, Matjaž Bolčina, Nika Grabar, Ernest Milčinovič, Teja Savelli, Jure Zavrtanik
Structural engineer: Primorje d.d.
Mechanical installations: Pinss d.o.o.
Electrical installations: Elita i.b. d.o.o.
Contractor: Primorje d.d.
Location: Nova Gorica, Slovenia
Area: 8200m²
Photographer: Peter Krapež

Residential Building C

In line with the Decree on the Location Plan for the site along the railway station in Nova Gorica, Phase 2.a of the construction encompasses the construction of three 11-storey tower blocks, beneath which the construction of an underground garage is foreseen on the entire site of this phase.

The site of residential building; Tower Block C is located in the future building complex of the residential area Majske Poljane along the railway station in Nova Gorica. Within this phase of construction, three tower blocks will be erected along Prvomajska Street.

Architectural Design

The architectural design of Tower Block C is derived from the competition design, as outlined in the study submitted to the competition in 2002 (which was subsequently awarded first prize in the competition) and in the adopted Decree on the Location Plan. The basic idea lies in the layering of two stories by displacing their floor plan – the elementary square of the floor plan is moved around the vertical core thus creating overhangs on which terraces are placed. The vertical line of the residential tower is cut up and transformed into horizontal layering, providing the tower with the necessary dynamics.

The common vertical building height with a ground floor and ten stories also includes a gallery on the ground floor, while the 10th floor houses duplex flats. The ground floor and gallery are intended for public use, for various service providers and office activities, and for grocery stores and other shops, while the upper floors are intended as flats (the tenth floor features two-storey duplex flats). The building has two basements intended for technical or maintenance facilities, storage rooms belonging to the flats and parking spaces for cars.

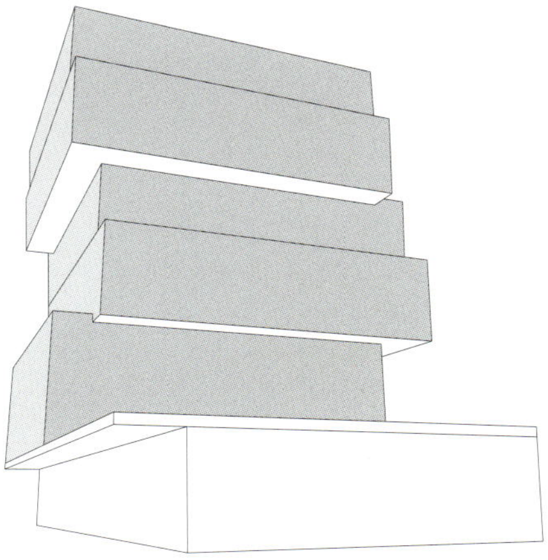

volume visualisation

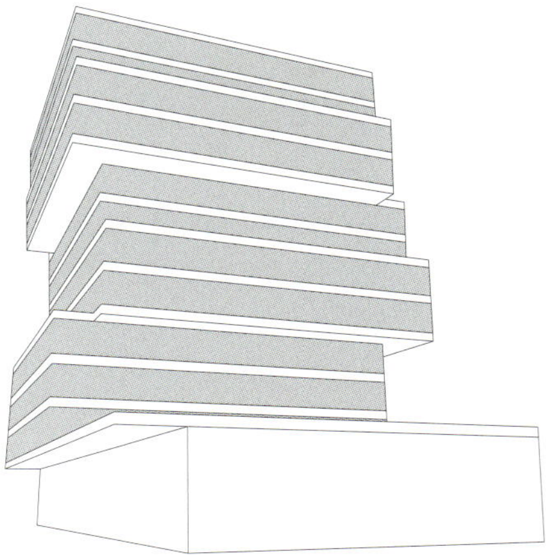

projection of plates

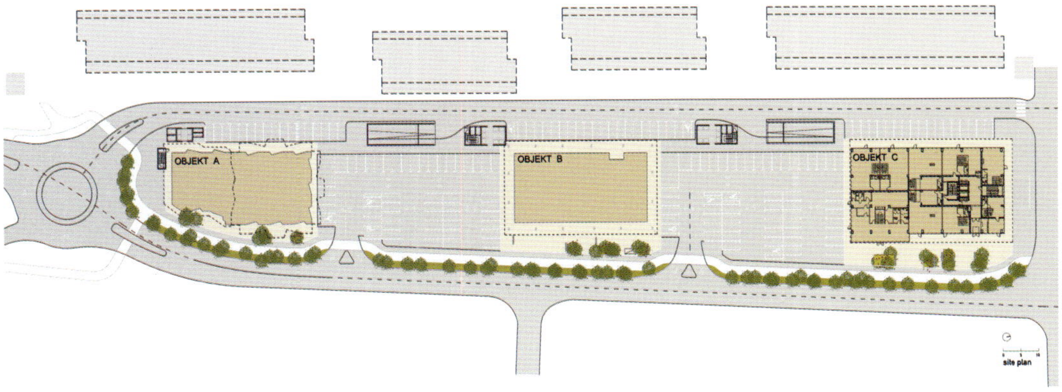

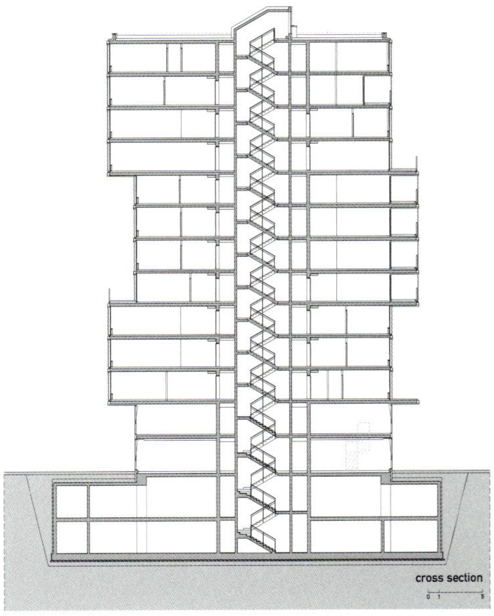

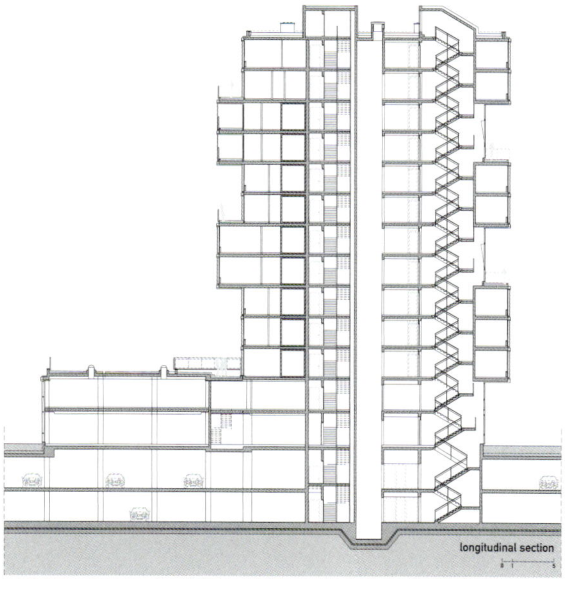

longitudinal section

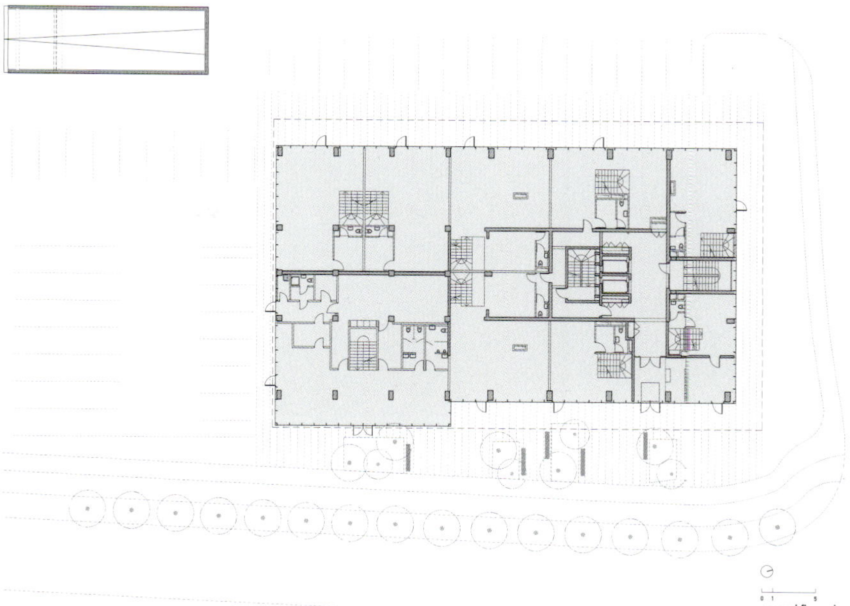

ground floor plan

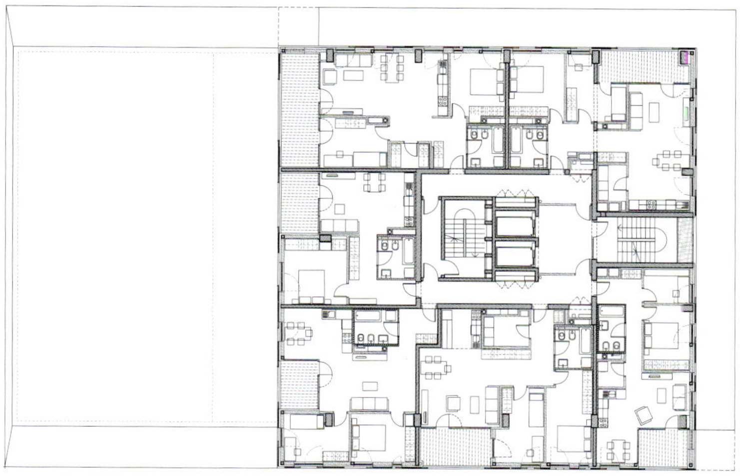

5th floor plan

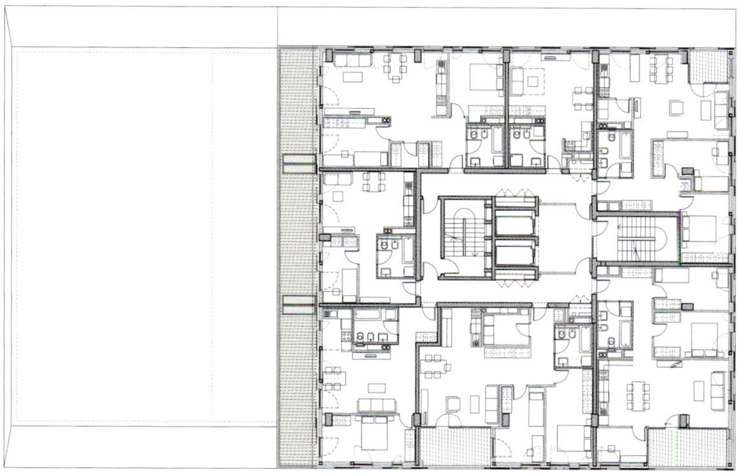

6th floor plan

Index

Stefano Boeri Architetti

Based in Milan, Stefano Boeri Architetti is committed to the research and practice of contemporary architecture and urbanism. Recently completed projects include the recovery of the Ex Arsenale in La Maddalena, Sardinia and the new RCS - Corriere della Sera Headquarters (work is now under way on the remaining buildings of the competition master plan). Projects currently under way include the design for a multifunctional building for the PACA Region on the Marseilles Waterfront (Center Régional de la Méditerranée) scheduled for inauguration for 2013, Marseilles European Cultural Capital; two eco compatible residential towers (il Bosco Verticale) in Milan are scheduled to begin construction at the beginning of 2010; the project for the Policlinic of Milan, a large scale renovation of one of the city's most prominent medical centers; research on new prototypes for social and sustainable housing; and the design the spaces for the new Italian private train transport company, NTV.

3+1 Architects

3+1 Architects is a Tallinn based practice committed to architectural design and urban planning as well as researching and analysing contemporary processes in both fields. The office was founded in 1994 by three architects – Markus Kaasik, Andres Ojari and Ilmar Valdur – all graduated from the Faculty of Architecture, Estonian Academy of Arts in the same year, 1992. All three are teaching architecture at the Academy of Arts as well.

Right from the start 3+1 architects have actively taken part in architectural competitions in Estonia as well as abroad.

3+1 Architects distinct strategy for competition projects as well as for other works is to cooperate with other architects outside their own. The variation of a project team is more project-centerd and flexible way of dealing with the ever-different design problems.

AART Architects

AART Architects is an international architecture company founded in 2000 with the vision to develop the Scandinavian architectural tradition by creating vibrant, sustainable physical surroundings. AART Architects has offices in Denmark and Norway and is a high-performance team of 50 young designers and technicians that deliver full-service consultants in the field of architecture. The projects of AART Architects are rooted in the essence of the site and characterized by a strong environmental and social awareness.

As architects we believe in the concept of "total design" and collaborate closely with clients, stakeholders and users to produce multi-disciplinary solutions that focus on the environmental, social and economic aspects of every project.

In other words, they believe that long-lasting architecture grows out of a profound understanding of the unique qualities of the specific site, cultural context and client's vision. Therefore they study the aesthetic, social and technological possibilities of every project in an entirely holistic way in the desire to develop the concept of sustainability and meet the needs of the present without compromising the ability of future generations to meet their own needs. A value based approach which permeates their design process and ensures a beneficial life-cycle economy.

Whether it's about complex masterplans, landmark buildings or inspiring interior design their prevalent goal is to create high-performance projects that set new standards of the sustainable architecture of tomorrow.

Since 2000 they have made their mark internationally with 1st prize competition projects such as Østfold Hospital and Østfold Science Center in Norway as well as notable Danish projects such as the Culture Yard in Elsinore, VUC Headquarter in Haderslev, Skagen Skipperskole in Skagen, Bikuben Student Residence in Copenhagen and the "Mies van der Rohe Award"—nominated Outdoor Center Sletten in Ry.

Zecc Architecten

Zecc is not an abbreviation, but it is a name. Zecc is inspired by the French word Sec: pure, simple, clear. Sec as a strong foundation, touched up with an extra 'c' and a spark of sensuality as expressed by the Z. Zecc equals clear & sparkling.

The strength of a tree equals the depth of its roots. Their architecture is the same. Zecc works from three ground principles: what they build is functional, sustainable and engages all senses at the same time. Uniting these three principles often seems an impossible assignment. However, this is the core of Zecc's strength. They call it Grounded Architecture. Architecture with a strong foundation. With deep roots. Since these bring forth the most wonderful ideas.

Zecc as a reliable and driven party. This side they call the "Zecc Firm". All employees have a constructing background and think about how it's made during the whole design process.

Besides the formal side they also know an informal side they call "the Zecc Family". In a young team they work with energy and enthusiasm.

Ian Bogle

"I'm an architect. Don't label me as a specialist. It's like cooking: I play with ingredients to create," says Ian Bogle who firmly believes that the design process should be generated by context, not the desire to impose a particular BFLS style.

After joining BFLS in 2002 as an associate director, Ian Bogle led two pioneering projects: Strata and Park House, the former now complete and the latter scheduled for completion in 2011.

Ian studied architecture at the Mackintosh school of Architecture, and subsequently worked for the Parr Partnership in Glasgow where he qualified as an architect in 1993. After joining Foster + Partners in 1995, he worked on projects including the Scottish Exhibition and Conference Center, the Dubai Cultural Center and Farnborough Business Park. As Project Director, he was responsible for London's Swiss Re Headquarters and the Bexley Business Academy, buildings which won and came second in the 2004 Stirling Prize respectively.

Ian is an active ambassador for BFLS abroad and oversees the practice's Prague office where he is responsible for a large masterplan at the airport and the mixed use urban masterplan at Zizkov. He also has a special interest in large Education projects such as the ELI Laser Research facility due on site in 2011

Marlies Rohmer

Architectural Office Marlies Rohmer is an Amsterdam (NL) based practice. Its field of operation and its twenty-five years' experience ranges from urban renewal and master planning, housing projects, school buildings and care centers to interior design.

With a staff of approximately fifteen employees. Architectural Office Marlies Rohmer is a network-based organization in which various disciplines are integrated and enrich one another. The design process is always driven by research into a variety of social and cultural phenomena. The firm is well equipped to carry projects through from start to finish, all stages of the process. Every assignment is approached as a unique project, demanding specific solutions. The results are a powerful and recognisable architecture with a strong detailing. Since 1999 Marlies Rohmer carries out a research into youth culture related to architecture and public space, resulting into the book 'Bouwen voor de NEXT GENERATION' (Building for the NEXT GENERATION, NAi Publishers, Rotterdam NL 2007). The firm got several awards, among which the Dutch School Building Award 2002 and 2008, the International School Building Award 2003 of the Buenos Aires International Forum of Educational Architecture, and the Golden Pyramid 2009.

KCAP Architects & Planners

KCAP Architects & Planners is a Dutch office for architecture, urban design and urban planning, founded by Kees Christiaanse in 1989. During the last 20 years KCAP has established itself as one of the leading international practices in the fields of architecture and urbanism.
Working on an international level from the start, KCAP has designed and realized projects in major countries and cities all over the world such as Hamburg, Zurich, Perm, London, Beijing and Amsterdam. The office gained international recognition with projects such as Kavel 25 housing block in The Hague (NL), Wijnhaveneiland and the Red Apple in Rotterdam (NL), GWL district in Amsterdam (NL), Hafencity in Hamburg (DE) and the Perm 2020 Strategic Masterplan in Russia. Recently KCAP won competitions including IAK office ensemble in Luxembourg, the Bay of Passaia in San Sebastian (ES), NEO Brussels (BE) and the office building on Ile de la Jatte in Paris (FR).

ECDM

Emmanuel Combarel and Dominique Marrec Architects (ECDM) is an office for architecture based in Paris. ECDM's work focuses on transformational and flexible concepts as well as on issues of sustainable development. The open programming provides an overall conceptual platform to define specific scenarios for specific projects. The method may be characterized as an integrative approach. In close collaboration with the client, associated ingeniers, consultants and future occupants, their work consists of bringing forth fundamental concepts and guiding a design response, appropriate for a specific time, place and program. The pertinent solution is as much concerned with functional legibility and technical appropriateness, as with architecture's significant and poetic content. The search for quality in design and construction is at the center of the firm's organization and is demonstrated through close collaboration between all disciplines. The team's philosophy is built around progressively acquired principles distilled from 12 years of experience in developing specific projects in all sectors of architectural practice including new construction, renovation, urbanism and design.

Thomas Christoffersen Bjarke Ingels

Thomas Christoffersen began his collaboration with Bjarke Ingels in 2001 when PLOT was first formed and is currently an Associate. Thomas has worked on every notable project from the VM Houses to one of our most recent and global developments, Astana National Library. Other accomplishments of his include the design for Iceland's National Bank and Stavanger Concert house in Norway.
In addition to his long standing participation in all things BIG, Thomas took a sabbatical year to work in New York City with WORK Architects and has also worked with Stan Allen, David Ling in NYC and Henning Larsen Architects.

Bjarke Ingels
Bjarke Ingels started BIG-Bjarke Ingels Group in 2005 after co-founding PLOT Architects in 2001 and working at OMA in Rotterdam. In 2004 he was awarded the Golden Lion at the Venice Biennale for the Stavanger Concert House, and the following year he received the Forum AID Award for the VM Houses. Since its completion, The Mountain has received numerous awards including the World Architecture Festival Housing Award, Forum Aid Award and the MIPIM

biq

The practice was founced in 1994 by Rick Wessels and Hans van der Heijden. Initially, the practice worked mostly on housing schemes. The workload has gradually expanded to offices, shops, medical and cultural buildings. In 1996 biq won the Europan-4 competition. In 2008, the Bluecoat arts center in Liverpool (GB) was completed. This project received many awards and was nominated for the Mies van der Rohe Award.
Hans van der Heijden
Hans van der Heijden is one of the founders of biq. In 1988 he graduated at TU Delft. His graduation project, a large scale urban redevelopment scheme, combined his interest for housing and his experience at the governmental building service and Mecanoo.
Rick Wessels
Rick Wessels is one of the founders of biq. He graduated at TU Delft in 1988 investigating recycling and demolition of dwellings. During his study, Rick worked at Bouwfonds and at the town council of Rotterdam. Rick's focus is on the internal organisation of the office and the process of projects.

HSH Hoyer Schindele

Hirschmüller BDA Architektur

Berlin is the starting point for the work of the architects HSH—Florian Hoyer, Harald Schindele and Markus Hirschmüller. Since establishing their office in 1997, they have been designing building projects from here domestically and internationally. HSH has worked across a broad spectrum from individual residential houses through office and residential quarters to the transformation of industry areas.
Beyond participating in national and international competitions and appraisals, the office mostly works on direct commissions. HSH Hoyer Schindele Hirschmüller present their work at numerous exhibitions locally and internationally, for example at the Architecture Biennale in Venice (2006), the Biennale for Architecture and Design "Fundacao Bienal de Sao Paulo"(2005/2003), and the 14th Architecture Biennale in Quito (2004).
Since 2005, all three partners have been members of the Bund Deutscher Architekten (BDA). There are currently 10 employees working for the office. HSH is supported by specialists in close interdisciplinary exchange, according to the particular challenges of each building project.

24H architecture

On 01.01.'01 Boris Zeisser and Maartje Lammers founded 24H architecture in Rotterdam, the Netherlands.
Maartje Lammers was born in 1963 and graduated from Technical University of Delft in 1988. Boris Zeisser was born 1968 and graduated from Technical University of Delft with honourable mention in 1995.
Before starting their own office Boris Zeisser and Maartje Lammers worked in several well-known architectural offices like (EEA) Erick van Egeraat associated architects, Mecanoo and the Office for Metropolitan Architecture (Rem Koolhaas).
In the past 10 years 24H architecture has been involved in various scale projects. One of the first projects was the summerhouse Dragspelhuset at Övre Gla, Sweden and the Ichthus Academy in Rotterdam. Various housing, exhibition and hotel projects have since then been realized.
The 16 person international team of 24H currently works on a broad scope of design tasks like schools, housing and hotel developments in Europe, Asia, Africa and the Americas.

Leen Borst
Mark Snitker

Leen Borst
1959 Born in Ter Aar, the Netherlands
Education
1988 Architecture, Delft University of Technology - cum laude
Employment
1988 – 1996 Lafour en Wijk Architects, Amsterdam
1997 – present SNITKER/BORST/ARCHITECTEN/
Lectures
1994 – 1995 guest lecturer Rotterdam Academy

Mark Snitker
1964 born in Heemstede, the Netherlands
Education
1990 Architecture, Delft University of Technology
Employment
1988 Rudy Uytenhaak, Amsterdam
1990 – 1996 Lafour en Wijk Architects, Amsterdam
1997 – present SNITKER/BORST/ARCHITECTEN/
Lectures
2001 guest lecturer Delft University of Technology
2007 guest lecturer Academy of Architecture, Amsterdam School of the Arts

Gerard Kruunenberg

1958 Born in amsterdam
1988 Graduate of the Academy of Architecture (2*), Amsterdam.
1989 Gerard Kruunenberg receives a Basic stipend by the Netherlands Foundation for Visual Arts (3*), Design and Architecture.
1991 – 1993 Partnership 'Atlas Architect - Bosch, Haslett, Kruunenberg'.
1994 – 2006 Partnership 'Kruunenberg Van der Erve Architects'.
September 2006 Kruunenberg Architects continues all actives of Kruunenberg Van der Erve Architects.
March 2009 Kruunenberg Architects and Change Architects share facilities and combine capacities when useful.
1991 – 2005 Visiting teacher and critic linked to the College of Advanced Technology, faculty of Architecture (1*), Amsterdam.
1993 – today Visiting teacher and critic linked to the Academy of Architecture (2*), Amsterdam.

Known in the Netherlands as:
1 Hogere Technische School, afdeling Bouwkunde
2 Academie van Bouwkunst
3 Fonds voor beeldende kunsten, vormgeving en bouwkunst (Fonds BKVB)

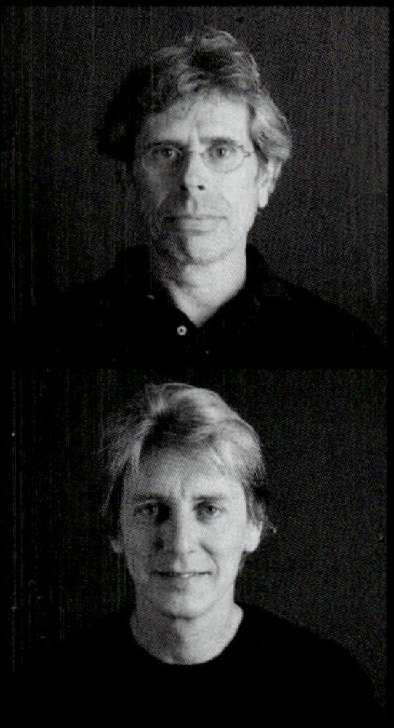

diederendirrix

diederendirrix is an architectural office with extensive experience in the fields of utilitarian building, housing, urban planning, residences and interior.
Because of a strong loyalty to the wishes and ideas of the client, diederendirrix looks for both an effective and an individualistic interpretation of the commission. Clarifying the complexity and the synthesis of contradictory interests constitute the core quality of our design approach. This research and interdisciplinary attitude results in inventive and striking buildings and architectural plans. The office's work is expressive, subtly sculptural. Buildings that are imagined from a clear convincing groundform. The architecture is not shocking or radically autonomous, but continually looks for the challenging relation between alienation and familiarity. Concepts such as simplicity, compactness, with maximum visual power and clear urban design characterize our work.

Sebastian Nagy
Ivan Matušík

Sebastian Nagy
1999 – 2003: Faculty of Architecture, Slovak Technical University Bratislava, department of public building
2003 – 2004: Royal Academy of Technology, School of Architecture, Stockholm / SWEDEN
2004: Berlage Institute, Rotterdam, Concrete Design Master Class led by Michael Speaks
2005: Master of Architecture, Faculty of Architecture, Slovak Technical University, Bratislava
2008: Authorized architect of Slovak Chamber of Architects

Ivan Matušík
1949 – 1953: Faculty of Architecture and Civil Engineering of the Slovak College of Technology in Bratislava
1968 – 1989: Led Bratislava's studio of the State Commercial Planning Office as director for 30 years
1976 – 1980: Worked as an external teacher at the Faculty of Architecture and Civil Engineering of the Slovak College of Technology in Bratislava
1994: Authorized architect of Slovak Chamber of Architects

Robert Potokar

Robert Potokar, born 1965
Graduated from the Faculty of Architecture in Ljubljana,1990.
Member of the Chamber for architecture and spatial planning in Slovenia.
Managing editor of the Piranesi review, since 1999.

Awards:
Piranesi Mention, Piran 1996, with: Vojteh Ravnikar, Maruša Zorec
Piranesi Mention, Piran 1998, with: Maruša Zorec, Ana Kučan
Plečnik medal, Ljubljana 2003
Piranesi Award, Piran 2004, with: Miha Klinar, Špela Kuhar, Blaž Medja, Uroš Pavasović
Slovenian Chamber of Architects award, Ljubljana 2007, with: Vojteh Ravnikar

Tony Owen

Tony Owen is the principle of Tony Owen Partners. Tony graduated from the University of NSW with the University Medal and the Board of Architects Prize. Tony studied advanced architectural design at Columbia University in New York where he was awarded the graduation design prize. At Columbia University, Tony worked with some of the leaders in contemporary design philosophy including Steven Holl, Bernard Tschumi and Greg Lynn. Here he explored contemporary approaches to design focusing on complex urban issues of density, sustainability and infrastructure as well as utilising current computer and catia technology in design generation.
Tony Owen has been a Design Director and senior designer in a number of leading Sydney firms such as Phillip Cox Architects, PTW and SJB. Tony has designed a number of prominent buildings and he has been the Australia winner of the CNBC Asia Pacific

Behnisch Architekten

Behnisch Architekten was founded in 1989 as a branch office of Behnisch & Partner and has since then developed, under the leadership of Stefan Behnisch, into a successful international practice. In 1991 it became economically independent and established its own partnership structure.
While Günter Behnisch ran his practice Behnisch & Partner until 2005, the branch office, often referred to as "Stadtbüro", operated under several names reflecting changing partnerships before it adopted, in 2005, the name "Behnisch Architekten". In 1999 a further office was founded in Los Angeles, followed by the Boston office in 2006, and the Munich office in 2009.
These four offices are run in different combinations by Stefan Behnisch, David Cook, Martin Haas, Robert Hösle, Christof Jantzen, Robert Matthew Noblett and Stefan Rappold. Stefan Behnisch shares his time and efforts among all four Behnisch offices who closely work together, but each are responsible for their own projects.
Over the years the practice has gained a reputation as a practice with in-depth experience in sustainable architecture, and many of its buildings received international awards.

Magén Arquitectos

Magén Arquitectos is an architectural practice founded by Jaime Magén (Zaragoza, 1974) and Francisco J. Magén (Zaragoza, 1980) based in Zaragoza since 2002.
Magén Arquitectos is interested in Architecture as a both technical and cultural fact, deeply focused on unique investigation and development of each project. This method of work allows people to manage different projects, in terms of size, scale and complexity.
The works design by Magén Arquitectos have won several awards, like the 1st Prize Bauwelt Preis 2007 or Mentions in 7th International Prize for Sustainable Architecture Fassa Bortolo and Ugo Rivolta European Architectural Award 2009, and have been published in many technical magazines, like C3 Korea, Arquitectura Viva, A10, Bauwelt or Detail. It has also been exhibited several times in Spain, Germany, Italy and France.

coco arquitectos

Jorge Martínez and Laura Sánchez established coco arquitectos in 2007. They are both architects graduated from Madrid Polytechnic University, Jorge in 2003 and Laura in 2007.
The office runs projects around Spain. They have been prize winners in several competitions, like: First prize for a Social Housing Building in Madrid (2007); First Prize for a Dancing and Music Pavilion in Madrid (2008); First prize for the logo design of the magazine "Arquitectos de Madrid" (2008). Their work has been selected to be exhibited in Madrid ReFresh (2008); Arquia Proxima Spain (2008); III International Conference on Architecture CONSTRUTEC, Madrid (2010); and published in different media.
They have just finished the construction of 168 Social Housing in Madrid, and are working on new projects.

Caramel Architekten

Martin Haller and Günter Katherl run the office haller+katherl since 1997 in vienna. From 1999 they work together with ulrich aspetsberger and build up caramel in 2001. caramel – günter katherl (*1965, vöcklabruck, upper austria), martin haller (*1966, mittelberg, vorarlberg) and ulrich aspetsberger (*1967, linz, upper austria) – have been working together since 2000 and with the help of their team have already completed a number of building projects of varying dimensions from vorarlberg to vienna.

AquiliAlberg

Founded in 2006, AquiliAlberg des-art-chitecture is a visionary and innovative international architectural and design practice based in Milan, Italy, with extensive experience in the fields of architecture, urbanism, infrastructure and design on different scale levels.
From 1999 they worked on various projects in Amsterdam, Rotterdam, Milan, Rome, London and Beijing, gaining experience in the fields of design and architecture @ Zaha Hadid, OMA (Rem Koolhaas), UNStudio (Ben Van Berkel & Caroline Bos), ONL (Kas Oosterhuis & Ilona Lenard) and Massimiliano Fuksas.
AquiliAlberg combine a multicultural approach with uncompromising implementation of high-quality design and architecture. The dynamic nature of the practice enables the exploration of new territories and the adaptation to future challenges using cutting edge digital design platforms and technological production techniques.

Kavellaris Urban Design

Kavellaris Urban Design (KUD) is an Architectural practice based in Melbourne, Australia. They are interested in architecture as an intellectual and philosophical construct founded on ideas that are materialised and translated into an architectural language; a critical dialectic which constructs their urbanism; a fragment of the built environment with meaning imbedded within every architectural intervention.
The core of their practice philosophy seeks to engage within the context and constraint for each individual project as areas of possibility and exploration in their design processes with conscientious environmental design as a foundation to deliver unique intelligent design responses for their client's needs and budget.
Their practices cover a broad and diverse range of projects and have completed work in Victoria, New South Wales, Queensland and South Australia ranging from small residential, retail, public buildings projects to large scale commercial developments and master planning.

CEBRA

CEBRA is a Danish architectural office with its background in the architectural circles of Aarhus. The office was founded 2001 by the architects Mikkel Frost, Carsten Primdahl and Kolja Nielsen – all graduates from the Aarhus School of Architecture.
In 2008 CEBRA received the most prestigious award in Scandinavia the 'Nykredits Architecture Award' and in 2006 CEBRA received the Golden Lion at the Venice Biennale 2006 for the project Co-evolution, which has been exhibited at the Biennale, São Paolo, Beijing, Manchester, Copenhagen and Venice 2006/2007. In 2008 CEBRA participated in the Biennale once again – this time curating the Danish pavilion at the Venice Biennale ECOTOPEDIA - Walk the Talk. The Bakkegaard School was nominated for the Mies van der Rohe award in 2006.
The office has worked with numerous architectural functions and scales but dwellings, sports facilities and school buildings have been predominant.

Josef Weichenberger

1964 Born in Salzburg
1990 Diploma / studies in Architecture in Innsbruck
1989 – 2002 Project partner at COOP HIMMELB AU Vienna, Lyon
1997 – 1998 Contracted Assistant at the University for Applied Arts, Institute a3 / Vienna
1998 – 2004 Assistant lecturer at the TU Vienna / Institute for Engineering and Design
2002 Foundation of JOSEF WEICHENBERGER_architects studio / Vienna

Igor Franić architectural engineer

Born in 1963 in Zagreb. Graduated from the Faculty of Architecture within the University of Zagreb. During his studies, he worked in Munich, and participated in the Congresses of European Students. After graduation, he lived and worked in London. He manages his own project bureau, and since 1996 he is employed in the Faculty of Architecture as an Assistant Professor. Member of the Executive Committee of the Association of Zagreb Architects during the 2001-2005 term. His project was given a special certificate on the 35th Zagreb Salon. Awarded on architectural competitions, participated in many exhibitions, both in Croatia and abroad; held a series of lectures on his work. Winner of yearly "Viktor Kovačić" Prize for the best architectural achievement in 2010, for his Museum of Contemporary Art in Zagreb.

x architekten

x architekten is a group of committed architects who develop conceptual positions on contemporary architecture through project-related work.
As a mathematical variable, x stands for openness. x demands the plural: the team with its flat hierarchy replaces the professional profile of the architect as lone fighter. The dynamics of a permanent work process oscillating between creativity and (self-) criticism permit the emergence of a quality standard above the capacities of each single member.
The activities of x architekten comprise the full range from housing and office construction, buildings for industry and commerce, shopping and entertainment projects to design and urbanism.

LOVE architecture and urbanism

love architecture and urbanism was founded in Graz (Austria) in 1997 and is being managed by three associates (Mark Jenewein, Herwig Kleinhapl, Bernhard Schönherr). Since its foundation, the team is heavily engaged in developing and realizing intelligent and innovative solutions for architectural and urban concepts.
The team has already realized projects in Austria, Japan, USA, South-Korea and Germany, and also won a number of competitions. They has been awarded several prizes and has been invited to take part in international competitions as well as Exhibitions and lectures.
Since 2003 the brandfield – built identity development company is being managed by the same owners. brandfield develops brand, marketing, experimental and identity strategies for real estates.

María José Pizarro Oscar Rueda

María José Pizarro
Graduates in Architecture from E.T.S. of Architecture in Madrid in 1995. Professor of Architectural Design at the School of Architecture of Madrid since 2001-02.

Oscar Rueda
Graduates in Architecture from E.T.S. of Architecture in Madrid in 1995. Professor of Architectural Design at the ETSA Alcalá de Henares and at the European University of Madrid since 2000-01.

Associate since 1998. Directors of the Department of Architecture of Editorial Rueda. They have published several articles and projects, and his work has been selected and exhibited in various national and international exhibitions.

OFIS architects

OFIS arhitekti (OFIS architects)
Based in Ljubljana formed by Rok Oman and Spela Videcnik (1998).

Rok Oman
(born 1970) studied architecture at the Ljubljana school of architecture (grad.1998) and at the architectural association in London (grad.2000).

Spela Videcnik
(born 1971) studied architecture at the Ljubljana school of architecture (grad.1997) and at the architectural association in London (grad.2000).

PLANT-Atelier Peter Kis

Established in May 1997, the studio deals with architecture, interior design and landscape planning. The quality of their work has been recognized by several Hungarian and international awards, competitions, publications and exhibitions. The studio was first established in the tower of the 'Pachydermi' House at the Budapest Zoo. Further on it is located in the penthouse of the Large Predators' House's monument building. According to some normative opinions, the succesive order of the two places shows a usual architect's career.

Biuro Projektów Architektonicznych

Biuro Projektów Architektonicznych (BPA) was founded by Jan Pallado and Aleksander Skupin in 1985 in Katowice, Poland, and currently employs 20 people. Office performs a variety of projects, both residential and services functions. Architects Pallado and Skupin have won many architectural competitions, their projects have been repeatedly awarded and exhibited in Poland and other countries and published in books and professional journals. Their recent project was highly commented by the judges of the World Architecture News Residential Awards 2011.

Keiichiro Sako

2004 to 2005 Visiting Scholar at Columbia University.
2004 to 2005 Overseas artist dispatched by Agency for Cultural Affairs.
2004 Established SAKO Architects
2004 Jointly presided over Asian Architects Associates
1996 to 2004 Worked with Riken Yamamoto & Field Shop
1996 Obtained a Masters Degree from the Tokyo Institute of Technology
1994 Graduated from Tokyo Institute of Technology
1970 Born in Fukuoka, Japan

Jun Mitsui, AIA, JIA

Principal of Jun Mitsui & Associates Architects and Pelli Clarke Pelli Architects Japan. After graduating from Tokyo University, Jun Mitsui worked in Japan until 1982. After he received his Master of Architecture from Yale University in 1984, he practiced at Cesar Pelli & Associates, Inc. until 1992. He returned to Tokyo in as a principal of Pelli Clarke Pelli Architects Japan and went on to found Jun Mitsui & Associates, Architects in 1995. Jun Mitsui is a member of AIA, JIA and Japan Architects Academy, and is a licensed architect in both Japan and the United States. He and his firm have been honored with numerous architectural awards.

Francis-Jones Morehen Thorp

Fjmt offers services in architecture, landscape architecture, master planning and interior design. With a staff of more than 75 personnel, the fjmt architectural team has a wealth of architectural and technical capability to assist them as required. This includes specialists in Landscape Architecture, Interior Design, Urban Design, 3D Visualisation (all tertiary qualified) and a team of administration/non-architectural staff including a practice manager, graphic designers and IT support.

Fjmt's team of professionals are proficient in, and aided by, the latest visualisation, communication and documentation technology. The practice has placed much emphasis on technology (computer hardware and software) to facilitate the design process. Accordingly, fjmt has the latest in CAD capability; designing and documenting using a fully transitioned 3D methodology (ArchiCAD) and the most powerful computer workstations and have an array of high-speed and large format colour printers.

Irina Cristea Grégoire Zündel

Irina Cristea
Born in 1966, Bucharest
Graduated from ENSA Strasbourg in 1995
Architectural studies in London South Bank University and in Buhcarest Ion Mincu institute
Since 2001 associate architect of A/ZC
Grégoire Zündel
Born in 1969, Colmar
Graduated from ENSA Strasbourg in 1995
Visiting lecturer at ENSA Strasbourg between 2004-2006
Since 2001 associate architect of A/ZC

Jakob+MacFarlane

Jakob+MacFarlane is an architectural firm based in Paris, France. Its work explores digital technology both as a conceptual consideration and as a means of fabrication, using new materials as a possibility to create a more flexible, responsive and immediate environment.

Their projects have been exhibited at the Victoria & Albert Museum (London, 2003), SFMOMA (San Francisco, 2004), the Museum of Architecture (Moscow, 2000), Artists Space (New York, 2003), Carnegie Melon (USA, 2001), the Mori Art Museum (Tokyo, 2004), at the MOCA in LA (Skin+Bones exhibition), the Pompidou Center, the Pavillon de l'Arsenal (Paris), and at the Orléans/ArchiLab International Architectural Exhibitions and Conferences (1999, 2001, 2003) and a personal exhibition at the FRAC of Orléans in 2005. Jakob + MacFarlane were among the architects selected by France for the 2002 Venice Architecture Biennale, and in the international 2004 selection as well as 2008.

Studio Kalamar

While the majority of its work focuses on office and public buildings, the practice's projects range from masterplans for cities to interior and furniture design.

In striving for design excellence in conjunction with technological innovation and functionality, the practice closely collaborates with clients and consultants from different fields. This active collaboration as well as the in-house supporting skills such as visualization, model making and graphic design enables the practice to offer clients full support from the concept stages to final projects without delays.

From 1996 to 2010, the practice's completed projects have totalled over 250000m² all built on schedule and within their respective budgets.

José Miguel Roldán Mercè Berengué

José Miguel Roldán graduated in 1988 in the Escola d'Arquitectura in Barcelona. At present he is Director for the Exchange Program for the Architecture Schools. He has been Deputy Professor of the School of Architecture of the Clemson University and of the Texas A&M University since 1999. He has been associated professor in the ETSAB (19888-2000) and professor Area of Projects in the ESARQ-UIC (1997-1999): He has also been a member of the Governing Board and Director of Educational Politics of the ESARQ-UIC.

Mercè Berengué graduated in the l'ETSAB-UPC in 1989. At present she presents the district of Barcelona in the Associació d'Arquitectes Urbanistes de Catalunya and she is member of the Executive Council of the Cooperativa d'Arquitectes of the COAC and of the PFC tribunal at the ETSAB. She has been, from 1998 to 2000, the directress and teacher of the Urbanism Area of the architecture school of the UIC. From 1990 to 1994 she collaborated with the department of the Architectural Projects of the ETSAB.

In 1988 they founded the Roldán+Berengué, arqts. office.

Fact Architects

Founder of Fact Architects:
Name: Peer Glandorff, architect MArch
Function: Architect Director
Residence and place of birth: Amsterdam
Born: 17 March 1970, Amsterdam
Education: Academie van Bouwkunst Amsterdam (Cum Laude)
Royal academy of Arts, Den Haag

Awards and Prizes:
Winner VKG architectuurprijs
Nomination IOC/IAKS AWARD (September 2011)
Nomination De Nederland's Bouwprijs 2011
Nomination Golden Piramide 2010
Nomination Amsterdam Architecture Prize 2010
Winner BNA JA Prijs 2006
Winner Future Hospitals 2005, second prize
Winner National Student Steel prize 2005, second prize
Nomination Archiprix 2005

Biuro Projektów Lewicki Latak

Biuro Projektów Lewicki Łatak, a company established by Kazimierz Łatak (b. 1962) and Piotr Lewicki (b. 1966) in March 1995, is a workshop specialising in architecture-building designs at various levels: urban planning, architecture, interior design. Biuro Projektów Lewicki Łatak is a laureate of numerous prizes and awards in architectural competitions.

Piotr Lewicki and Kazimierz Łatak, who graduated from the Faculty of Architecture of Cracow University of Technology in 1990, have been working together since 1988.

Brooks + Scarpa

Brooks + Scarpa (formerly known as Pugh + Scarpa) has grown to a firm of over 20 dedicated professionals. Angela Brooks, AIA, LEED AP, and Lawrence Scarpa, FAIA are the sole principals and the firm is consciously structured to ensure their participation in each project.